MW00846584

Accelerator Physics

Editors: F. Bonaudi
C. W. Fabjan

Springer
Berlin
Heidelberg
New York
Barcelona
Budapest
Hong Kong
London
Milan
Paris
Santa Clara
Singapore
Tokyo

Levi Schächter

Beam-Wave Interaction in Periodic and Quasi-Periodic Structures

With 109 Figures

Springer

Dr. Levi Schächter
Department of Electrical Engineering
Technion – Israel Institute of Technology
Haifa 32000, Israel

Editors:

Professor F. Bonaudi
Professor C. W. Fabjan
CERN, Div. PPE
CH-1211 Genève 23, Switzerland

Library of Congress Cataloging-in-Publication Data

Schächter, Levi.
Beam-wave interaction in periodic and quasi-periodic structures / Levi Schächter.
p. cm. -- (Accelerator physics)
Includes bibliographical references and index.
ISBN 3-540-61568-7 (hardcover)
1. Electron waves. 2. Electromagnetic waves. 3. Lasers. I. Title. II. Series.
QC793.5.E622S33 1996
537.5'6--dc20 96-35940
CIP

ISBN 3-540-61568-7 Springer-Verlag Berlin Heidelberg New York

Typesetting: Camera ready copy from the author using a Springer TEX macro package
Cover design: *design & production* GmbH, Heidelberg

SPIN 10520442 55/3144 – 5 4 3 2 1 0 – Printed on acid-free paper

To my parents

Preface

During the past seven years I have been involved in the investigation of high-power microwave sources for accelerator and radar applications. As for many others before me, the starting point of this book was a collection of notes on theoretical topics out of the material I had been working on. The notes were the core of a course for graduate students at Cornell University. When I started to prepare these notes it seemed a fairly straight-forward and not very time-consuming task since I had most of the material well organized. Today, three years after the preparation of the first notes, I can only wonder how naive this thought was.

Most of my work was oriented towards analytic and quasi-analytic techniques for the investigation of the interaction of an electron beam with electromagnetic waves. These topics are presented in Chaps. 4 and 6. However, for a systematic elaboration of these topics it was necessary to provide some general background, therefore parts of what are today Chaps. 2, 3, and 5 were prepared. Related topics of acceleration concepts were also prepared to some extent but I ran out of time and the material (Chap. 8) was not delivered. In the meantime, various sections of this book were taught at the Technion–Israel Institute of Technology and Ben-Gurion University. In the last version I included a discussion on free electron lasers (Chap. 7).

In this book I present primarily models of the various concepts of beam–wave interactions with emphasis on analytic and quasi-analytic methods. These have to be the basis of any numerical simulation which is today the elementary design tool of any advanced radiation source. A model is an idealization of what we conceive as the real system. In practice, this idealization manifests itself as mathematical approximations that if not revealed and/or evaluated, may lead to a non-realistic analysis and therefore to wrong conclusions. Therefore I believe the reader should be given all reasonable details for tracing each step in the analysis. She or he has to decide whether to skip a certain step because it is trivial or because the implications to the model are obvious. For this reason detailed analysis is used in most of the cases.

Many people have contributed directly or indirectly to this book and I wish to thank them all. The collaboration with Prof. J.A. Nation of Cornell University, his encouragement and critique, had a crucial impact on the material presented here and he deserves my sincere gratitude. I benefited from enlightening discussions with Prof. Norman M. Kroll of the University of California at San Diego. I wish to thank Dr. G. S. Kerslick and Mr. J. D. Ivers who contributed hours and days in discussions on many of the topics presented here. Special acknowledgement is deserved by those who were, or still are, graduate students and carried out the experiments that, in turn, were either the cause or the result of the concepts developed and presented here. They are: Dr. D.A. Shiffler, Dr. T.J. Davis, Dr. E. Kuang, Dr. S. Naqvi and Mr. D. Flechtner. Most of this research was supported by the United States Department of Energy. During the last three years I also benefited from the support of the Bi-National United States - Israel Science Foundation.

Last but by no means least I want to thank my wife Tal, and my children Michal and Roy who endured the long hours of work with patience and love.

Haifa, August 1996 Levi Schächter

Table of Contents

1. Introduction

In recent years there has been a fast growth in the demand for communication systems. Radios, phones or a TV sets have for a long time been an integral part of the everyday life but there are many others, such as wireless phones, which are climbing very rapidly to the top place held by the former three. It is sufficient to think what benefit an interactive TV combined with a personal computer would be to the user, in order to have an idea of the demands on the communication networks. The fact that within this framework, audio as well as video information will be transmitted in both directions, pushes the frequency bandwidth, and thus the operating frequency, to new limits. In most of these systems the information is carried by radiation at microwave or millimeter wave frequencies. Although for portable devices the solid state technology has a substantial advantage over the vacuum electronics due to its compact systems, ground stations will still play an important role in long distance transmission. Therefore they will require new vacuum tubes which will operate at frequencies and power levels much higher than those existing today.

This latter category of radiation sources also plays a crucial role in systems which are not necessarily for communication or information. For example, radars driven by microwave sources enable the control of air traffic or facilitate the weather forecast. At a smaller scale, in our household, microwave oven driven by 2.45 GHz magnetron became an integral part of our life.

Two major scientific programs rely on microwave or millimeter waves as part of their operation. Millimeter waves are used to heat up electrons which in turn raise the temperature of the hydrogen-based plasma in order to facilitate fusion for future power plants. Modern particle accelerators rely on the acceleration experienced by a relativistic particle as it moves in the presence of a wave which propagates at the speed of light. In fact the analysis presented in this book relies on the experience gained from a research program whose goal is to develop high power microwave radiation using distributed interaction in quasi-periodic structures for particles acceleration. Another effort, which is now in its early stages, involves the investigation of ceramics-processing by means of microwave radiation heating.

Future plans may present great new challenges for the designer of radiation sources. High-power radiation sources may contribute to repairing the ozone layer which is so vital to life on earth. Another possible future application is to launch, in low orbit, loads which would then be used to construct the international space station. The latter approach may have a substantial advantage over chemical rockets in which the weight of the load is a small fraction of the total rocket since in the case of electromagnetic propulsion the load is the majority of the weight.

The heart of all the applications mentioned above and many others, is the radiation source which can be of many kinds and a few of which are briefly overviewed in Sect 1.2. In all cases the radiation is generated by converting kinetic energy from electrons. With the exception of solid state devices, these electrons form a beam which propagates in vacuum where it interacts with electromagnetic waves in the presence of an auxiliary structure. The development of these sources started at the beginning of the century with the magnetron, followed by the klystron in the thirties, the traveling wave tube in the late forties, the gyrotron in the early sixties and the free electron laser in the mid seventies. These are a small fraction of the devices which have been developed during the years and have played a crucial role in defense, communication and research. Over the years, with the better understanding of their operation principles, their performance improved and with it, the demand of the systems' designers. Consequently, the regime these tubes operate has broadened. For example, the first traveling wave tubes operated in a continuous mode with currents of the order of mA's and beam voltages of kV's whereas today, in addition to these kind of tubes, one can find high-power devices which are driven by kA's beams (and sometime tens of kA's) with voltages on the order of 1MV – in a pulse mode. The six order of magnitude increase in current and three in the voltage, correspond to an increase of nine orders of magnitude in the power level. Consequently, a whole new variety of conceptual and technological problems evolve. Obviously, power is not the only design parameter. Frequency, bandwidth, tunability, stability and repetition rate are only a few of the considerations which should be taken into account while designing a radio frequency generator or amplifier.

In this text we present a detailed description and analysis of the concepts involved in the interaction of electromagnetic waves and electrons. Since we intend to present a thorough analysis, within the limits of a reasonably sized volume, we chose a small fraction of the existing devices. In particular we discuss various aspects of the interaction in periodic or quasi-periodic structures. Reviews of other devices can be found in recent literature which we will refer to later in this chapter. But first, we shall discuss some basic concepts of electron-wave interaction.

1.1 Single-Particle Interaction

On its own, an electron cannot transfer energy via a linear process to a monochromatic electromagnetic wave in vacuum if the interaction extends over a very long region. Non-linear processes may facilitate energy exchange in vacuum, but this kind of mechanism is rarely used since most systems require a linear response at the output. Therefore throughout this text we shall consider primarily linear processes and in this introductory chapter we shall limit the discussion to *single*-particle schemes. *Collective* effects, where the current is sufficiently high to affect the electromagnetic field, are discussed in Chaps. 4, 6 and 7.

1.1.1 Infinite Length of Interaction

Far away from its source, in vacuum, an electromagnetic wave forms a plane wave which is characterized by a wavenumber whose magnitude equals the angular frequency, ω, of the source divided by $c = 299,792,458\,\mathrm{m\,s^{-1}}$, the phase velocity of the plane wave in vacuum, and its direction of propagation is perpendicular to both the electric and magnetic field. For the sake of simplicity let us assume that such a wave propagates in the z direction and the component of the electric field is parallel to the x axis i.e.,

$$E_x(z,t) = E_0 \cos\left[\omega\left(t - \frac{z}{c}\right)\right] . \tag{1.1.1}$$

If a charged particle moves at v parallel to z axis, then the electric field this charge experiences (neglecting the effect of the charge on the wave) is given by

$$E_x(z(t),t) = E_0 \cos\left[\omega\left(t - \frac{z(t)}{c}\right)\right] . \tag{1.1.2}$$

A crude estimate for the particle's trajectory is

$$z(t) \simeq vt\,, \tag{1.1.3}$$

therefore if the charge moves in the presence of this wave from $t \to -\infty$ to $t \to \infty$ then the average electric field it experiences is *zero*,

$$\int_{-\infty}^{\infty} \mathrm{d}t \cos\left[\omega t\left(1 - \frac{v}{c}\right)\right] = 0\,, \tag{1.1.4}$$

even if the particle is highly relativistic [Pantell (1981)]. The lack of interaction can be illustrated in a clearer way by superimposing the dispersion relation of the wave and the particle on the same diagram. Firstly, the relation between energy and momentum for an electron is given by

$$E = c\sqrt{p^2 + (mc)^2}\,, \tag{1.1.5}$$

where $m = 9.1094 \times 10^{-31}$Kg is the rest mass of the electron. Secondly, the corresponding relation for a photon in free space is

$$E = cp\,. \tag{1.1.6}$$

For the interaction to take place the electron has to change its initial state, subscript i, denoted by $(E_\mathrm{i}, p_\mathrm{i})$ along the dispersion relation to the final, subscript f, denoted by $(E_\mathrm{f}, p_\mathrm{f})$ in such a way that the resulting photon in case of emission or absorbed photon for absorption, has exactly the same difference of energy and momentum i.e.,

$$E_\mathrm{i} = E_\mathrm{f} + E_\mathrm{ph}\,, \tag{1.1.7}$$

and

$$p_\mathrm{i} = p_\mathrm{f} + p_\mathrm{ph}\,. \tag{1.1.8}$$

In the case of vacuum this is impossible, as can be shown by substituting (1.1.5-6) in (1.1.7-8). We can also reach the same conclusion by examining Fig. 1.1. The expression, $E = cp$, which describes the photon's dispersion relation, is parallel to the *asymptote* of the electron's dispersion relation. Thus, if we start from one point on the latter, a line parallel to $E = cp$ will never intersect (1.1.5) again. In other words, energy and momentum can not be conserved simultaneously in vacuum.

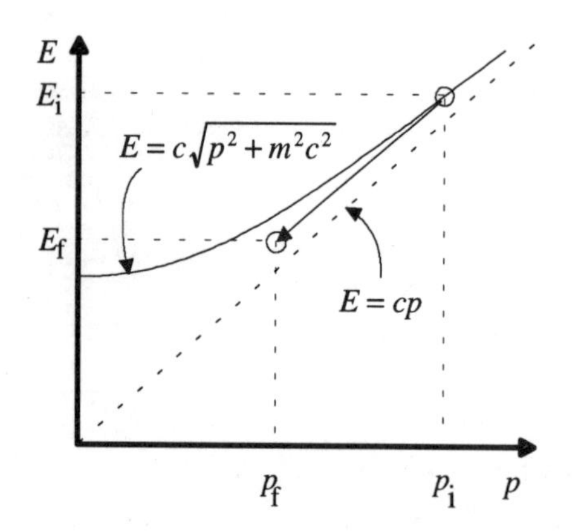

Fig. 1.1. The dispersion relation of a free electron, $E = \sqrt{(pc)^2 + (mc^2)^2}$, and an electromagnetic plane wave in vacuum, $E = pc$, are described on the same diagram. The dispersion relation of the wave is also the asymptote of the dispersion relation of the electron. Consequently, it cannot change its state along a line parallel to the asymptote. In other words, the energy and momentum laws cannot be satisfied simultaneously

1.1.2 Finite Length of Interaction

If we go back to (1.1.4) we observe that if the electron spends only a finite time in the interaction region then it can experience a net electric field. Let us denote by $-T$ the time the electron enters the interaction region and by T the exit time. The average electric field experienced by the electron (subject to the same assumptions indicated above) is

$$\langle E\rangle = E_0 \frac{1}{2T}\int_{-T}^{T} \mathrm{d}t \cos\left[\omega t\left(1-\frac{v}{c}\right)\right],$$
$$= E_0 \mathrm{sinc}\left[\omega T\left(1-\frac{v}{c}\right)\right]; \tag{1.1.9}$$

here $\mathrm{sinc}(x) = \sin(x)/x$. That is to say that if the time the electron spends in the interaction region, as measured in its frame of reference, is small on the scale of the radiation period $T_0 = 2\pi/\omega$ then the net electric field it experiences, is not zero. From the perspective of the conservation laws, the interaction is possible since although the energy conservation remains unchanged i.e.,

$$E_\mathrm{i} = E_\mathrm{f} + \hbar\omega, \tag{1.1.10}$$

the constraint on momentum conservation was released somewhat and it reads

$$\left|p_\mathrm{i} - p_\mathrm{f} - \hbar\frac{\omega}{c}\right| < \frac{\hbar}{cT}, \tag{1.1.11}$$

which clearly is less stringent than in (1.1.8) as also illustrated in Fig. 1.2; $\hbar = 1.05457\times10^{-34}$ J sec is the Planck constant. The operation of the klystron relies on the interaction of an electron with a wave in a region which is shorter than the radiation wavelength.

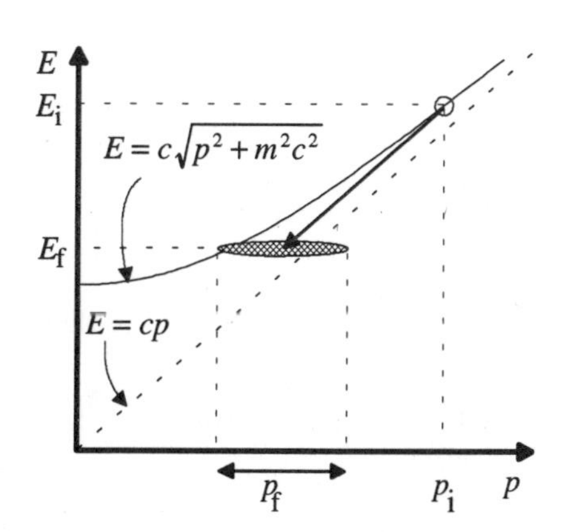

Fig. 1.2. The dispersion relation of a free electron, $E = \sqrt{(pc)^2 + (mc^2)^2}$, and an electromagnetic plane wave in vacuum, $E = pc$, are described on the same diagram. The constraint on the momentum conservation is less stringent because the interaction occurs in a finite length

1.1.3 Finite Length Pulse

Another case where energy transfer is possible in vacuum is when the pulse duration is short. In order to examine this case we consider, instead of a periodic wave whose duration is infinite, a short pulse of a typical duration τ. In order to visualize the configuration, consider a field given by

$$E_x(z,t) = E_0 \mathrm{e}^{-(t-z/c)^2/\tau^2}. \tag{1.1.12}$$

A particle following the same trajectory as in (1.1.3) will clearly experience an average electric field which is non-zero even when the interaction duration is infinite. This is possible since the spectrum of the radiation field is broad – in contrast to Sect. 1.1.1 where it was peaked – therefore again the constraint of the conservation laws is less stringent:

$$|E_{\rm i} - E_{\rm f}| < \frac{\hbar}{T}, \tag{1.1.13}$$

and

$$|p_{\rm i} - p_{\rm f}| < \frac{\hbar}{cT}. \tag{1.1.14}$$

For schematics of this mechanism see Fig. 1.3. In Chap. 8 we shall discuss various wake field acceleration schemes which rely in part on this concept. It should be also pointed out that in this section we consider primarily the kinematics of the interaction and we pay no attention to the dynamics. In other words, we examined whether the conservation laws can be satisfied without details regarding the field configuration. As will be shown later, the efficiency of energy conversion is strongly dependent on the ability to preserve one of the major component of the elecrtric field parallel to the trajectory of the charged particle.

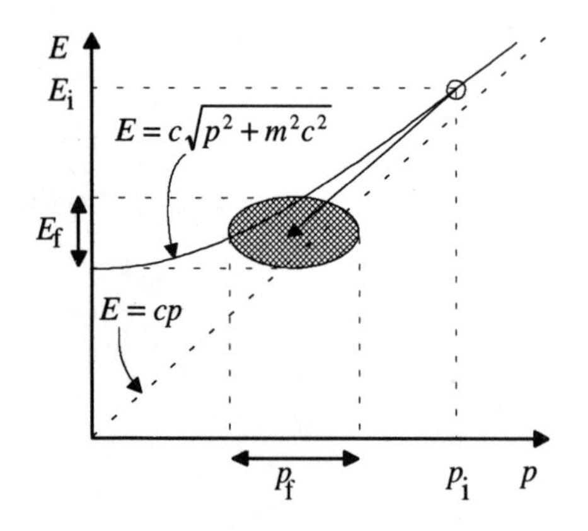

Fig. 1.3. The interaction of an electron with an electromagnetic pulse is possible since both the energy and momentum constraints are less stringent

1.1.4 Cerenkov Interaction

It was previously indicated that since the dispersion relation of the photon is parallel to the asymptote of the electron's dispersion relation, the interaction is not possible in an infinite domain. However, it is possible to change the slope of the photon, namely to change its phase velocity – see Fig. 1.4. The easiest way to do so is by "loading" the medium where the wave propagates with a material whose dielectric coefficient is larger than one. Denoting the refraction coefficient by n, the dispersion relation of the photon is given by

$$E_{\mathrm{ph}} = \frac{c}{n} p_{\mathrm{ph}} \,, \tag{1.1.15}$$

while the dispersion relation of the electron remains unchanged. Substituting in the expressions for the energy and the momentum conservation laws we find that the condition for the interaction to occur is

$$\frac{c}{n} = v \,, \tag{1.1.16}$$

where it was assumed that the electron's recoil is relatively small i.e., $\hbar\omega/mc^2 \ll 1$. The result in (1.1.16) indicates that for the interaction to occur, the phase velocity of a plane wave in the medium has to equal the velocity of the particle. This is the so-called Cerenkov condition in the 1D case. Although dielectric loading is conceptually simple, it is not always practical because of electric charges which accumulate on the surface and of a relatively low breakdown threshold which is critical in high-power devices. For these reasons the phase velocity is typically slowed down using metallic structures with periodic boundaries. The operation of traveling wave tubes (or backward wave oscillators) relies on this concept and it will be discussed extensively in Chaps. 4, 5 and 6.

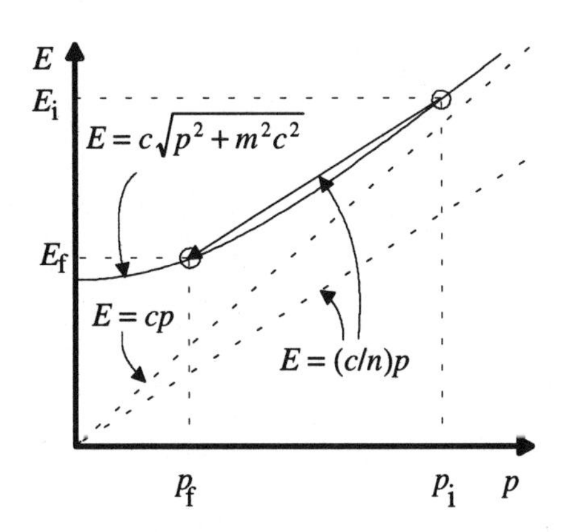

Fig. 1.4. The interaction of an electron with an electromagnetic wave whose phase velocity is smaller than c is possible

1.1.5 Compton Scattering: Static Fields

It is not only a structure with periodic boundaries which facilitates the interaction between electrons and electromagnetic waves but also periodic fields. For example, if a magneto-static field of periodicity L is applied on the electron in the interaction region, then this field serves as a momentum "reservoir" which can supply momentum quanta of $n\hbar(2\pi/L)$ where $n = 0, \pm1, \pm2, \ldots$; see Fig. 1.5. The energy conservation law remains unchanged i.e.,

$$E_{\mathrm{i}} = E_{\mathrm{f}} + E_{\mathrm{ph}} \,, \tag{1.1.17}$$

but the momentum is balanced by the applied static field

$$p_{\rm i} = p_{\rm f} + p_{\rm ph} + \hbar\frac{2\pi}{L}n\,. \tag{1.1.18}$$

For a relativistic particle ($\beta \simeq 1$) and when the electron's recoil is assumed to be small, these two expressions determine the so-called resonance condition which reads

$$\omega \simeq 2\gamma^2\left(\frac{2\pi c}{L}n\right), \tag{1.1.19}$$

where $\gamma \equiv [1-(v/c)^2]^{-1/2}$. Note that the frequency of the emitted photon depends on the velocity of the electron which means that by varying the velocity we can change the operating frequency. A radiation source which possesses this feature is a tunable source. Identical result is achieved if we assume a periodic electrostatic field and both field configurations are employed in free electron lasers discussed in Chap. 7.

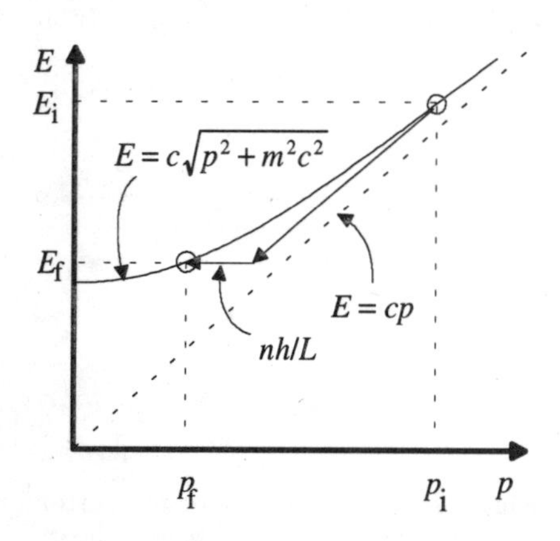

Fig. 1.5. The interaction of an electron with an electromagnetic wave in a periodic system whose periodicity is L

1.1.6 Compton Scattering: Dynamic Fields

Static electric or magnetic field can be conceived as limiting cases of a dynamic field of zero or vanishingly small frequency and we indicated above that they facilitate the interaction between an electron and a wave. Consequently we may expect that the interaction of an electron with a wave will occur in the presence of another wave. Indeed, if we have an initial wave of frequency ω_1 and the emitted wave is at a frequency ω_2 the conservation laws read

$$E_{\rm i} + \hbar\omega_1 = E_{\rm f} + \hbar\omega_2\,, \tag{1.1.20}$$

and

$$p_{\rm i} = p_{\rm f} + \hbar\frac{\omega_1}{c} + \hbar\frac{\omega_2}{c}\,. \tag{1.1.21}$$

Following the same procedure as above we find that the ratio between the frequencies of the two waves is

$$\frac{\omega_2}{\omega_1} \simeq 4\gamma^2 , \tag{1.1.22}$$

which is by a factor of 2 larger than in the static case. Figure 1.6 illustrates this process.

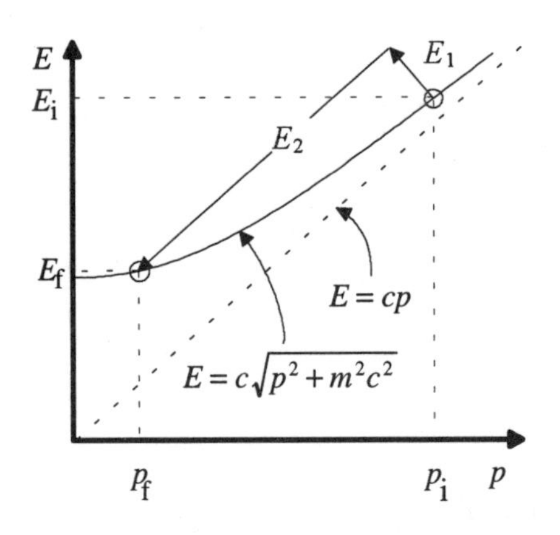

Fig. 1.6. The interaction of an electron with an electromagnetic wave in the presence of another electromagnetic wave

1.1.7 Uniform Magnetic Field

A periodic magnetic field can provide quanta of momentum necessary to satisfy the conservation law. It does not affect the average energy of the particle. An opposite situation occurs when the electron moves in a uniform magnetic field (B): there is no change in the momentum of the particle whereas its energy is given by

$$E = c\sqrt{p^2 + (mc)^2 - 2n\hbar eB} , \tag{1.1.23}$$

where $e = 1.6022 \times 10^{-19}$ C is the charge of the electron and $n = 0, \pm 1, \pm 2....$

For most practical purposes the energy associated with the magnetic field is much smaller than the energy of the electron therefore we can approximate

$$E_i - n_1\hbar\frac{ec^2B}{E_i} = E_f - n_2\hbar\frac{ec^2B}{E_f} + E_{ph} , \tag{1.1.24}$$

and the momentum conservation remains unchanged i.e.,

$$p_i = p_f + p_{ph} . \tag{1.1.25}$$

From these two equations we find that the frequency of the emitted photon is

$$\omega = 2\gamma \frac{eB}{m} = 2\gamma^2 \left(\frac{eB}{m\gamma} \right) . \qquad (1.1.26)$$

The last term is known as the relativistic cyclotron angular frequency, $\omega_{c,rel} \equiv eB/m\gamma$. Figure 1.7 illustrates schematically this type of interaction. It indicates that the dispersion line of the electron is split by the magnetic field in many lines (index n) and the interaction is possible since the electron can move from one line to another. Gyrotron's operation relies on this mechanism and it will be discussed briefly in the next section.

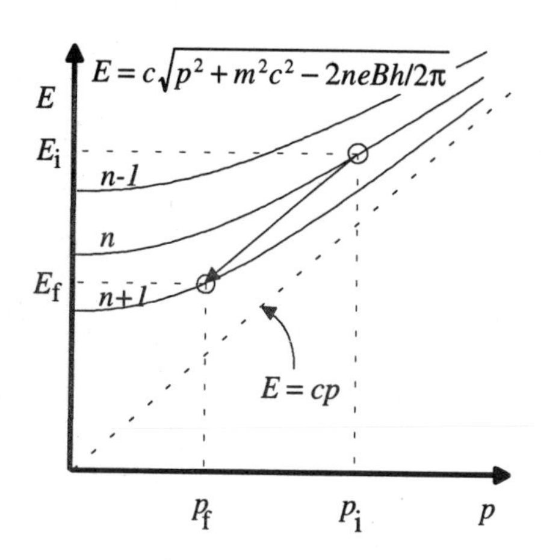

Fig. 1.7. The interaction of an electron with an electromagnetic wave in the presence of a uniform magnetic field

1.1.8 Synchronism Condition

All the processes in which the interaction of an electron with a monochromatic wave extends to large regions, have one thing in common: the velocity of the electron has to equal the *effective* phase velocity of the pondermotive wave along the electron's main trajectory, namely

$$v = v_{ph,eff} . \qquad (1.1.27)$$

Here by pondermotive wave we mean the effective wave along the longitudinal trajectory of the particle which accounts for transverse or longitudinal oscillation. In the Cerenkov case we indicated that the phase velocity is c/n and there is no transverse motion therefore the condition for interaction implies $n = c/v$ where n is the refraction coefficient of the medium. In the presence of a periodic static field the wavenumber of the pondermotive wave is $\omega/c + 2\pi/L$ therefore

$$v_{ph,eff} = \frac{\omega}{\omega/c + 2\pi/L} , \qquad (1.1.28)$$

and for a dynamic field

$$v_{\mathrm{ph,eff}} = \frac{\omega_2 - \omega_1}{k_2 + k_1}, \tag{1.1.29}$$

where $k_{1,2} = \omega_{1,2}/c$ are the wavenumbers of the two waves involved [see Chap. 7]. Finally, in a uniform magnetic field only the effective frequency varies

$$v_{\mathrm{ph,eff}} = \frac{\omega - \omega_{\mathrm{c,rel}}}{k}. \tag{1.1.30}$$

The reader can check now that within the framework of this formulation we obtain (1.1.19) from (1.1.28), in the case of the dynamic field we have from (1.1.29) the $4\gamma^2$ term as in (1.1.22) and finally (1.1.30) leads to the gyrotron's operation frequency presented in (1.1.26).

1.2 Radiation Sources: Brief Overview

There are numerous types of radiation sources driven by electron beams. Our purpose in this section is to continue the general discussion from the previous section and briefly describe the operation principles of one "member" of each class of what we consider the main classes of radiation sources. A few comments on experimental work will be made and for further details the reader is referred to recent review studies. The discussion continues with the classification of the major radiation sources according to several criteria which we found to be instructive.

1.2.1 The Klystron

The klystron was one of the first radiation sources to be developed. It is a device in which the interaction between the particle and the wave is localized to the close vicinity of a gap of a cavity, as illustrated in Fig. 1.8. Electrons move along a drift tube and its geometry is chosen in such a way that at the frequency of interest it does not allow the electromagnetic wave to propagate. The latter is confined to cavities attached to the drift tube. The wave which feeds the first cavity modulates the velocity of the otherwise uniform beam. This means that after the cavity, half of the electrons have a velocity larger than the average beam velocity whereas the second half has a smaller velocity. According to the change in the (non-relativistic) velocity of the electrons the beam becomes bunched down the stream since accelerated electrons from one period of the electromagnetic wave catch up with the slow electrons from the previous period. When this bunch enters the gap of another cavity it may generate radiation very efficiently. In practice, several intermediary cavities are necessary to achieve good modulation.

A modern klystron, of the type which is now under intensive research operates at 11.4 GHz. It is driven by 440 kV, 500 A beam [Caryotakis (1994)] and

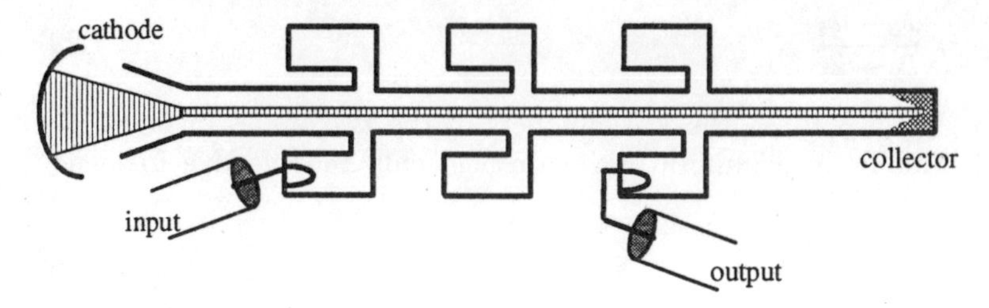

Fig. 1.8. The basic configuration of a klystron: the first cavity bunches the beam, the second amplifies the modulation and the third extracts power from the beam and converts it into radiation power

the goal is to generate power levels of the order of 100 MW in a 1.5 μsec pulse for the Next Linear Collider (NLC) developed at Stanford Linear Accelerator Center (SLAC). The operation of a klystron driven by a relativistic electron beam is different from that described above for a non-relativistic beam. If we were to use the same implementation in the case of a relativistic beam, then the distance the beam has to propagate in order to become bunched is prohibitively long since the change in velocity is relatively small. What comes to our aid in the relativistic case is the fact that the current is much higher than in the non-relativistic case and when bunching the beam, we generate the, so-called, space-charge waves (discussed in detail in Chaps. 3, 4, 6 and 7). Fortunately, the velocity modulation from the input cavity translates in a density modulation in a quarter period of the plasma wave number (defined in Chap. 3) which is proportional to the square root of the current. Therefore, if the current is sufficiently high, then the distance between two cavities again becomes reasonable.

For efficient modulation of the beam, the quality factor of the cavities has to be high and therefore in general the klystron is not a tunable device. In high power devices the choice of geometry is a trade-off between a small cavity gap required for good modulation and a large gap required to sustain the large electric field in the gap associated with high power levels. Finally, at relatively low frequencies ($\sim$1 GHz) where the geometry is sufficiently large such that a large amount of current can be injected before reaching the limiting current (to be discussed in Chap. 3), it was shown by Friedman (1985) experimentally and by Lau (1990) theoretically, that annular beams have some advantages in generating multi-gigawatt pulses – see also Serlin (1994). In this case the current carried by the beam is very high ($>$5 kA).

1.2.2 The Traveling Wave Tube

The traveling wave tube (TWT) is a Cerenkov device, namely the phase velocity of the interacting wave is smaller than c and the interaction is distributed along many wavelengths. Generally speaking, as the beam and the wave advance, the beam gets modulated by the electric field of the wave and in turn, the modulated beam increases the amplitude of the electric field. In this process both the beam modulation and the radiation field grow exponentially in space. The coupling between the wave and the beam is determined by the interaction impedance which is a measure of the electric field acting on the electrons (E) for a given total electromagnetic power (P) flowing in the system:

$$Z_{\text{int}} = \frac{E^2}{2k^2 P}, \tag{1.2.1}$$

where k is the wave-number. This definition introduced first by Pierce (1950) is the basis of his theory of the TWT which is in very good agreement with experiments in uniform low power devices. The concept of using space-charge waves in order to generate high power microwave radiation with traveling wave structures was first introduced by Nation (1970).

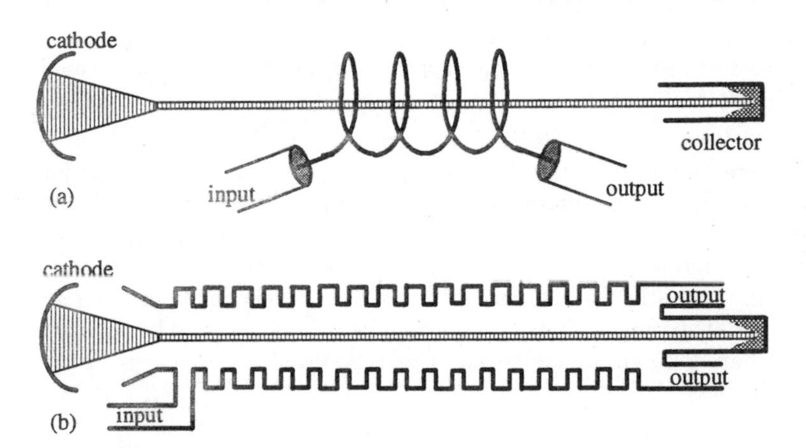

Fig. 1.9a,b. The basic configuration of a traveling wave amplifier: (**a**) helix and (**b**) coupled cavity structure

The TWT can be designed to be a very broad band device and it can occur in various configurations: helix, disk-loaded waveguide (coupled cavities), dielectric loaded waveguide, gratings, dielectrically coated metal and others. Several of these configurations are illustrated in Fig. 1.9a–e. Whenever the electromagnetic wave can propagate parallel to the beam, it means that a wave can also propagate in the opposite direction. Therefore the input is not isolated from the output, and in amplifiers, this problem can be detrimental. At the same time this is the basis for the design of an oscillator.

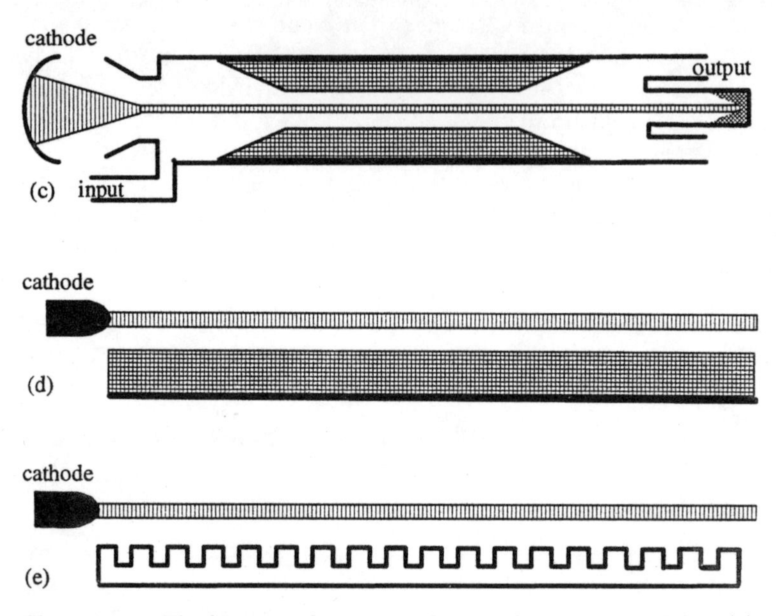

Fig. 1.9c–e. The basic configuration of a traveling wave amplifier: (**c**) based on a dielectric loaded waveguide (**d**) based on an open dielectric structure and (**e**) based on an open periodic structure

In the interaction process the electron oscillates primarily along the major axis (z direction) and the interaction is with the parallel component of the electric field. Correspondingly, the interaction occurs here with the transverse magnetic (TM) mode. This device will be extensively treated in this text since it has the potential for generating radiation at a very high efficiency as discussed in Chap. 6. For this purpose the definition in (1.2.1) will be changed slightly in Chap. 2 to account for spatial variation which may occur in the interaction region.

1.2.3 The Gyrotron

The gyrotron relies on the interaction between an annular beam, gyrating around the axis of symmetry due to an applied uniform magnetic field, and a transverse electric (TE) mode. The concept of generating coherent radiation from electrons gyrating in a magnetic field was proposed independently by three different researchers in the late fifties, Twiss (1958), Schneider (1959) and Gaponov (1959), and it has attracted substantial attention due to its potential to generate millimeter and submillimeter radiation.

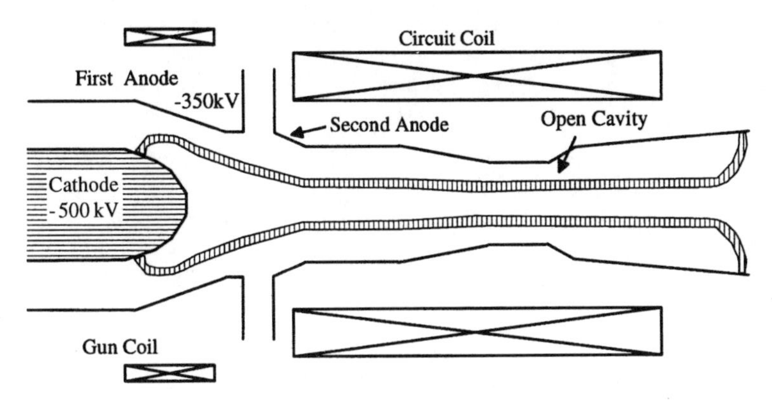

Fig. 1.10. The basic configuration of a gyrotron. The magnetic insulated gun (MIG) generates electrons which spin azimuthally therefore they are suitable for interaction with a transverse electric (TE) mode

In this device electrons move in the azimuthal direction and they get bunched by the corresponding azimuthal electric field. As in the case of the TWT the bunches act back on the field and amplify it. In contrast to traveling wave tubes or klystrons in which the beam typically interacts with the lowest mode, in the gyrotron the interaction is with high modes therefore various suppression techniques are employed in order to obtain coherent operation with a single mode.

The operation frequency is determined by the applied magnetic field, the energy of the electrons and, in cases of high mode operation, also by the radius of the waveguide:

$$\omega = \omega_c \gamma + \gamma\beta\sqrt{\omega_c^2 + \omega_{co}^2}\,, \tag{1.2.2}$$

where $\beta = v/c$, $\omega_c = eB/m$ and ω_{co} is the cutoff frequency of the mode. The operating frequency in this case can reach very high values: for a magnetic field of 1T and $\gamma \simeq 2.5$ the operation frequency is of the order of 150 GHz or higher according to the mode with which the electrons interact.

Since the interaction of the electrons is with an azimuthal electric field, it is necessary to provide the electrons with maximum momentum in this direction. The parameter which is used as a measure of the injected momentum is the ratio of the transverse to longitudinal momentum $\alpha \equiv v_\perp / v_z$. This transverse motion is acquired by the electrons in the gun region as can be deduced from the schematics illustrated in Fig. 1.10. In relativistic devices this ratio is typically smaller than unity whereas in non-relativistic devices it can be somewhat larger than one.

Beam location is also very important. In the TWT case the interaction is with the lowest symmetric TM mode. Specifically the electrons usually form a pencil beam and they interact with the longitudinal electric field which has a maximum on axis. We indicated that gyrotrons operate with high TE modes and the higher the mode, the higher the number of nulls the azimuthal electric field has along the radial direction. Between each two nulls there is a peak value of this field. It is crucial to have the annular beam on one of these peaks for an efficient interaction to take place.

Reviews of gyrotrons have been given by Flyagin, Gaponov, Petelin and Yulpatov (1977) and Hirshfield and Granatstein (1977). An instructive overview of gyrotron theory was published by Baird (1987) and the experimental results were reviewed by Granatstein (1987). More recent work on gyrotrons can be found in the Special Issue of IEEE Trans. of Plasma Science Vol. 22.

1.2.4 The Free Electron Laser

The free electron laser (FEL) will be discussed in detail in Chap. 7. As the gyrotron, it is a fast-wave device in the sense that the interacting electromagnetic wave has a phase velocity larger or equal to c but instead of a uniform magnetic field it has a periodic magnetic field. The "conventional" free electron laser has a magnetic field perpendicular to the main component of the beam velocity. As a result, the electrons undergo a helical motion which is suitable for interaction with either a TE or a TEM mode. The oscillation of electrons is in the transverse direction but the bunching is longitudinal and in this last regard the process is similar to the one in the traveling wave tube. However, its major advantage is the fact that it does not require a metallic (or other type of) structure for the interaction to take place. Consequently, it has the potential to either generate very high power at which the contact of radiation with metallic walls would create very serious problems, or produce radiation at UV, XUV or X-ray where there are no other coherent radiation sources. Figure 1.11 illustrates the basic configuration.

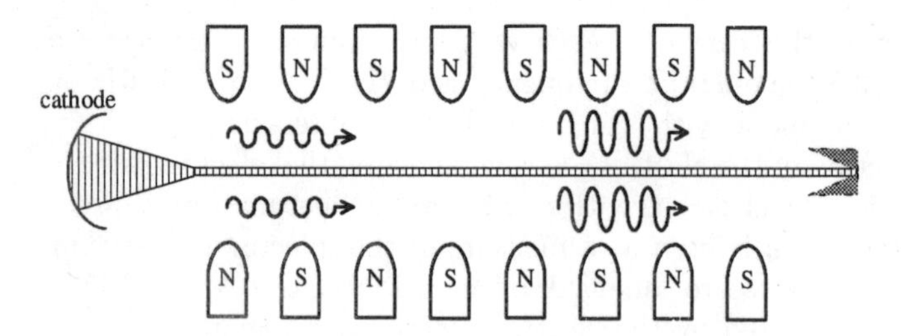

Fig. 1.11. The schematic of a free electron laser

1.2.5 The Magnetron

The magnetron was invented at the beginning of the century but because of its complexity there is no analytical model, as yet, which can describe its operation adequately as a whole. In recent years great progress has been made in the understanding of the various processes with the aid of particle in cell (PIC) codes. Its operation combines potential and kinetic energy conversion. Figure 1.12 illustrates the basic configuration. Electrons are generated on the cathode (inner surface) and since a perpendicular magnetic field is applied they form a flow which rotates azimuthally. The magnetic field and the voltage applied on the anode are chosen in such a way that, in equilibrium, the average velocity of the electrons equals the phase velocity of the wave supported by the periodic structure at the frequency of interest.

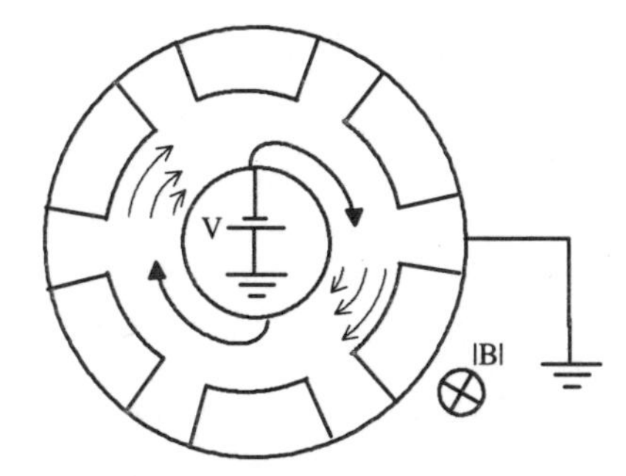

Fig. 1.12. The schematic of a magnetron

A simplistic picture of the interaction can be conceived in the following way: electrons which lose energy to the wave via the Cerenkov type interaction, move in upward trajectories – closer to the anode. Consequently, two processes occur. Firstly, the closer the electron is to the periodic surface the stronger the radiation field and therefore the deceleration is larger, causing a further motion upwards. Secondly, as it moves upwards its (dc) potential energy varies. Again, this is converted into electromagnetic energy.

Two major differences between the magnetron and other radiation sources mentioned above, are evident: *(i)* in the magnetron the beam generation, acceleration and collection occur all in the same region where the interaction takes place. *(ii)* The potential energy associated with the presence of the charge in the gap plays an important role in the interaction; the other device where this is important is the vircator which will be briefly discussed next.

1.2.6 The Vircator

The vircator takes advantage of the fact that the amount of current generated by a given voltage which can be injected into a grounded metallic waveguide, is limited. This limiting current is developed in Sect. 3.4.3. Any current injected above this limit is reflected, but on average there is a finite amount of charge in the waveguide. See Fig. 1.13. This charge forms what is called a virtual cathode (i.e. negative potential) which can be conceived as the reason for the reflection of the electrons. These oscillate between the real and the virtual cathode at a frequency which is directly related to the electrons' density (plasma frequency). A review of the vircator's theory has been given by Sullivan, Walsh and Coutsias (1987). More recently Alyokhin (1994) presented a review of the studies of this device.

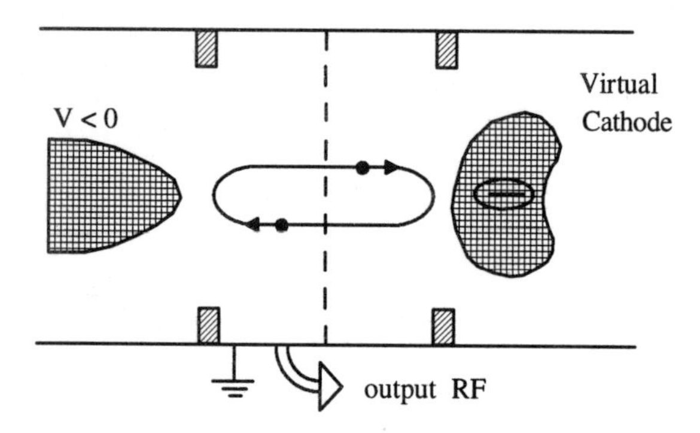

Fig. 1.13. The schematic of a vircator

1.2.7 The Magnicon

The magnicon or its earlier version the gyrocon, utilizes pencil beams which are deflected in a cavity by the electromagnetic field. They take advantage of the fact that typically the transverse dimension of the beam is small on the scale of the wavelength therefore they form very good bunches provided that a proper extraction is designed. Figure 1.14 illustrates, schematically, an axial gyrocon. A first cavity deflects the beam causing a conical motion. A dc deflecting system forces the beam to enter perpendicularly into a waveguide which forms a closed loop. The wave supported by this loop interacts with the beam. Each single electron has two velocity components: one is longitudinal and the other is azimuthal. These correspond to the components of a transverse magnetic mode supported by the waveguide.

The magnicon is an even simpler version of this device in the sense that the rf deflection is the same but the static deflection system is a uniform magnetic field. The rf converter is a simple TM_{110} cavity – see Fig. 1.15. Conversion of energy is primarily from the transverse motion since the longitudinal force (at least sufficiently close to the axis) is zero as can be checked for the TM mode

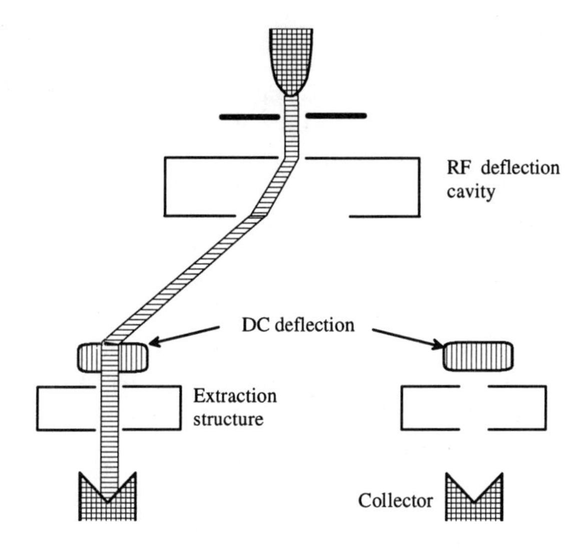

Fig. 1.14. The schematic of a gyrocon

and assuming the wave frequency coincides with the cyclotron frequency. Detailed analysis and experimental results of both devices were presented by Nezhevenko (1994).

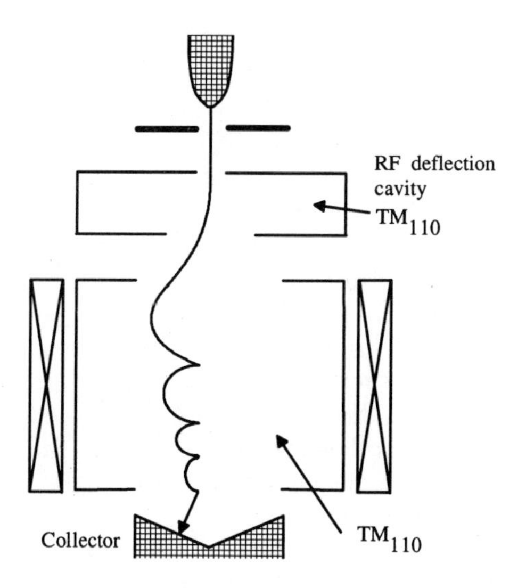

Fig. 1.15. The schematic of a magnicon

1.2.8 Classification Criteria

The variety of operation principles and consequently of devices does not allow to define a single criterion for their classification. We shall start our discussion from the trivial observation that any radiation source consists of at least two components, namely the beam and the wave. From this starting point, we shall consider various subdivisions. With regard to the *wave*: the first question we should ask is whether it is guided or confined by metallic walls as is the case in most sources, or if it can propagate in free space as is the case in a few of the free electron laser schemes. If it is guided, then the next question is whether its phase velocity is smaller or greater than c. The first category is that of slow-wave devices and its main members are the traveling wave tube and the backward wave oscillator (BWO). The second category, that of fast-wave devices, consists of the gyrotron, cyclotron auto-resonance maser (CARM) and the free electron laser. Among the slow-wave structures there is room for an additional subdivision since there are closed and open slow-wave structures. Although the great majority of today's systems rely on closed structures, the continuous demand for high frequency sources will probably in future increase the number of devices which have open structures as their main component; primarily because of the limited number of modes which may develop.

Still in the context of the electromagnetic wave, the various sources can be classified according to the interacting mode. In TWTs the interaction is always with the TM mode whereas in gyrotrons the interaction is always with the TE mode. FELs, on the other hand, may interact with either TE or TEM mode. Combinations of TE and TM modes (hybrid) are, in general destructive – as happens in acceleration sections where the hybrid mode (HE_{11}) causes beam breakup. However, this effect can be utilized for constructive purposes in particular when a highly relativistic beam has to be dumped to the wall. Beam break-up and hybrid modes are discussed in Chap. 8.

Even in two devices in which the interaction is with the same mode, say TM mode such as in the case of TWT and klystron, we realize that there is room for additional sub-division regarding the character of the interaction. In the TWT (as in FEL, gyrotron and magnetron) the interaction is distributed and it occurs over many wavelengths. On the other hand, in a klystron the interaction is localized and it is limited to the close vicinity of the cavity's gap – which is typically a fraction of the wavelength.

The electromagnetic structure determines whether there are reflections in the system and, as we shall see in Chaps. 4 and 6, these determine if the system operates as an amplifier or an oscillator. In the case of metallic periodic structures the feedback can be designed to be part of the electromagnetic characteristics of the structure as happens in the case of the backward wave oscillator (BWO). Furthermore, the transverse dimension of the structure determines the number of electromagnetic modes in the structure. In most cases the geometry is chosen such that a single mode is supported at a given frequency but there are cases where frequency or power impose large geometry therefore the system becomes a multi-mode device. This is the case for gyrotrons and also a few Cerenkov devices, e.g., Bugaev (1990).

If we examine the sources from the point of view of the *electron beam* there are also many possible classifications. High-power devices utilize typically relativistic beams and devices like the free electron laser or vircator have meaningful operations primarily in this regime. Others like the TWT, gyrotron and klystron can operate either with relativistic or non-relativistic electrons. Relativistic beams in many cases are associated not only with high voltages (>200 kV) but also high currents (>250 A) which implies high power levels (>50 MW). These can be sustained for relatively short periods of time; typically of the order of 1μsec or shorter. In many of the cases of interest several such pulses are fired per second and this is referred to as the repetition rate. For example, the design for the next linear collider (NLC) considers klystrons which are driven by electron pulses which are 1.5μsec long with a repetition rate of 180Hz. At the other extreme, there are continuous wave (CW) sources such as magnetrons, gyrotrons and TWTs which operate at high average power (>1 kW). Repetitive pulse sources, such as the one driven by the 50 MW beam mentioned above, may provide a maximum average power (assuming 50% rf efficiency conversion) of 2.5 W if the pulse duration is 100 nsec and the repetition rate is 1 Hz. At 1μsec and 10 Hz the average power goes up to 250 W.

Without exception the beam has to be guided, otherwise the electrons blow apart and they are of no use for energy exchange. In most cases the beam is guided by a uniform magnetic field and in a small fraction by a permanent periodic magnetic field. In the gyrotron or the free electron laser this field plays a crucial role in the interaction process itself. Furthermore, in cross-field devices the uniform magnetic field is accompanied by a perpendicular electric field which also contributes to the interaction.

Beam quality, which is associated with fluctuations in the energy around the average value, is another classification criterion. This topic is addressed in Sect. 3.4.5 and it is of particular interest in accelerators and in free electron lasers. In the former because the electrons have to travel very long distances and ultimately have to be focused with great precision, therefore both the transverse and longitudinal momentum are important. In free electron lasers this parameter is important as we go up in frequency and in fact beam quality is one of the major limitations of today's free electron lasers – at least with regard to optical or shorter wavelengths. Energy conversion efficiency brings us to another possible way of classification, based on whether the initial beam is uniform or pre-bunched. In the latter case, the efficiency of energy extraction can be very high. There are basically two ways to pre-bunch a beam: either in a two (or more) stage system as in a klystron or to form the bunches at the same place where the electrons are generated, namely to produce bunches in which all electrons have the same velocity. In this regard the way the electrons are generated is critical and may have a dramatic impact on the performance of the device.

Finally, the amount of current injected into the system can also be used for classification of sources. SLAC klystrons, for example, operate well below the limiting current whereas the relativistic klystron amplifier (RKA) developed at NRL by Friedman (1985) operates close to the limiting current. At the extreme, the vircator operates well above the limiting current.

1.3 Choice and Organization of the Material

With such a variety of interaction schemes and sources we owe the reader an explanation of how the material was selected, and why we chose to present one topic, whereas another, which might be as important, was left out. The principle which directed us was to have a coherent and thorough presentation of the beam-wave interaction in a few modern devices with most, if not all, the mathematical details associated with the models which explain their operation.

From the very beginning it was clear that one could not meet the requirement of detailed presentation and encompass the whole variety of sources and interaction schemes, discussed above, in one reasonably-sized volume. In addition, in the last years we were actively involved in the development of high-power, high-efficiency traveling wave amplifiers. These two facts have biased the choice of presentation towards the interaction in *periodic* and *quasi-periodic structures.* This includes traveling wave tube, free electron laser and linear accelerators, but excludes the gyrotron, CARM, vircator and magnicon. We had a dilemma with the magnetron since in principle it belongs to the periodic structure category but, to the best of our knowledge, the main developments in recent years were the result of particle-in-cell (PIC) simulations which are beyond our scope in this book since we concentrate primarily on *analytical methods.*

The book can be divided into three parts. The first includes Chaps. 2 and 3 which present some of the elementary concepts in the electromagnetic theory and electrons' dynamics which are relevant to beam-wave interaction. The second part includes Chaps. 4, 5 and 6. It addresses the interaction in periodic (and quasi-periodic) metallic structures. The third part (Chaps. 7 and 8) presents two additional devices in which similar phenomena occur but in different configurations or regimes of operation, namely the free electron laser and the linear accelerator. Let us now review the motivation and the content of the various chapters.

Chapter 2. After we discuss Maxwell equations in general we present three simple homogeneous solutions corresponding to the TEM, TM and TE modes (Sect. 2.3). When the current density is present it is useful to use Green's function method for solution of the electromagnetic field. This method is discussed in Sect. 2.4 and is accompanied by two examples which illustrate the Cerenkov radiation in free space and in a waveguide. In Sect. 2.5 several finite length effects are considered; in particular, reflection effects and transients in a cavity.

Chapter 3. All topics considered throughout the text rely on classical mechanics (Sect. 3.1) and without exception they are consistent with the special theory of relativity (Sect. 3.2), therefore the fundamentals of these two theories are summarized. Beyond reviewing the fundamental concepts of relativistic classical dynamics, we consider, in Sect. 3.3, some of the methods of electron generation and discuss the Child-Langmuir law which draws a limit on the maximum current achievable when applying a voltage on a cathode from which electrons are emitted. After electrons are generated, they are typically guided by magnetic fields and waveguides to the interaction region. In Sect. 3.4 we present some basics of beam propagation in free space with uniform or periodic magnetic field and in a uniform and disk-loaded waveguide. The section concludes with the basic measures of beam quality: emittance and brightness. The last section is dedicated to space-charge waves. After introducing the basic concepts of fast and slow space-charge waves, we consider two instabilities which can develop when these waves are present. One is the resistive wall instability and the other is the two-beam instability. The section and chapter conclude with a discussion on the interference of two space charge waves and their interaction with a cavity as it happens in relativistic klystrons.

In order to understand the motivation behind the second part of this book (Chaps. 4, 5 and 6), we shall next review the progress in the research on the interaction in traveling wave structures. The first experiments on high power TWT, performed by J. A. Nation at Cornell University [Shiffler (1989)], indicated that 100 MW of power at 8.76 GHz can be achieved before the system oscillates. Although no rf breakdown was observed, the fact that the input is no longer isolated from the output, allows waves to be reflected backwards and this feedback causes the system to oscillate. In order to isolate the input from the output the TWT was split in two sections separated by a sever (waveguide made of lossy material which operates below cutoff). The second set of experiments on a two-stage high power TWT indicated that power levels in excess of 400 MW are achievable with no indication of rf breakdown [Shiffler (1991)]. In this case, however, the spectrum of output frequencies was 300 MHz wide and a significant amount of power (up to 50%) was measured in asymmetric sidebands. The latter observation was investigated theoretically [Schächter (1991)] and it was concluded that it is a result of amplified noise at frequencies selected by the interference of the two waves bouncing between the ends of the last stage. In fact we have shown [Schächter and Nation (1992)] that what we call amplifier and oscillator are the two extremes of possible operation regimes and any practical device operates somewhere in between, according to the degree of control we have on the reflection process.

In order to eliminate the reflections effect it was suggested to design a low group velocity structure such that the time it would take the first reflection to reach the input section of the structure is longer than the electrons' pulse duration. As a result, the reflected wave has no electrons to interact with. This method is called the transit-time isolation method and it was successfully demonstrated [Kuang (1993)] experimentally when power levels of 200 MW were achieved at 9 GHz. The spectrum of the output signal was less than 50 MHz wide and the passband of the periodic structure was less than 200 MHz – for this reason we called it the narrow band structure. The 200 MW power levels generated with this structure were accompanied by gradients larger than 200 MV/m and no rf breakdown was observed experimentally. However, for any further increase in the power levels it is necessary to increase the volume of the last two or three cells in order to minimize the electric field on the metallic surface. The system then becomes quasi-periodic. In order to envision the process in a clearer way, let us assume that 80% efficiency is required from our source. If the initial beam is not highly relativistic, which is the case in most systems, such an efficiency implies a dramatic change in the geometry of the structure over a short distance. Specifically, for a 500 keV beam, the initial velocity is $v \sim 0.86\,c$, thus an 80% efficiency would require a phase velocity of $0.55\,c$ at the output. This corresponds to a 36% change in the phase velocity and a similar change would be required in the geometry which is not at all an adiabatic change when it occurs in one period of the wave.

Based on our experience the main problems of a quasi-periodic traveling wave extraction structure are: *(i)* minimize the reflections primarily at the output end of the structure in order to maintain a clean spectrum and to avoid oscillations and *(ii)* taper the output section to avoid breakdown and *(iii)* compensate for the decrease in the velocity of the electrons. All these topics are discussed in detail in Chap. 4. In order to optimize these conflicting requirements we have developed an analytical technique which permits us to design a quasi-periodic structure. This technique is presented in Chap. 6.

Chapter 4. In this chapter we investigate the fundamentals of beam-wave interaction in a distributed slow-wave structure. A dielectric loaded waveguide was chosen as the basic model in the first sections because it enables us to illustrate the essence of the interaction without the complications associated with complex boundary conditions. In the first section we present part of Pierce's theory for the traveling wave amplifier applied to dielectric loaded structure and extended to the relativistic regime. The interaction for a semi-infinitely long system is formulated in terms of the interaction impedance introduced in Chap. 2. Finite length effects are considered in the second section where we first examine the operation of an oscillator. The macro-particle approach is described in Sect. 4.3 where the beam dynamics, instead of being considered in the framework of the hydrodynamic approximation i.e. as a single fluid flow, is represented by a large number of clusters of electrons. This formalism enables us to examine the interaction in phase-space either in the linear regime of operation or close to saturation. It also permits investigation of tapered structures and analysis of the interaction of pre-bunched beams in tapered structures. In Sect.4.4 we extend the concept of the interaction impedance to a complex quantity and formulate the equations adequately. This impedance is allowed to vary in space and no prior assumption of the form of the electromagnetic field is made, but it is assumed that the effect of the interaction is local. The chapter concludes with a further extension of the macro-particle approach formalism to include the effect of reflections. This framework combines the formulations of an amplifier and an oscillator and permits us to quantify and illustrate the operation of a realistic device which is neither an ideal amplifier nor an ideal oscillator.

Chapter 5. This chapter presents various characteristics of periodic structures with emphasis on those aspects relevant to interaction with electrons. In the first section we present the basic theorem of periodic structures, namely the Floquet theorem. Following this theorem we bring an investigation of closed periodic structures in Sect. 5.2 and open structures in the third. Smith-Purcell effect is considered as a particular case of a Green's function calculation for an open structure and a simple scattering problem is also considered. The chapter concludes with an example of a simple transient solution in a periodic structure which is of importance in accelerators where wake-fields left behind one bunch may affect trailing bunches.

Chapter 6. This chapter deals with metallic quasi-periodic structures which are required in order to maintain an interacting bunch in resonance with the wave when high efficiency is required. Non-adiabatic change of geometry dictates a wide spatial spectrum, in which case the formulation of the interaction in terms of a single wave with a varying amplitude and phase is inadequate. In fact, the electromagnetic field cannot be expressed in a simple (analytic) form if substantial geometric variations occur from one cell to another. To be more specific: in uniform or weakly tapered structures the beam-wave interaction is analyzed assuming that the general functional form of the electromagnetic wave is known i.e., $A(z)\cos[\omega t - kz - \phi(z)]$ and the beam affects the amplitude $A(z)$ and the phase, $\phi(z)$. Furthermore, it is assumed that the variation due to the interaction is small on the scale of one wavelength of the radiation. Both assumptions are not acceptable in the case of a structure designed for high efficiency interaction. In order to overcome this difficulty and others, we present, in Chap. 6, an analytic technique which has been developed in order to design and analyze quasi-periodic metallic structures of the type discussed in Chap. 5. The method relies on a model which consists of a cylindrical waveguide to which a number of pill-box cavities and radial arms are attached. In principle the number of cavities and arms is arbitrary. In Sect. 6.1 we examine the homogeneous electromagnetic characteristic of quasi-periodic structures. The technique is further developed to include Green's function formulation in Sect. 6.2 followed by the investigation of space-charge waves (Sect. 6.3) within the framework of the (linear) hydrodynamic approximation for the beam dynamics. In Sect. 6.4 the method is further generalized to include effects of large deviations from the initial average velocity of the electrons by formulating the beam-wave interaction in the framework of the macro-particle dynamics.

In the third part of this book we consider the beam-wave interaction in periodic and quasi-periodic structures different from that in the second part, namely free electron lasers and particle accelerators. These two topics have been extensively discussed in literature and they are the subject of many articles, books and conferences. Therefore our approach in this part combines our approach of detailed analysis used in the previous chapters with a general discussion of alternative concepts and configurations.

Chapter 7. This chapter deals with the principles of free electron laser. In the first section we consider the spontaneous emission as an electron traverses an ideal wiggler. It is followed by the investigation of coherent interaction in the low-gain Compton regime and Sect. 7.3 deals with the high-gain Compton regime which includes cold and warm beam operations. The macro-particle approach is introduced in Sect. 7.4 and we conclude the chapter with a brief overview of the various alternative schemes of free electron lasers.

Chapter 8. One of the important systems where beam-wave interaction in periodic structures plays a crucial role is the particle accelerator which is discussed in this chapter. It has basically two parts. In the first part (Sect. 8.1) we discuss in detail the basics of the linear accelerator (linac) concepts with particular emphasis on the beam-wave interaction. The discussion is limited to a linear accelerator of the type *operational* today at SLAC (Stanford Linear Accelerator Center) whose basic concepts are applicable to what is today conceived as the Next Linear Collider. The second part (Sects. 8.2–6) is a collection of brief reviews of different alternative schemes of acceleration which are in their early stages of research. In these sections the discussion is in general limited to the basic concepts and the figure of merit which characterizes their application i.e. the achievable gradient.

2. Elementary Electromagnetic Phenomena

All the effects discussed in this text rely on the presence of electric, magnetic or electro-magnetic fields in the system. It is therefore natural to discuss first the governing equations and some basic electromagnetic phenomena. In this regard, "elementary" in the title of this chapter refers to subjects related to beam-wave interaction and not necessarily to undergraduate-level topics though a few really elementary concepts are discussed in Sects. 2.1 and 2.2.

In many of the devices of interest the waves are guided by waveguides, therefore a short discussion on TEM, TM and TE modes in various configurations is presented in Sect. 2.3. When source terms are present (current and charge density) it is particularly useful to use Green's function method for solution of the electromagnetic field. This method is discussed in Sect. 2.4 and is accompanied by two examples which illustrate the Cerenkov radiation in free space and in a waveguide. In Sect. 2.5 several finite length effects are considered, in particular, reflection effects and transients in a cavity.

2.1 Maxwell's Equations

The basis for the analysis of all electro-magnetic phenomena are Maxwell's equations which relate the electric ($\mathbf{E}$) and magnetic ($\mathbf{H}$) field, the electric ($\mathbf{D}$) and magnetic ($\mathbf{B}$) induction with the current ($\mathbf{J}$) and charge (ϱ) densities:

$$\nabla \times \mathbf{E}(\mathbf{r},t) + \frac{\partial}{\partial t}\mathbf{B}(\mathbf{r},t) = 0\,, \tag{2.1.1}$$

$$\nabla \times \mathbf{H}(\mathbf{r},t) - \frac{\partial}{\partial t}\mathbf{D}(\mathbf{r},t) = \mathbf{J}(\mathbf{r},t)\,, \tag{2.1.2}$$

$$\nabla \cdot \mathbf{D}(\mathbf{r},t) = \varrho(\mathbf{r},t)\,, \tag{2.1.3}$$

$$\nabla \cdot \mathbf{B}(\mathbf{r},t) = 0\,. \tag{2.1.4}$$

This set of equations determines the electromagnetic field at any point in space and in time provided that the source terms (ϱ and $\mathbf{J}$), are known. In addition, the boundary conditions have to be determined together with the constitutive relations of the medium, i.e., the relation between the inductions ($\mathbf{B}$ and $\mathbf{D}$) and the field ($\mathbf{H}$ and $\mathbf{E}$).

2.1.1 Constitutive Relations

Matter reacts to the presence of an electromagnetic field. The constitutive relations characterize this reaction in terms of the dynamics of charges at the microscopic level. In general these relations are non-linear and they couple together all the components of the electromagnetic field. In many cases the constitutive relations are linear and even then the relations are not necessarily simple since they can form a tensor. A particular case of constitutive relations is the linear and scalar case in which

$$\mathbf{B}(\mathbf{r},t) = \mu_0\mu_\mathrm{r}\mathbf{H}(\mathbf{r},t)\,, \tag{2.1.5}$$

$$\mathbf{D}(\mathbf{r},t) = \varepsilon_0\varepsilon_\mathrm{r}\mathbf{E}(\mathbf{r},t)\,, \tag{2.1.6}$$

and

$$\mathbf{J}(\mathbf{r},t) = \sigma\mathbf{E}(\mathbf{r},t)\,; \tag{2.1.7}$$

here $\varepsilon_0 = 8.85\times10^{-12}$ farad/m and $\mu_0 = 4\pi\times10^{-7}$ henry/m are the vacuum permittivity and permeability respectively. The relative dielectric coefficient ε_r and its permeability counterpart μ_r characterize the material. Expression (2.1.7) is Ohm's law, which relates the current density and the electric field in metals; σ is the conductivity of the material and it is measured in units of $\Omega^{-1}\mathrm{m}^{-1}$. In vacuum obviously $\varepsilon_\mathrm{r}\equiv1$, $\mu_\mathrm{r}\equiv1$ and $\sigma=0$, i.e.,

$$\nabla\times\mathbf{E}(\mathbf{r},t) + \frac{\partial}{\partial t}\mu_0\mathbf{H}(\mathbf{r},t) = 0\,, \tag{2.1.8}$$

$$\nabla\times\mathbf{H}(\mathbf{r},t) - \frac{\partial}{\partial t}\varepsilon_0\mathbf{E}(\mathbf{r},t) = \mathbf{J}(\mathbf{r},t)\,, \tag{2.1.9}$$

$$\nabla\cdot\varepsilon_0\mathbf{E}(\mathbf{r},t) = \varrho(\mathbf{r},t)\,, \tag{2.1.10}$$

$$\nabla\cdot\mu_0\mathbf{H}(\mathbf{r},t) = 0\,. \tag{2.1.11}$$

Assuming that we know the source terms (ϱ and $\mathbf{J}$) it is sufficient to use the first two equations (2.1.8–9) in conjunction with the charge conservation,

$$\nabla\cdot\mathbf{J}(\mathbf{r},t) + \frac{\partial}{\partial t}\varrho(\mathbf{r},t) = 0\,, \tag{2.1.12}$$

in order to solve the electromagnetic field. This statement can be examined by applying $\nabla\cdot$ on both (2.1.8) and (2.1.9). Since any vector function $\mathbf{V}$ satisfies $\nabla\cdot(\nabla\times\mathbf{V})\equiv0$, one obtains (2.1.11) from (2.1.8) and (2.1.10) from (2.1.9).

For an electron beam the constitutive relations cannot in general be expressed as the simple relations in (2.1.5–7). Instead, they are solutions of the electrons' equations of motion as will be shown. In this chapter we shall consider only linear constitutive relations and we present solutions of Maxwell's equations for known sources. These will be either a stationary dipole oscillating at a known frequency or a charged particle moving at a constant velocity.

2.1.2 Boundary Conditions

In cases of sharp discontinuities in the properties of the medium one finds it convenient to define the boundary conditions associated with such a jump. Consider two regions (subscripts 1 and 2) separated by a surface which is locally characterized by its normal $\mathbf{n}$. The boundary condition associated with (2.1.1) is given by

$$\mathbf{n} \times (\mathbf{E}_1 - \mathbf{E}_2) = 0 . \tag{2.1.13}$$

Similarly, from the integral form of (2.1.2) we conclude that

$$\mathbf{n} \times (\mathbf{H}_1 - \mathbf{H}_2) = \mathbf{J}_\mathrm{s} , \tag{2.1.14}$$

from (2.1.3)

$$\mathbf{n} \cdot (\mathbf{D}_1 - \mathbf{D}_2) = \varrho_\mathrm{s} , \tag{2.1.15}$$

and finally, from the integral form of (2.1.4) we can deduce that

$$\mathbf{n} \cdot (\mathbf{B}_1 - \mathbf{B}_2) = 0 . \tag{2.1.16}$$

Here $\mathbf{J}_\mathrm{s}$ is the surface current density and ϱ_s is the surface charge density.

Equation (2.1.13) indicates that the the tangential component of the electric field, at any time, has to be continuous at the transition between two discontinuities. In a similar way, the tangential component of the magnetic field can be discontinuous only if there is a surface current($\mathbf{J}_\mathrm{s}$) which can account for the field discontinuity [see (2.1.14)]. The other two expressions indicate that any discontinuity in the normal component of the electric induction is due to surface charge and the normal component of the magnetic induction is always continuous. As in the case of the Maxwell's equations, it is sufficient to use the first two sets of boundary conditions since the latter two are then automatically satisfied. An outcome of the boundary conditions as formulated above is that at the surface of an ideal metal ($\sigma \to \infty$) the tangential electric field vanishes. This can be readily understood based on the fact that the electric field is zero in the metal and the tangential electric field has to be continuous.

2.1.3 Poynting's Theorem

The energy conservation associated with the electromagnetic field can be deduced from Maxwell's equations by multiplying (scalarly) (2.1.1) by $\mathbf{H}$, (2.1.2) by $\mathbf{E}$ and subtracting the latter from the former. In a linear medium, the result reads

$$\nabla \cdot \mathbf{S} + \frac{\partial}{\partial t}\left[\frac{1}{2}\varepsilon_0\varepsilon_\mathrm{r}\mathbf{E}\cdot\mathbf{E} + \frac{1}{2}\mu_0\mu_\mathrm{r}\mathbf{H}\cdot\mathbf{H}\right] = -\mathbf{J}\cdot\mathbf{E} , \tag{2.1.17}$$

where

$$\mathbf{S}(\mathbf{r},t) \equiv \mathbf{E}(\mathbf{r},t) \times \mathbf{H}(\mathbf{r},t) \tag{2.1.18}$$

is the instantaneous Poynting vector which represents the energy flux (power per unit surface) in the vector direction. The second term,

$$W(\mathbf{r},t) \equiv \frac{1}{2}\varepsilon_0\varepsilon_r\mathbf{E}(\mathbf{r},t)\cdot\mathbf{E}(\mathbf{r},t) + \frac{1}{2}\mu_0\mu_r\mathbf{H}(\mathbf{r},t)\cdot\mathbf{H}(\mathbf{r},t), \tag{2.1.19}$$

represents the instantaneous energy density stored in the electric and magnetic field respectively. And the right-hand side term (in (2.1.17)) represents the coupling between the electromagnetic field and the sources in the system.

Gauss's theorem can be used to formulate Poynting's theorem in its integral form. We integrate over a volume V whose boundary is denoted by $\mathbf{a}$; the result is

$$\int_V \mathrm{d}V\nabla\cdot\mathbf{S} \equiv \int_a \mathrm{d}\mathbf{a}\cdot\mathbf{S} = -\frac{\mathrm{d}}{\mathrm{d}t}\mathrm{w}(t) - \int_V \mathrm{d}V\mathbf{J}\cdot\mathbf{E}, \tag{2.1.20}$$

where for a linear medium

$$\mathrm{w}(t) \equiv \int_V \mathrm{d}V\left[\frac{1}{2}\varepsilon_0\varepsilon_r\mathbf{E}\cdot\mathbf{E} + \frac{1}{2}\mu_0\mu_r\mathbf{H}\cdot\mathbf{H}\right], \tag{2.1.21}$$

is the total energy stored in the volume V.

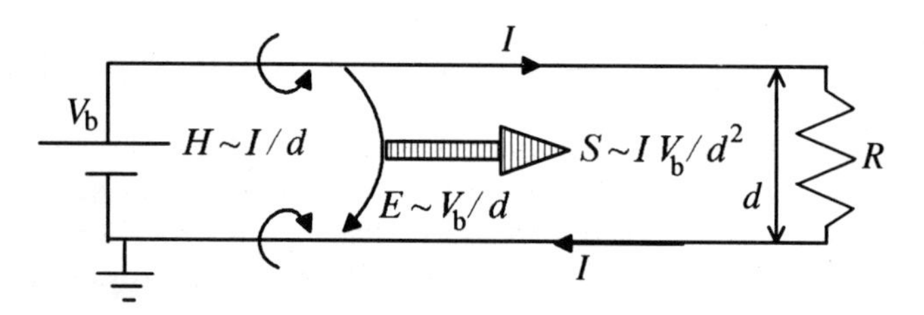

Fig. 2.1. Energy flow in a simple circuit. The power flows in the air and is guided by the wires

One important point to emphasize at this stage is that the electromagnetic power is carried by the field and not by the metallic boundaries; the latter only guide the energy flow. This is important since we shall later discuss propagation of electromagnetic waves and electron beams of hundreds of megawatts and all this power propagates in vacuum (assuming lossless walls). To illustrate the process let us consider an elementary electric circuit consisting of a battery, two parallel and lossless wires, and a resistor at the end – as illustrated in Fig. 2.1. Firstly, we examine the left term of (2.1.20): the voltage V_b is determined by the battery whereas the current by the resistor (R) namely, $I = V_b/R$. Since the distance between the two wires is d, the typical electric field between the two wires is V_b/d; furthermore, the azimuthal magnetic field generated by one wire at the location of the other is of the order of I/d. Consequently, the Poynting vector is parallel to the wires and it is proportional to the product of the two field components $S \propto IV_\mathrm{b}/d^2$. The power which propagates from the battery towards the resistor is proportional to Poynting vector and the effective area of propagation. In this over-simplified model the only length parameter is d, therefore this area is expected to be quadratic in d which entails that the power is proportional to $V_\mathrm{b} * I$ or V_b^2/R. In the right-hand side of (2.1.20) the first term is identically zero since there are no time variations and the second term can be readily calculated to show that the power dissipated in the resistor is V_b^2/R. For further discussion see Chap. 11 in the book by Haus and Melcher (1989).

2.1.4 Steady-State Regime

In many cases of interest all the components of the electromagnetic field oscillate at a single angular frequency (ω) thus all components have the following functional form

$$F(\mathbf{r},t) = f(\mathbf{r}) \cos\Big[\omega t + \psi(\mathbf{r})\Big] . \tag{2.1.22}$$

It is convenient to omit the time dependence and represent the function $F(\mathbf{r},t)$ using a complex notation, namely we introduce the imaginary number $j \equiv \sqrt{-1}$ and utilize the fact that $\mathrm{e}^{j\xi} \equiv \cos(\xi) + j\sin(\xi)$ or $\cos(\xi) \equiv \frac{1}{2}(\mathrm{e}^{j\xi} + \mathrm{e}^{-j\xi})$ and $\sin(\xi) \equiv \frac{1}{2j}(\mathrm{e}^{j\xi} - \mathrm{e}^{-j\xi})$ hence

$$F(\mathbf{r},t) = \frac{1}{2}\Big[f(\mathbf{r})\,\mathrm{e}^{j\psi(\mathbf{r})}\,\mathrm{e}^{j\omega t} + f(\mathbf{r})\,\mathrm{e}^{-j\psi(\mathbf{r})}\,\mathrm{e}^{-j\omega t}\Big] . \tag{2.1.23}$$

With this notation it is convenient to define

$$\bar{F}(\mathbf{r},\omega) \equiv f(\mathbf{r})\,\mathrm{e}^{j\psi(\mathbf{r})} , \tag{2.1.24}$$

which permits us to use this function instead of $F(\mathbf{r},t)$ and consequently,

$$F(\mathbf{r},t) = \mathrm{Re}\left[\bar{F}(\mathbf{r},\omega)\mathrm{e}^{j\omega t}\right] ; \tag{2.1.25}$$

$\bar{F}(\mathbf{r},\omega)$ is called the phasor associated with the function $F(\mathbf{r},t)$. To illustrate the use of this notation, Maxwell's equations read

$$\nabla \times \bar{\mathbf{E}} + j\omega\bar{\mathbf{B}} = 0 , \tag{2.1.26}$$

$$\nabla \times \bar{\mathbf{H}} - j\omega\bar{\mathbf{D}} = \bar{\mathbf{J}} , \tag{2.1.27}$$

$$\nabla \cdot \bar{\mathbf{D}} = \bar{\varrho} , \tag{2.1.28}$$

$$\nabla \cdot \bar{\mathbf{B}} = 0 . \tag{2.1.29}$$

The main advantage of this notation is now evident since the differential operator $\partial/\partial t$ was replaced by a simple algebric operator $j\omega$.

2.1.5 Complex Poynting's Theorem

The phasor notation, as introduced above, cannot be directly applied to Poynting's theorem since all quantities are quadratic in the electromagnetic field. In principle we have two options: *(i)* transform the field components to the time domain and then substitute in Poynting's theorem as defined in (2.1.17) – abandoning in the process the phasor notation. *(ii)* Limit the information to the average energy and power – but preserving the phasor notation. Since in the former case there is no real advantage to the phasor notation, we shall next pursue the latter option.

When we consider the product of two oscillating quantities we have

$$A_1 \cos(\omega t + \psi_1) A_2 \cos(\omega t + \psi_2) = \frac{1}{2}\left[\bar{A}_1 e^{j\omega t} + \bar{A}_1^* e^{-j\omega t}\right] \frac{1}{2}\left[\bar{A}_2 e^{j\omega t} + \bar{A}_2^* e^{-j\omega t}\right] . \tag{2.1.30}$$

The *average* of the product of these two oscillating functions corresponds to the non-oscillating term in the expression above i.e.,

$$\frac{1}{4}\left[\bar{A}_1 \bar{A}_2^* + \bar{A}_1^* \bar{A}_2\right] = \frac{1}{2} A_1 A_2 \cos(\psi_1 - \psi_2) . \tag{2.1.31}$$

We shall now use this fact in order to formulate the complex Poynting's theorem. We multiply (scalarly) (2.1.26) by the complex conjugate of the magnetic field phasor ($\bar{\mathbf{H}}^*$). From the product we subtract the complex conjugate of (2.1.27) multiplied by the electric field; the result reads

$$\nabla \cdot \bar{\mathbf{S}} + 2j\omega \left[\bar{W}_{\mathrm{M}} - \bar{W}_{\mathrm{E}}\right] = -\frac{1}{2} \bar{\mathbf{E}} \cdot \bar{\mathbf{J}}^* , \tag{2.1.32}$$

where

$$\bar{\mathbf{S}} = \frac{1}{2} \bar{\mathbf{E}} \times \bar{\mathbf{H}}^* , \tag{2.1.33}$$

is the complex Poynting vector,

$$\bar{W}_{\mathrm{M}} = \frac{1}{4} \mu_0 \mu_{\mathrm{r}} \bar{\mathbf{H}} \cdot \bar{\mathbf{H}}^* , \tag{2.1.34}$$

is the magnetic energy density and

$$\bar{W}_{\mathrm{E}} = \frac{1}{4} \varepsilon_0 \varepsilon_{\mathrm{r}} \bar{\mathbf{E}} \cdot \bar{\mathbf{E}}^* , \tag{2.1.35}$$

is the electric energy density, both averaged in time.

Energy conversion is associated with the real part of the Poynting vector whereas the imaginary part is associated with electro-magnetic energy stored in the system. Throughout the text we shall omit the bar from the phasor quantities, except if ambiguities may occur.

2.1.6 Potentials

It is convenient, instead of solving a couple of first order differential equations, to solve a single second-order differential equation. For this purpose we take advantage of the fact that the divergence of the magnetic induction is zero ($\nabla \cdot \mathbf{B} = 0$) and introduce the magnetic vector potential $\mathbf{A}$ which determines the magnetic induction through

$$\mathbf{B} = \nabla \times \mathbf{A}, \tag{2.1.36}$$

and by virtue of which the former equation becomes an identity. Substituting this definition in (2.1.26) we obtain

$$\nabla \times (\mathbf{E} + j\omega \mathbf{A}) = 0. \tag{2.1.37}$$

Using the fact that $\nabla \times (\nabla \Phi) \equiv 0$ we conclude that

$$\mathbf{E} = -j\omega \mathbf{A} - \nabla \Phi, \tag{2.1.38}$$

where Φ is the scalar electric potential.

The two potentials we have just introduced satisfy, in a cartesian coordinate system and in a linear medium ($\mu_{\rm r} = 1$ and $\varepsilon_{\rm r} > 1$), the non-homogeneous wave equation:

$$\left[\nabla^2 + \varepsilon_{\rm r}\frac{\omega^2}{c^2}\right] \mathbf{A} = -\mu_0 \mathbf{J}, \tag{2.1.39}$$

and

$$\left[\nabla^2 + \varepsilon_{\rm r}\frac{\omega^2}{c^2}\right] \Phi = -\frac{1}{\varepsilon_0 \varepsilon_{\rm r}}\varrho, \tag{2.1.40}$$

provided that the divergence of the vector function $\mathbf{A}$ is chosen to be

$$\nabla \cdot \mathbf{A} + j\omega \frac{\varepsilon_{\rm r}}{c^2}\Phi = 0. \tag{2.1.41}$$

This is the so-called Lorentz gauge. Consistency with the special theory of relativity dictates this choice over the well-known Coulomb gauge; $c \equiv 1/\sqrt{\mu_0 \varepsilon_0}$ is the phase velocity of a plane electromagnetic wave in vacuum.

2.2 Simple Wave Phenomena

The wave equation developed in the previous section for the scalar electric potential and the three components of the magnetic vector potential will be next solved for several simple cases. A few of the examples presented here will be used later in this text to develop models which in turn enable the investigation of complex structures. In the context of these examples we formulate the radiation condition.

2.2.1 Simple Propagating Waves

With the source terms, constitutive relations and boundary conditions determined, one can proceed towards solution of a few simple wave phenomena. For simplicity we shall consider a scalar function $\psi(\mathbf{r})$ which oscillates at an angular frequency ω (i.e., we assume a steady-state regime of the form $\mathrm{e}^{j\omega t}$) and which is a solution of

$$\left[\nabla^2 + \frac{\omega^2}{c^2}\right]\psi(\mathbf{r}) = 0\,. \tag{2.2.1}$$

As a first stage, waves propagating in one dimension are examined. In a cartesian system (x, y, z) we consider a system in which all variations are only in the z direction ($\partial/\partial x \sim 0$ and $\partial/\partial y \sim 0$), and the homogeneous wave equation reads

$$\left[\frac{\mathrm{d}^2}{\mathrm{d}z^2} + \frac{\omega^2}{c^2}\right]\psi(z) = 0\,. \tag{2.2.2}$$

A second order differential equation, has two solutions:

$$\psi(z) = A_+\mathrm{e}^{-j(\omega/c)z} + A_-\mathrm{e}^{j(\omega/c)z}\,; \tag{2.2.3}$$

these represent plane waves since the phase is constant in the plane defined by z =const. The first term corresponds to a wave propagating in the z direction and the second propagates in the opposite direction.

In a cylindrical coordinate system (r, ϕ, z), ignoring azimuthal and longitudinal variations ($\partial^2/\partial\phi^2 \sim 0$ and $\partial^2/\partial z^2 \sim 0$), the wave equation reads

$$\left[\frac{1}{r}\frac{\mathrm{d}}{\mathrm{d}r}r\frac{\mathrm{d}}{\mathrm{d}r} + \frac{\omega^2}{c^2}\right]\psi(r) = 0\,. \tag{2.2.4}$$

Its solution is

$$\psi(r) = A_+\mathrm{H}_0^{(2)}\left(\frac{\omega}{c}r\right) + A_-\mathrm{H}_0^{(1)}\left(\frac{\omega}{c}r\right), \tag{2.2.5}$$

where $\mathrm{H}_0^{(1)}(\xi)$ and $\mathrm{H}_0^{(2)}(\xi)$ are the zero order Hankel function of the first and second kind; they are related to Bessel functions of the first and second kind by $\mathrm{H}_0^{(1)}(x) \equiv \mathrm{J}_0(x) + j\mathrm{Y}_0(x)$ and $\mathrm{H}_0^{(2)}(x) \equiv \mathrm{J}_0(x) - j\mathrm{Y}_0(x)$. As in the previous case, the first term represents the wave which is propagating from the axis outwards and the second term describes a wave propagating inwards. For completeness we shall also present the solution in a spherical coordinate system (r, ϕ, θ). Ignoring all angular variations the wave equation is given by

$$\left[\frac{1}{r}\frac{\mathrm{d}^2}{\mathrm{d}r^2}r + \frac{\omega^2}{c^2}\right]\psi(r) = 0\,, \tag{2.2.6}$$

and its solution is

$$\psi(r) = A_+\frac{1}{(\omega/c)r}\mathrm{e}^{-j(\omega/c)r} + A_-\frac{1}{(\omega/c)r}\mathrm{e}^{j(\omega/c)r}\,, \tag{2.2.7}$$

where the first term represents a spherical wave propagating outwards (from the center out) whereas the second represents an inward flow.

2.2.2 The Radiation Condition

From the pure mathematical point of view, the two waves in each one of the solutions of above are a direct result of the fact that the wave equation is a second order differential equation. However, in absence of obstacles, our daily experience dictates a wave which propagates from the source outwards; this implies that in all three cases there are no "advanced" waves i.e., $A_- \equiv 0$. This is one possible interpretation of the so-called the *radiation condition* and it can be considered an additional boundary condition which is a byproduct of the causality constraint imposed on the solutions of the wave equation.

This formulation relies on the simple solutions presented above, however the general trend is valid for more complex solutions. In the case of cylindrical azimuthally non-symmetric waves, the radiation condition implies for a solution $\psi(r, \phi, z,)$ that the limit

$$\left[\psi(r, \phi, z)\, \mathrm{e}^{j(\omega/c)r}\, r^{1/2}\right]_{r\to\infty}\,, \tag{2.2.8}$$

is finite and it is r independent. In a similar way, for spherical waves described by a function $\psi(r, \phi, \theta)$, the limit

$$\left[\psi(r, \phi, \theta)\, \mathrm{e}^{j(\omega/c)r} r\right]_{r\to\infty}\,, \tag{2.2.9}$$

is finite and r independent.

Advanced solutions of the wave equation have been used by Wheeler and Feynmann (1945) in order to explain the source of the so-called *radiation reaction force*. It is well known that electromagnetic power is emitted by a particle when it is accelerated. This power is emitted from the particle outwards and comes at the expense of its kinetic energy. Since this change in the kinetic energy of the particle can be conceived as an effective force this is also referred to as the radiation reaction force. However, this force is negligible for all regimes discussed throughout this text therefore will not be included in the electron dynamics considered in the next chapter.

2.2.3 Evanescent Waves

So far we have presented only waves which vary and propagate in one dimension, namely solutions of the wave equation either in a cartesian, cylindrical or spherical system of coordinate. We increase now the complexity of the analysis and examine waves which vary in two dimensions. First consider a cartesian coordinate system in which we ignore variations in the y direction. The wave equation in this case reads

$$\left[\frac{\partial^2}{\partial x^2} + \frac{\partial^2}{\partial z^2} + \frac{\omega^2}{c^2}\right] \psi(x, z, \omega) = 0 , \tag{2.2.10}$$

and the formal solution is given by

$$\psi(x, z, \omega) = \mathrm{e}^{-jkz} \left[A_+ \mathrm{e}^{-\sqrt{k^2-(\omega/c)^2}\, x} + A_- \mathrm{e}^{\sqrt{k^2-(\omega/c)^2}\, x}\right] . \tag{2.2.11}$$

However in the half-plane defined by $x > 0$ the solution is

$$\psi(x, z, \omega) = A_+ \, \mathrm{e}^{-jkz} \mathrm{e}^{-\sqrt{k^2-(\omega/c)^2}\, x} , \tag{2.2.12}$$

since otherwise the solution diverges at $x \to \infty$. For $|k|c > \omega$ the wave decays exponentially in the x direction. This is an evanescent wave: it propagates in one direction and decays exponentially in another. In the opposite case, for $|k|c < \omega$, the wave propagates at an angle $\theta = \cos^{-1}(kc/\omega)$ relative to the z axis.

It is instructive to examine (2.2.12) in the time domain. Assuming zero phase for A_+ then

$$\psi(x, z, t) = A_+ \cos(\omega t - kz) \mathrm{e}^{-\sqrt{k^2-(\omega/c)^2}\, x} . \tag{2.2.13}$$

Based on this solution it is convenient to introduce the concept of phase velocity: this is the velocity at which an imaginary observer has to move in order to measure a constant phase ($\omega t - kz = \mathrm{const}$). Explicitly it reads

$$v_{\mathrm{ph}} \equiv \frac{\omega}{k} . \tag{2.2.14}$$

With this definition, we observe that in a two dimensional case, an evanescent wave is characterized by a phase velocity smaller than c.

2.2.4 Waves of a Moving Charge

Evanescent waves play an important role in the interaction process of particles and waves. The simplest manifestation of their role is the representation of the spectrum of a moving charge in the laboratory frame of reference. For this purpose we shall examine now the waves associated with a point charge (e) moving in the z direction at a constant velocity v_0 in vacuum; no boundaries are involved and the system is azimuthally symmetric ($\partial/\partial\phi = 0$). The current distribution in this case is given by

$$\mathbf{J}(\mathbf{r}, t) = -ev_0 \frac{1}{2\pi r} \delta(r)\, \delta(z - v_0 t)\, \mathbf{1}_z , \tag{2.2.15}$$

where $\mathbf{1}_z$ is a unit vector in the z direction. This current distribution excites the z component of the magnetic vector potential which in turn satisfies

$$\left[\frac{1}{r}\frac{\partial}{\partial r} r \frac{\partial}{\partial r} + \frac{\partial^2}{\partial z^2} - \frac{1}{c^2}\frac{\partial^2}{\partial t^2}\right] A_z(r, z, t) = -\mu_0 J_z(r, z, t) ; \tag{2.2.16}$$

its solution is assumed to have the form

$$A_z(r, z, t) = \int_{-\infty}^{\infty} \mathrm{d}\omega \mathrm{e}^{j\omega t} \int_{-\infty}^{\infty} \mathrm{d}k\, \mathrm{e}^{-jkz}\, a_z(r, k, \omega) , \tag{2.2.17}$$

where $a_z(r, k, \omega)$ satisfies

$$\left[\frac{1}{r}\frac{\mathrm{d}}{\mathrm{d}r} r \frac{\mathrm{d}}{\mathrm{d}r} - \Gamma^2\right] a_z(r, k, \omega) = \frac{ev_0\mu_0}{(2\pi)^2 r} \delta(r)\, \delta(\omega - kv_0) , \tag{2.2.18}$$

and

$$\Gamma^2 = k^2 - \frac{\omega^2}{c^2} . \tag{2.2.19}$$

Off-axis the solution of this equation is

$$a_z(r, k, \omega) = A_+(k, \omega)\, \mathrm{K}_0(\Gamma r) , \tag{2.2.20}$$

where $\mathrm{K}_0(\xi)$ is the zero order modified Bessel function of the second kind. In order to determine the amplitude A_+ there are two ways to proceed: (*i*) calculate the azimuthal magnetic field and then impose the boundary conditions at $r = 0$. An alternative way is to (*ii*) integrate (2.2.18) from $r = 0$ to $r = \delta \to 0$. At this point we shall prefer the latter primarily because this approach will be utilized extensively in Chaps. 5 and 6. Since for small arguments the modified Bessel function behaves as $\mathrm{K}_0(\xi) \simeq -\ln(\xi)$ [see Abramowitz and Stegun (1968) p. 375] then

$$A_+(k, \omega) = -\frac{ev_0\mu_0}{(2\pi)^2} \delta(\omega - kv_0) . \tag{2.2.21}$$

Substituting this result in (2.2.17,20) we obtain

$$A_z(r,z,t) = -\frac{e\mu_0}{(2\pi)^2}\int_{-\infty}^{\infty} d\omega\, e^{j\omega(t-z/v_0)}\, K_0\left(\frac{\omega}{c}r\frac{1}{\gamma\beta}\right), \tag{2.2.22}$$

where $\beta = v_0/c$ and $\gamma = [1-\beta^2]^{-1/2}$. Using the Lorentz gauge one can determine the scalar electric potential

$$\Phi(r,z,t) = -\frac{e}{(2\pi)^2\varepsilon_0 v_0}\int_{-\infty}^{\infty} d\omega e^{j\omega(t-z/v_0)} K_0\left(\frac{\omega}{c}r\frac{1}{\gamma\beta}\right). \tag{2.2.23}$$

This expression indicates that the field associated with a moving charge is a superposition of cylindrical evanescent waves [for large arguments the modified Bessel function decays exponentially following $K_0(\xi) \simeq e^{-\xi}\sqrt{\pi/2\xi}$ Abramowitz and Stegun (1968) p. 378]. There is no electromagnetic *average* power emitted by this particle in the radial direction however, this average power is non-zero in the direction parallel to the particle's motion. When scattered by periodic structures, the evanescent waves can be "converted" into propagating waves as we shall see when the Smith-Purcell effect will be discussed in Chap. 5.

2.3 Guided Waves

In all the solutions presented above, no boundaries were involved, while in many of the topics to be considered, the electromagnetic wave is guided by a metallic structure. In addition to the injection of electromagnetic power into the system, metallic structures facilitate the interaction process itself and ultimately, they allow extraction of the power out of the system. As we shall later see the three processes are inter-dependent but at this stage only the simplest configurations will be considered.

2.3.1 Transverse Electromagnetic Mode

The simplest mode which may develop when two metallic surfaces are present is the transverse electro-magnetic (TEM) mode. In conjunction with the electromagnetic field generated by a moving charge let us consider a radial transmission line which consists of two parallel lossless plates; the distance between the plates is denoted by d and it is much smaller than the (vacuum) wavelength i.e., $\lambda(\equiv 2\pi c/\omega) \gg d$. Subject to this condition, we ignore the longitudinal variations ($\partial^2/\partial z^2 \simeq 0$) therefore for an azimuthally symmetric system the wave equation reads

$$\left[\frac{1}{r}\frac{d}{dr}r\frac{d}{dr} + \frac{\omega^2}{c^2}\right] A_z(r,\omega) = -\mu_0 J_z(r,\omega). \tag{2.3.1}$$

A dipole which oscillates at an angular frequency, ω, is located on axis and is represented by the following current density

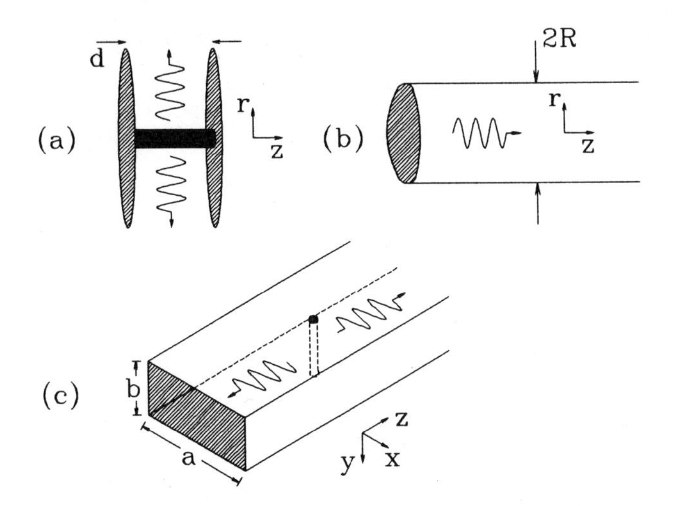

Fig. 2.2. (**a**) Propagation of transverse electro-magnetic mode in a radial transmission line $\lambda \gg d$. (**b**) Propagation of a transverse magnetic (TM) mode in a circular waveguide – see Sect. 2.3.2. (**c**) Propagation of transverse electric (TE) mode in rectangular waveguide – Sect. 2.3.4

$$J_z(r,\omega) = I\frac{1}{2\pi r}\delta(r). \tag{2.3.2}$$

Figure 2.2a illustrates schematically the system under consideration. A solution of the homogeneous wave equation which satisfies the radiation condition is given by

$$A_z(r,\omega) = A_+ \mathrm{H}_0^{(2)}\left(\frac{\omega}{c}r\right), \tag{2.3.3}$$

and A_+ is determined by the discontinuity at $r = 0$. Integrating (2.3.1) in the close vicinity of $r = 0$,

$$\left[r\frac{\mathrm{d}}{\mathrm{d}r}A_z(r,\omega)\right]_{r=0+} = -\frac{\mu_0}{2\pi}I, \tag{2.3.4}$$

and using the expression for Hankel function for small arguments i.e., $\mathrm{H}_0^{(2)}(x) \simeq -j\ln(x)2/\pi$ [Abramowitz and Stegun (1968) p. 360], we obtain $A_+ = -jI\mu_0/4$.

The longitudinal component of the electric field and the azimuthal counterpart of the magnetic field are

$$E_z(r,\omega) = -j\omega A_z(r,\omega) = -j\omega A_+ \mathrm{H}_0^{(2)}\left(\frac{\omega}{c}r\right),$$

$$H_\phi(r,\omega) = -\frac{1}{\mu_0}\frac{\mathrm{d}}{\mathrm{d}r}A_z(r,\omega) = \frac{1}{\mu_0}\frac{\omega}{c}A_+ \mathrm{H}_1^{(2)}\left(\frac{\omega}{c}r\right), \tag{2.3.5}$$

respectively. With these two components, the radial component of the Poynting vector is

$$S_r(r) = -\frac{1}{2}E_z(r)H_\phi^*(r)\,, \tag{2.3.6}$$

and the total power radiated is

$$P = \mathrm{Re}\left[2\pi r d S_r(r)\right] = \frac{1}{8}\left(\frac{\omega}{c}d\right)\eta_0 I^2\,. \tag{2.3.7}$$

In the last expression, we used the asymptotic approximation for large arguments of Hankel function i.e., $\mathrm{H}_0^{(2)}(x) \simeq \mathrm{e}^{-jx}\sqrt{2/\pi x}$ [see Abramowitz and Stegun (1968) p. 364]. The impedance associated with the radiation process is

$$R_{\mathrm{rad,TEM}} \equiv \frac{P}{\frac{1}{2}I^2} = \frac{1}{4}\eta_0\left(\frac{\omega}{c}d\right)\,; \tag{2.3.8}$$

in this expression $\eta_0 \equiv \sqrt{\mu_0/\varepsilon_0}$ is the vacuum impedance of a plane wave. At 9 GHz and for $d = 5$mm the impedance is 90 Ω which is 5 times larger (for the same parameters) than the radiation impedance in free space defined as $R_{\mathrm{rad}} = \eta_0\left(\omega d/c\right)^2/6\pi$ i.e., 18 Ω. The radiation impedance is a measure, extensively used in antenna theory, which represents the effect of the surroundings on the radiation emitted by a source. This can be a single dipole or a distribution of dipoles as is the case for a bunched beam, therefore this impedance can be used as a direct measure of the power extracted from electrons.

2.3.2 Transverse Magnetic Mode

Transverse magnetic (TM) modes can develop in the radial system discussed previously and their characteristics will be further investigated in Chap. 4, in the context of periodic structures. Here we review the characteristics of these modes for a circular cylindrical waveguide of radius R filled with a dielectric material of relative permittivity ε_{r}; the relative permeability is taken to unity ($\mu_r = 1$). The walls of the waveguide are assumed to be made of an ideal conducting material ($\sigma \to \infty$) therefore the tangential electric field at the walls vanishes. To this configuration, we attach a cylindrical system of coordinates (r, ϕ, z) – see Fig. 2.2b and the waves are assumed to be excited by an azimuthally symmetric source thus we may take $\partial/\partial\phi = 0$.

The electromagnetic field in the waveguide has two contributions. One is from the z component of the magnetic vector potential

$$A_z(r, z, \omega) = \sum_{s=1}^{\infty} A_s \mathrm{J}_0\left(p_s \frac{r}{R}\right)\mathrm{e}^{-\Gamma_s z}\,, \tag{2.3.9}$$

where

$$\Gamma_s^2 = \frac{p_s^2}{R^2} - \varepsilon_\mathrm{r}\frac{\omega^2}{c^2}\,, \tag{2.3.10}$$

$\mathrm{J}_0(\xi)$ is the zero order Bessel function of the first kind and p_s are the zeros of this function ($p_1 = 2.4048$, $p_2 = 5.52\ldots$). The second, is from the scalar electric potential Φ

$$\Phi(r,z,\omega) = \sum_{s=1}^{\infty} \Phi_s \mathrm{J}_0\left(p_s \frac{r}{R}\right) \mathrm{e}^{-\Gamma_s z}\,. \tag{2.3.11}$$

Lorentz gauge [(2.1.41)] correlates the two amplitudes, namely

$$\Phi_s = \frac{c^2 \Gamma_s}{j\omega\varepsilon_\mathrm{r}} A_s\,. \tag{2.3.12}$$

In this solution the waves are assumed to propagate from the source without obstacles thus no reflected waves were included.

The three non-trivial components of the electromagnetic field are: the azimuthal magnetic field

$$\begin{aligned} H_\phi(r,z,\omega) &= -\frac{1}{\mu_0}\frac{\partial}{\partial r} A_z(r,z,\omega) \\ &= \frac{1}{\mu_0}\sum_{s=1}^{\infty} A_s \frac{p_s}{R} \mathrm{J}_1\left(p_s \frac{r}{R}\right) \mathrm{e}^{-\Gamma_s z}\,, \end{aligned} \tag{2.3.13}$$

the radial electric field

$$\begin{aligned} E_r(r,z,\omega) &= -\frac{\partial}{\partial r}\Phi(r,z,\omega) \\ &= \sum_{s=1}^{\infty} A_s \frac{c^2\Gamma_s}{j\omega\varepsilon_\mathrm{r}} \frac{p_s}{R} \mathrm{J}_1\left(p_s \frac{r}{R}\right) \mathrm{e}^{-\Gamma_s z}\,, \end{aligned} \tag{2.3.14}$$

and the longitudinal electric field

$$\begin{aligned} E_z(r,z,\omega) &= -\frac{\partial}{\partial z}\Phi(r,z,\omega) - j\omega A_z(r,z,\omega) \\ &= \sum_{s=1}^{\infty} A_s \frac{c^2}{j\omega\varepsilon_\mathrm{r}} \left(\frac{p_s}{R}\right)^2 \mathrm{J}_0\left(p_s \frac{r}{R}\right) \mathrm{e}^{-\Gamma_s z}\,. \end{aligned} \tag{2.3.15}$$

With the electromagnetic field determined the average magnetic and electric energy per unit length can be calculated. These are given by

$$W_{\rm M} = \frac{1}{4}\mu_0 \int_0^R {\rm d}r(2\pi r)|H_\phi|^2$$
$$= \frac{\pi}{2\mu_0}\sum_{s=1}^{\infty}|A_s|^2 \frac{p_s^2}{R^2}\left[\frac{R^2}{2}{\rm J}_1^2(p_s)\right] {\rm e}^{-(\Gamma_s+\Gamma_s^*)z} ,$$

$$W_{\rm E} = \frac{1}{4}\varepsilon_0\varepsilon_{\rm r} \int_0^R {\rm d}r(2\pi r)[|E_r|^2 + |E_z|^2]$$
$$= \frac{\pi}{2}\varepsilon_0\varepsilon_{\rm r}\sum_{s=1}^{\infty}|A_s|^2 \left[\frac{c^4}{\omega^2\varepsilon_{\rm r}^2}\frac{p_s^2}{R^2}\right]\left[\Gamma_s\Gamma_s^* + \frac{p_s^2}{R^2}\right]$$
$$\times \left[\frac{R^2}{2}{\rm J}_1^2(p_s)\right] {\rm e}^{-(\Gamma_s+\Gamma_s^*)z} . \tag{2.3.16}$$

In these expressions the orthogonality of the Bessel functions was used i.e.,

$$\int_0^R {\rm d}r\, r\, {\rm J}_0\left(p_s\frac{r}{R}\right){\rm J}_0\left(p_{s'}\frac{r}{R}\right) = \frac{1}{2}R^2{\rm J}_1^2(p_s)\delta_{s,s'} . \tag{2.3.17}$$

In a similar way we can determine the total average power which flows in the waveguide:

$$P = {\rm Re}\left[2\pi \int_0^R {\rm d}r r\frac{1}{2}E_r H_\phi^*\right]$$
$$= \frac{\pi}{\mu_0}\sum_{s=1}^{\infty}|A_s|^2 \frac{p_s^2}{R^2}\left[\frac{R^2}{2}{\rm J}_1^2(p_s)\right] {\rm Re}\left[{\rm e}^{-(\Gamma_s+\Gamma_s^*)z}\frac{c^2\Gamma_s}{j\omega\varepsilon_{\rm r}}\right] . \tag{2.3.18}$$

According to this expression we observe that power is carried along the waveguide only by the propagating modes namely those which satisfy

$$\Gamma_s^2 = \frac{p_s^2}{R^2} - \varepsilon_{\rm r}\frac{\omega^2}{c^2} < 0 . \tag{2.3.19}$$

The remainder are below cut-off and they do not carry any (real) power. The situation is different when reflections are present.

2.3.3 Velocities and Impedances

Energy Velocity. In the context of power flow presented above it is convenient to define several parameters which help to characterize the interaction of waves and electron beams in various configurations. The energy velocity is a measure of the power flow in the system relative to the total energy stored per unit length namely,

$$v_{\rm en} = \frac{P}{W_{\rm M} + W_{\rm E}} . \tag{2.3.20}$$

In a circular cylindrical waveguide with a single propagating mode ($s = 1$), the energy velocity reads

$$v_{\rm en} = c\frac{1}{\varepsilon_{\rm r}}\sqrt{\varepsilon_{\rm r} - \left(\frac{p_1 c}{\omega R}\right)^2} \,. \tag{2.3.21}$$

From the definition of the energy velocity (2.3.20) it is evident that whenever more than one mode propagates in the waveguide the energy velocity is dependent on the *relative* amplitudes of the various modes. Another point which should be emphasized since it will be encountered again later in this text is the fact that even if only one mode propagates but there is a substantial amount of energy stored in the higher modes, the energy velocity will be much slower than indicated by the expression in (2.3.21).

Phase Velocity. A general definition of this quantity was introduced in Sect. 2.2.3 [(2.2.14)]. In a cylindrical waveguide with no dielectric, the phase velocity is always larger than c. However if $\varepsilon_{\rm r} > 1 + (p_1 c/\omega R)^2$ the phase velocity is smaller than c. In fact for high frequencies ($\omega R/c \gg p_1$) the phase velocity is determined entirely by the medium: $v_{\rm ph} \sim c/\sqrt{\varepsilon_{\rm r}}$.

Group Velocity. This is a kinematic quantity indicative of the propagation of a spectrum of waves. To envision the meaning of the group velocity, imagine that a system is fed by two adjacent frequencies $\omega_1 = \omega + \Delta\omega$, $\omega_2 = \omega - \Delta\omega$ and the waves which develop have the form

$$f(z,t) = \cos(\omega_1 t - K_1 z) + \cos(\omega_2 t - K_2 z) \,, \tag{2.3.22}$$

where $K_1 = k + \Delta k$, $K_2 = k - \Delta k$ and $k = \sqrt{(\omega/c)^2 - (p_1/R)^2}$. Explicitly we can now write the expression in (2.3.22) as

$$f(z,t) = 2\cos(\Delta\omega t - \Delta k z)\cos(\omega t - kz) \,. \tag{2.3.23}$$

Assuming that $|\Delta\omega| \ll \omega$, we can consider the first trigonometric function as a slow varying amplitude. As such, we can ask what has to be the velocity of an observer in order to experience a constant *amplitude* i.e., $\Delta\omega\delta t - \Delta k\delta z = 0$; in this case, the answer will be $v_{\rm gr} \equiv \frac{\Delta\omega}{\Delta k}$ or ar the limit of $\Delta\omega \to 0$

$$v_{\rm gr} \equiv \frac{\partial\omega}{\partial k} \,. \tag{2.3.24}$$

If the dielectric coefficient is not frequency dependent, the group velocity of a propagating TM mode is $v_{\rm gr} = c^2 k/\omega\varepsilon_{\rm r}$ and it satisfies

$$v_{\rm gr} v_{\rm ph} = \frac{c^2}{\varepsilon_{\rm r}} \,. \tag{2.3.25}$$

Although this relation is valid only for uniformly filled waveguide it provides information about the general trend in the variation of the group velocity as the (effective) dielectric coefficient changes in partially loaded systems.

Characteristic Impedance. There are several kinds of impedances which can be defined. Two of which will be defined here and a third one will be defined in Chap. 8. The first is basically oriented towards the propagation of the electromagnetic mode in the structure and this is the characteristic impedance which is the ratio between the two transverse components of the field, E_r and H_ϕ, it reads

$$Z_{\mathrm{ch}} \equiv \frac{E_r}{H_\phi} = \eta_0 \frac{c\Gamma_s}{j\omega\varepsilon_{\mathrm{r}}} . \tag{2.3.26}$$

Interaction Impedance. The second impedance is indicative of the electric field which a thin pencil or annular beam experiences as it traverses the waveguide. For this purpose we define the effective longitudinal electric field in the region where the electron beam will be injected. For a pencil beam $(0 \le r \le R_{\mathrm{b}})$ this is given by

$$|E(z)|^2 \equiv \frac{2}{R_{\mathrm{b}}^2} \int_0^{R_{\mathrm{b}}} \mathrm{d}r\, r \left|E_z(r,z,\omega)\right|^2 , \tag{2.3.27}$$

whereas for an annular beam $(R_{\mathrm{b}} - \Delta/2 \le r \le R_{\mathrm{b}} + \Delta/2)$ it reads

$$|E(z)|^2 \equiv \frac{1}{\Delta R_{\mathrm{b}}} \int_{R_{\mathrm{b}}-\Delta/2}^{R_{\mathrm{b}}+\Delta/2} \mathrm{d}r\, r \left|E_z(r,z,\omega)\right|^2 . \tag{2.3.28}$$

For either one of the cases we define the interaction impedance as

$$Z_{\mathrm{int}} \equiv \frac{1}{2} |E(z)|^2 \pi R^2 \frac{1}{P(z)} . \tag{2.3.29}$$

Note that although we are motivated by the presence of a beam of electrons, all the quantities in the definition of the interaction impedance are "cold" quantities namely, they do not account for the presence of the beam. It should be pointed out that the definition introduced here differs from Pierce's (1957) definition, $Z_{\mathrm{int}} = |E|^2/2k^2P$ by the factor k^2 which was replaced by the inverse of the area where the wave propagates, $1/\pi R^2$. This definition is in particular useful in tapered structures where the internal radius of the system is maintained constant but the other geometric parameters may vary in space such that the phase velocity may vary.

For our particular system the interaction impedance reads

$$Z_{\mathrm{int}} = \eta_0 \left[\frac{p_1}{\varepsilon_{\mathrm{r}}} \frac{c}{\omega R}\right]^2 \frac{\mathrm{J}_0^2(p_1 R_{\mathrm{b}}/R) + \mathrm{J}_1^2(p_1 R_{\mathrm{b}}/R)}{\mathrm{J}_1^2(p_1)} \frac{1}{\beta_{\mathrm{en}}} ; \tag{2.3.30}$$

here $\beta_{\mathrm{en}} = v_{\mathrm{en}}/c$ is the normalized energy velocity which in many cases is equal or close to the group velocity (in this particular case it is equal). One may expect to achieve maximum efficiency when the longitudinal electric field $[E(z)]$ experienced by the electron is maximum. Therefore according to the definition in (2.3.29), from the point of view of the beam-wave interaction, the purpose should be to design a structure with the highest interaction impedance. According to (2.3.30) there are three possibilities: (*i*) operate at low frequency, which in many cases is not desirable, (*ii*) have a structure with small radius which might be acceptable or (*iii*) design a structure with low energy (group) velocity. It should be pointed out that these three possibilities are interdependent since for example, the energy velocity depends both on frequency and radius. One possibility to design a low group velocity structure is to have a small radius.

Interaction Dielectric Coefficient. This quantity is indicative of the total average electromagnetic energy stored per unit length and the longitudinal component of the electric field experienced by a thin annular/pencil beam:

$$\varepsilon_{\rm int} \equiv W(z)\left[\frac{1}{2}\varepsilon_0|E(z)|^2\pi R^2\right]^{-1} . \qquad (2.3.31)$$

In our particular case it reads

$$\varepsilon_{\rm int} = \left[\frac{\varepsilon_{\rm r}}{p_1}\frac{\omega}{c}R\right]^2 \frac{{\rm J}_1^2(p_1)}{{\rm J}_0^2(p_1 R_{\rm b}/R)+{\rm J}_1^2(p_1 R_{\rm b}/R)} . \qquad (2.3.32)$$

Note that according to the definitions of the interaction impedance (2.3.29) and the effective dielectric coefficient (2.3.31) their product is inversely proportional to the energy velocity:

$$Z_{\rm int}\varepsilon_{\rm int} = \eta_0\frac{1}{\beta_{\rm en}} . \qquad (2.3.33)$$

Since the definitions above (2.3.29) and (2.3.31) are general, as long as there is only one dominant mode in the system, the result in the last expression is also general.

2.3.4 Transverse Electric Mode

In many cases, electromagnetic power is transferred along a waveguide in the transverse electric (TE) mode due to its low loss [Ramo, Whinnery and Van Duzer (1965) p. 424]. In many devices, power is extracted using rectangular waveguides, therefore we shall consider next the characteristics of such a waveguide. In Sect. 2.3.1 we examined the radiation emitted from a dipole oscillating in azimuthally symmetric radial transmission line. In this geometry the main mode generated was the transverse electro-magnetic (TEM) mode. In this section we consider the same problem in a rectangular waveguide whose wide dimension is a and the narrow one is b – see Fig. 2.2c. Variations along the narrow dimension are neglected ($\partial/\partial y \sim 0$). A dipole is located in the center of the waveguide and it prescribes a current density which is given by

$$J_y(x,z,\omega) = I\delta\left(x - \frac{a}{2}\right)\delta(z)\,. \tag{2.3.34}$$

It excites the transverse electric field $E_y(x,z,\omega)$ which satisfies

$$\left[\frac{\partial^2}{\partial x^2} + \frac{\partial^2}{\partial z^2} + \frac{\omega^2}{c^2}\right] E_y(x,z,\omega) = j\omega\mu_0 J_y(x,z,\omega)\,, \tag{2.3.35}$$

subject to the boundary conditions: $E_y(x = 0,z,\omega) = 0$ and $E_y(x = a,z,\omega) = 0$. The solution can be represented as a superposition of trigonometric functions i.e.,

$$E_y(x,z,\omega) = \sum_{n=1}^{\infty} E_n(z,\omega)\sin\left(\frac{\pi n}{a}x\right)\,, \tag{2.3.36}$$

where $E_n(z,\omega)$ satisfies

$$\begin{aligned}\left[\frac{\mathrm{d}^2}{\mathrm{d}z^2} - \left(\frac{\pi n}{a}\right)^2 + \frac{\omega^2}{c^2}\right] E_n(z,\omega) &= j\omega\mu_0 I \sin\left(\frac{\pi}{2}n\right)\frac{2}{a}\delta(z) \\ &\equiv I_n\delta(z)\,.\end{aligned} \tag{2.3.37}$$

For $z > 0$ the solution of this equation is

$$E_n(z>0) = A_+\,\mathrm{e}^{-\Gamma_n z}\,, \tag{2.3.38}$$

and for $z < 0$

$$E_n(z<0) = A_-\,\mathrm{e}^{\Gamma_n z}\,, \tag{2.3.39}$$

where $\Gamma_n^2 = (\pi n/a)^2 - (\omega/c)^2$. The transverse electric field has to be continuous at $z = 0$ thus

$$A_+ = A_-\,, \tag{2.3.40}$$

whereas its derivative is discontinuous. The discontinuity is determined by the Dirac delta function in (2.3.37) therefore by integrating the latter we obtain

$$\left[\frac{\mathrm{d}}{\mathrm{d}z}E_n(z)\right]_{z=0+} - \left[\frac{\mathrm{d}}{\mathrm{d}z}E_n(z)\right]_{z=0-} = I_n\,, \tag{2.3.41}$$

hence

$$-\Gamma_n A_+ - \Gamma_n A_- = I_n\,. \tag{2.3.42}$$

From (2.3.40,42) we conclude that the transverse electric field reads

$$E_y(x,z,\omega) = -\sum_{n=1}^{\infty} \frac{I_n}{2\Gamma_n} \, \mathrm{e}^{-\Gamma_n|z|} \sin\left(\frac{\pi n}{a}x\right) . \tag{2.3.43}$$

As in Sect. 2.3.1 we shall next calculate the power generated by the current distribution in (2.3.34). For this purpose the transverse magnetic field is calculated since it is the only component which contributes to the longitudinal component of the Poynting vector; H_x for $z > 0$ reads

$$H_x(x,z>0,\omega) = -\sum_{n=1}^{\infty} \frac{\Gamma_n}{j\omega\mu_0} \frac{I_n}{2\Gamma_n} \, \mathrm{e}^{-\Gamma_n z} \sin\left(\frac{\pi n}{a}x\right) . \tag{2.3.44}$$

Before proceeding note that similar to the transverse magnetic mode, the phase velocity (for $\omega > \pi nc/a$ and $\varepsilon_\mathrm{r} = 1$) is always larger than c. Nevertheless, the characteristic impedance (in vacuum) of the nth propagating mode,

$$\begin{aligned} Z_{\mathrm{ch,TE}} &\equiv \frac{E_y}{H_x} \\ &= \frac{j\omega\mu_0}{\Gamma_n} , \end{aligned} \tag{2.3.45}$$

is always larger than the vacuum impedance (η_0), in contrast to the TM mode, where the characteristic impedance is always smaller than η_0.

Now we can direct our attention to the power flow: the average power which flows in the positive z direction, assuming a single mode above cut-off, is given by

$$P_+ = \frac{1}{2} \frac{I_1}{2\sqrt{(\omega/c)^2 - (\pi/a)^2}} \frac{I_1^*}{2\omega\mu_0} \frac{1}{2} ab . \tag{2.3.46}$$

The radiation impedance is determined by the power emitted in both directions divided by $\frac{1}{2}|I|^2$ and it reads

$$R_{\mathrm{rad,TE}} \equiv \frac{P_+ + P_-}{\frac{1}{2}|I|^2} = \eta_0 \frac{\omega b/c}{\sqrt{(\omega a/c)^2 - \pi^2}} = \frac{b}{a} Z_{\mathrm{ch,TE}} . \tag{2.3.47}$$

At 9 GHz, and for $a = 2.5$ cm, $b = 0.5$ cm, this impedance is 100 Ω which is close to that calculated in the case of the radial transmission line as calculated in Sect. 2.3.1.

2.4 Green's Scalar Theorem

Green's function is a useful tool for calculation of electromagnetic field generated by a distributed source (particles) subject to the boundary conditions imposed by the structure. The logic behind the method presented below is the following: instead of solving for an arbitrary source we solve for a point source and by virtue of the linearity of Maxwell's equations, the field at a given location is a superposition of all the point sources which constitute the real source.

Let us assume that we have to solve the non-homogeneous wave equation:

$$\left[\nabla^2 + \frac{\omega^2}{c^2}\right]\psi(\mathbf{r}) = -s(\mathbf{r}), \tag{2.4.1}$$

where $s(\mathbf{r})$ is an arbitrary source which is assumed to be known. Instead of solving this equation let us assume for the moment that we know how to solve a simpler problem namely,

$$\left[\nabla^2 + \frac{\omega^2}{c^2}\right]G(\mathbf{r}|\mathbf{r}') = -\delta(\mathbf{r}-\mathbf{r}'), \tag{2.4.2}$$

where the coefficient of the Dirac delta function on the right-hand side was chosen such that the result of the integration over the entire space is unity. We can then multiply (2.4.1) by $G(\mathbf{r}|\mathbf{r}')$ and (2.4.2) by $\psi(\mathbf{r})$, subtract the two results to obtain

$$G(\mathbf{r}|\mathbf{r}')\,\nabla^2\psi(\mathbf{r}) - \psi(\mathbf{r})\,\nabla^2 G(\mathbf{r}|\mathbf{r}') = -G(\mathbf{r}|\mathbf{r}')s(\mathbf{r}) + \psi(\mathbf{r})\,\delta(\mathbf{r}-\mathbf{r}'). \tag{2.4.3}$$

Integrating over the entire space we have

$$\begin{aligned}\int_V \mathrm{d}V'\,\nabla\cdot\left[G(\mathbf{r}|\mathbf{r}')\,\nabla\psi(\mathbf{r}) - \psi(\mathbf{r})\,\nabla G(\mathbf{r}|\mathbf{r}')\right] \\ = -\int_V \mathrm{d}V' G(\mathbf{r}|\mathbf{r}')s(\mathbf{r}) + \psi(\mathbf{r}),\end{aligned} \tag{2.4.4}$$

which can be further simplified using Gauss theorem to read

$$\begin{aligned}\psi(\mathbf{r}) = &\int_V \mathrm{d}V' G(\mathbf{r}|\mathbf{r}')s(\mathbf{r}') \\ &+ \int_a \mathrm{d}\mathbf{a}'\cdot\left[G(\mathbf{r}|\mathbf{r}')\,\nabla\psi(\mathbf{r}') - \psi(\mathbf{r}')\,\nabla G(\mathbf{r}|\mathbf{r}')\right];\end{aligned} \tag{2.4.5}$$

$\int_a$ is a surface integral which encloses the volume V. This is the scalar Green's theorem. In free space or for zero boundary conditions on a, Green's theorem reads

$$\psi(\mathbf{r}) = \int_V \mathrm{d}V' G(\mathbf{r}|\mathbf{r}')s(\mathbf{r}'). \tag{2.4.6}$$

Next we shall employ Green's theorem for the calculation of the Cerenkov effect in two cases: firstly, in a boundless system and secondly in a waveguide.

2.4.1 Cerenkov Radiation in the Boundless Case

Let us examine the electromagnetic field generated by a charge (e) as it moves in a medium which is characterized by a dielectric coefficient larger than unity, $\varepsilon_{\mathrm{r}} > 1$; its velocity is v_0. This is the case when a charged particle traverses a gas. If the medium is a solid, such as glass, where the mean free path of the electron may be small one should consider the field generated the particle as it moves in a vacuum channel bored in the solid. We shall consider here the former case.

A current density described by the same expression as in (2.2.15) drives the system and for an azimuthally symmetric medium the wave equation is

$$\left[\frac{1}{r}\frac{\partial}{\partial r}r\frac{\partial}{\partial r} + \frac{\partial^2}{\partial z^2} - \varepsilon_{\mathrm{r}}\frac{1}{c^2}\frac{\partial^2}{\partial t^2}\right] A_z(r,z,t) = -\mu_0 J_z(r,z,t)\,; \tag{2.4.7}$$

the other two components of the magnetic vector potential are zero and the electric scalar potential can be determined using Lorentz gauge. The time Fourier transform of the magnetic vector potential is defined by

$$A_z(r,z,t) = \int_{-\infty}^{\infty} \mathrm{d}\omega \mathrm{e}^{j\omega t} A_z(r,z,\omega)\,, \tag{2.4.8}$$

where $A_z(r,z,\omega)$ satisfies

$$\left[\frac{1}{r}\frac{\partial}{\partial r}r\frac{\partial}{\partial r} + \frac{\partial^2}{\partial z^2} + \varepsilon_{\mathrm{r}}\frac{\omega^2}{c^2}\right] A_z(r,z,\omega) = -\mu_0 J_z(r,z,\omega)\,, \tag{2.4.9}$$

and the time Fourier transform of the current density in (2.2.15) is

$$J_z(r,z,\omega) = -\frac{e}{(2\pi)^2 r}\,\delta(r)\,\mathrm{e}^{-j(\omega/v_0)z}\,. \tag{2.4.10}$$

Green's function associated with this problem is a solution of

$$\left[\frac{1}{r}\frac{\partial}{\partial r}r\frac{\partial}{\partial r} + \frac{\partial^2}{\partial z^2} + \varepsilon_{\mathrm{r}}\frac{\omega^2}{c^2}\right] G(r,z|r',z') = \frac{-1}{2\pi r}\,\delta(r-r')\,\delta(z-z') \tag{2.4.11}$$

which can be represented by

$$G(r,z|r',z') = \int_{-\infty}^{\infty} \mathrm{d}k\, g(r|r';k)\,\mathrm{e}^{-jk(z-z')}\,, \tag{2.4.12}$$

and $g(r|r';k)$ satisfies

$$\left[\frac{1}{r}\frac{\mathrm{d}}{\mathrm{d}r}r\frac{\mathrm{d}}{\mathrm{d}r} - \Gamma^2\right] g(r|r';k) = -\frac{1}{(2\pi)^2 r}\,\delta(r-r')\,, \tag{2.4.13}$$

where

$$\Gamma^2 = k^2 - \varepsilon_{\rm r}\frac{\omega^2}{c^2}\,. \tag{2.4.14}$$

The solution of this equation for $r > r' > 0$ is

$$g(r|r' < r; k) = F_1(r')\mathrm{K}_0(\Gamma r)\,, \tag{2.4.15}$$

and for $r' > r > 0$ it reads

$$g(r < r'|r'; k) = F_2(r')\mathrm{I}_0(\Gamma r)\,. \tag{2.4.16}$$

The function $g(r|r'; k)$ has to be continuous at $r = r'$ i.e.,

$$F_1(r')\mathrm{K}_0(\Gamma r') = F_2(r')\mathrm{I}_0(\Gamma r')\,, \tag{2.4.17}$$

whereas its derivative is discontinuous at the same location. To determine the discontinuity we integrate (2.4.13)

$$\left[r\frac{\mathrm{d}}{\mathrm{d}r}g(r|r')\right]_{r=r'+0} - \left[r\frac{\mathrm{d}}{\mathrm{d}r}g(r|r')\right]_{r=r'-0} = -\frac{1}{(2\pi)^2}\,, \tag{2.4.18}$$

hence

$$-r'F_1(r')\Gamma\mathrm{K}_1(\Gamma r') - r'F_2(r')\Gamma\mathrm{I}_1(\Gamma r') = -\frac{1}{(2\pi)^2}\,. \tag{2.4.19}$$

From (2.4.17,19) and using the fact that $\mathrm{K}_0(\xi)\mathrm{I}_1(\xi) + \mathrm{K}_1(\xi)\mathrm{I}_0(\xi) = 1/\xi$ [see Abramowitz and Stegun (1968) p. 375] we can finally obtain

$$g(r|r'; k) = \frac{1}{(2\pi)^2}\begin{cases} \mathrm{I}_0(\Gamma r)\mathrm{K}_0(\Gamma r') & \text{for } 0 \le r \le r' < \infty\,, \\ \mathrm{K}_0(\Gamma r)\mathrm{I}_0(\Gamma r') & \text{for } 0 \le r' \le r < \infty\,. \end{cases} \tag{2.4.20}$$

This expression together with (2.4.12) determine Green's function in a boundless space.

With this function, Green's theorem (2.4.6) and the current density as given in (2.4.10), we can determine the magnetic vector potential. It reads

$$A_z(r, z, \omega) = -\frac{e\mu_0}{(2\pi)^2}\mathrm{K}_0\left(\frac{\omega}{c}r\sqrt{\beta^{-2} - n^2}\right)\mathrm{e}^{-j(\omega/v_0)z}\,, \tag{2.4.21}$$

where $n \equiv \sqrt{\varepsilon_{\rm r}}$ is the refractive index of the medium. If we examine this solution far away from the source and use the asymptotic value for large arguments $\left[(\omega/c)r\left|\sqrt{\beta^2 - n^2}\right| \gg 1\right]$ of the modified Bessel function, the magnetic vector potential reads

$$A_z(r, z, \omega) \propto \mathrm{e}^{-(\omega/c)r\sqrt{\beta^{-2}-n^2}}\,\mathrm{e}^{-j(\omega/v_0)z}\,. \tag{2.4.22}$$

If n is smaller than $1/\beta$ the field decays exponentially in the radial direction since, as in vacuum, this is an *evanescent* wave.

When the velocity of the particle, $v_0 = \beta c$, is larger than the phase velocity of a plane wave in the medium (c/n) i.e., $\beta > 1/n$, the expression above represents a *propagating* wave – this radiation is called Cerenkov radiation. The emitted wave is not parallel to the electron's trajectory but it propagates at an angle θ relative to this direction (z axis) given by

$$\begin{aligned} k_z &= \frac{\omega}{c} n \cos\theta\,, \\ &= \frac{\omega}{c}\frac{1}{\beta}\,. \end{aligned} \tag{2.4.23}$$

This determines what it is known as the Cerenkov radiation angle, θ_c

$$\theta_c = \cos^{-1}\left(\frac{1}{n\beta}\right)\,. \tag{2.4.24}$$

Since the phase velocity of the wave is smaller than that of the particle, clearly, the radiation lags behind the particle. This fact will become evident in the next subsection.

2.4.2 Cerenkov Radiation in a Cylindrical Waveguide

In this subsection we consider the electromagnetic field associated with the symmetric transverse magnetic (TM) mode in a dielectric filled waveguide. As in the previous subsection, the source of this field is a particle moving at a velocity v_0, however, the main difference is that the solution has a constraint since on the waveguide's wall ($r = R$) the tangential electric field vanishes. Therefore, we shall calculate the Green function in the frequency domain subject to the condition $G(r = R, z|r', z') = 0$. We assume a solution of the form

$$G(r, z|r', z') = \sum_{s=1}^{\infty} G_s(z|r', z')\, \mathrm{J}_0\left(p_s \frac{r}{R}\right)\,, \tag{2.4.25}$$

substitute in (2.4.11) and use the orthogonality of the Bessel functions we find that

$$G_s(z|r', z') = \mathrm{J}_0\left(p_s \frac{r'}{R}\right) \frac{1}{\frac{1}{2}R^2 \mathrm{J}_1^2(p_s)}\, g_s(z|z')\,, \tag{2.4.26}$$

where $g_s(z|z')$ satisfies

$$\left[\frac{\mathrm{d}^2}{\mathrm{d}z^2} - \Gamma_s^2\right] g_s(z|z') = -\frac{1}{2\pi}\,\delta(z - z')\,, \tag{2.4.27}$$

and

$$\Gamma_s^2 = \frac{p_s^2}{R^2} - \varepsilon_{\rm r}\frac{\omega^2}{c^2}\,. \tag{2.4.28}$$

For $z > z'$ the solution of (2.4.27) is

$$g_s(z|z') = A_+ {\rm e}^{-\Gamma_s(z-z')}\,, \tag{2.4.29}$$

and for $z < z'$ the solution is

$$g_s(z|z') = A_- {\rm e}^{\Gamma_s(z-z')}\,. \tag{2.4.30}$$

Green's function is continuous at $z = z'$ i.e.,

$$A_+ = A_-\,, \tag{2.4.31}$$

and its first derivative is discontinuous. The discontinuity is determined by integrating (2.4.27) from $z = z' - 0$ to $z = z' + 0$ i.e.,

$$\left[\frac{\rm d}{{\rm d}z}g_s(z|z')\right]_{z=z'+0} - \left[\frac{\rm d}{{\rm d}z}g_s(z|z')\right]_{z=z'-0} = -\frac{1}{2\pi}\,. \tag{2.4.32}$$

Substituting the two solutions introduced above, and using (2.4.31) we obtain

$$g_s(z|z') = \frac{1}{4\pi\Gamma_s}\,{\rm e}^{-\Gamma_s|z-z'|}\,. \tag{2.4.33}$$

Finally, the explicit expression for the Green's function corresponding to azimuthally symmetric TM modes in a circular waveguide is given by

$$G(r,z|r',z') = \sum_{s=1}^{\infty}\frac{{\rm J}_0(p_s r/R)\,{\rm J}_0(p_s r'/R)}{\frac{1}{2}R^2{\rm J}_1^2(p_s)}\,\frac{1}{4\pi\Gamma_s}\,{\rm e}^{-\Gamma_s|z-z'|}\,. \tag{2.4.34}$$

In this expression it was tacitly assumed that $\omega > 0$ and Γ_s [defined in (2.4.28)] is non-zero.

With Green's function established, we can calculate the magnetic vector potential as generated by the current distribution described in (2.4.10); the result is

$$\begin{aligned} A_z(r,z,\omega) &= 2\pi\mu_0\int_0^R {\rm d}r\,r\int_{-\infty}^{\infty}{\rm d}z' G(r,z|r',z')J_z(r',z') \\ &= -\frac{e\mu_0}{8\pi^2}\sum_{s=1}^{\infty}\frac{{\rm J}_0(p_s r/R)}{\frac{1}{2}R^2{\rm J}_1^2(p_s)}\,\frac{2}{\Gamma_s^2+\omega^2/v_0^2}\,{\rm e}^{-j(\omega/v_0)z}\,. \end{aligned} \tag{2.4.35}$$

It will be instructive to examine this expression in the time domain; the Fourier transform is

$$\begin{aligned} A_z(r,z,t) &= -\frac{e}{2\pi^2\varepsilon_0 R^2}\,\frac{\beta^2}{1-n^2\beta^2} \\ &\quad\times\sum_{s=1}^{\infty}\frac{{\rm J}_0(p_s r/R)}{{\rm J}_1^2(p_s)}\int_{-\infty}^{\infty}{\rm d}\omega\frac{{\rm e}^{j\omega(t-z/v_0)}}{\omega^2+\Omega_s^2}\,, \end{aligned} \tag{2.4.36}$$

where

$$\Omega_s^2 = \left(\frac{p_s c}{R}\right)^2 \frac{\beta^2}{1-n^2\beta^2}\,. \tag{2.4.37}$$

The problem has been now simplified to the evaluation of the integral

$$F_s(\tau = t - z/v_0) \equiv \int_{-\infty}^{\infty} \mathrm{d}\omega \frac{\mathrm{e}^{j\omega\tau}}{\omega^2 + \Omega_s^2}\,, \tag{2.4.38}$$

which in turn is equivalent to the solution of the following differential equation

$$\left[\frac{\mathrm{d}^2}{\mathrm{d}\tau^2} - \Omega_s^2\right] F_s(\tau) = -2\pi\,\delta(\tau)\,. \tag{2.4.39}$$

If the particle's velocity is smaller than the phase velocity of a plane wave in the medium ($n\beta < 1$) then $\Omega_s^2 > 0$ and the solution for $\tau > 0$ is

$$F_s(\tau > 0) = A_+ \mathrm{e}^{-\Omega_s \tau}\,, \tag{2.4.40}$$

or

$$F_s(\tau < 0) = A_- \mathrm{e}^{\Omega_s \tau}\,. \tag{2.4.41}$$

As previously, in the case of Green's function, $F_s(\tau)$ has to be continuous at $\tau = 0$ and its derivative is discontinuous:

$$\left(\frac{\mathrm{d}}{\mathrm{d}\tau} F_s(\tau)\right)_{\tau=0+} - \left(\frac{\mathrm{d}}{\mathrm{d}\tau} F_s(\tau)\right)_{\tau=0-} = -2\pi\,. \tag{2.4.42}$$

When the velocity of the particle is smaller than c/n (i.e., $n\beta < 1$) the characteristic frequency Ω_s is real, therefore

$$F_s(\tau) = \frac{\pi}{\Omega_s} \mathrm{e}^{-\Omega_s|\tau|}\,, \tag{2.4.43}$$

and

$$A_z(r, z, t) = -\frac{e}{2\pi\varepsilon_0 R^2} \frac{\beta^2}{1-\beta^2 n^2} \sum_{s=1}^{\infty} \frac{\mathrm{J}_0(p_s r/R)}{\mathrm{J}_1^2(p_s)\Omega_s} \mathrm{e}^{-\Omega_s|t - z/v_0|}\,. \tag{2.4.44}$$

This expression represents a superposition of evanescent modes attached to the particle. It is important to emphasize that since the phase velocity of a plane wave in the medium is larger than the velocity of the particle there is an electromagnetic field in front ($\tau < 0$) of the particle. The situation is different in the opposite case, $\beta > 1/n$, since $\Omega_s^2 < 0$. In this case the waves are slower than the particle and there is no electromagnetic field in front of the particle i.e.,

$$F_s(\tau < 0) = 0\,. \tag{2.4.45}$$

By virtue of the continuity at $\tau = 0$ we have for $\tau > 0$

$$F_s(\tau > 0) = A_+ \sin(|\Omega_s|\tau) . \tag{2.4.46}$$

Substituting these two expressions in (2.4.42) we obtain

$$F_s(\tau) = -\frac{2\pi}{|\Omega_s|} \sin(|\Omega_s|\tau)\, h(\tau) , \tag{2.4.47}$$

and the magnetic vector potential reads

$$\begin{aligned} A_z(r,z,t) = & -\frac{e}{\pi\varepsilon_0 R^2}\frac{\beta^2}{n^2\beta^2-1} \\ & \times \sum_{s=1}\frac{\mathrm{J}_0(p_s r/R)}{\mathrm{J}_1^2(p_s)|\Omega_s|}\sin\left[|\Omega_s|\left(t-\frac{z}{v_0}\right)\right] h\left(t-\frac{z}{v_0}\right) , \end{aligned} \tag{2.4.48}$$

where $h(\xi)$ is the Heaviside step function. This expression indicates that when the velocity of the particle is larger than c/n, there is an entire superposition of *propagating* waves traveling behind the particle. Furthermore, all the waves have the same phase velocity which is identical with the velocity of the particle, v_0. It is important to bear in mind that this result was obtained after tacitly assuming that ε_r is frequency independent which generally is not the case, therefore the summation is limited to a finite number of modes. The modes which contribute are determined by the Cerenkov condition $n(\omega = \Omega_s)\beta > 1$.

After we established the magnetic vector potential, let us now calculate the average power which trails behind the particle. Firstly, the azimuthal magnetic field is given by

$$\begin{aligned} H_\phi(r,z,t) &= -\frac{1}{\mu_0}\frac{\partial}{\partial r}A_z(r,z,t) \\ &= \frac{1}{\mu_0}\sum_{s=1} A_s \frac{p_s}{R}\mathrm{J}_1\left(p_s\frac{r}{R}\right) \\ &\quad \times \sin\left[|\Omega_s|\left(t-\frac{z}{v_0}\right)\right] h\left(t-\frac{z}{v_0}\right) , \end{aligned} \tag{2.4.49}$$

where

$$A_s = -\frac{e}{\pi\varepsilon_0 R^2}\frac{\beta^2}{n^2\beta^2-1}\frac{1}{\mathrm{J}_1^2(p_s)|\Omega_s|} . \tag{2.4.50}$$

Secondly, the radial electric field is determined by the electric scalar potential which in turn is calculated using the Lorentz gauge and it reads

$$\begin{aligned} E_r(r,z,t) &= -\frac{\partial}{\partial r}\Phi(r,z,t) \\ &= \frac{c^2}{\varepsilon_\mathrm{r} v_0}\sum_{s=1} A_s \frac{p_s}{R}\mathrm{J}_1\left(p_s\frac{r}{R}\right) \\ &\quad \times \sin\left[|\Omega_s|\left(t-\frac{z}{v_0}\right)\right] h\left(t-\frac{z}{v_0}\right) . \end{aligned} \tag{2.4.51}$$

With these expressions we can calculate the average electromagnetic power trailing the particle. It is given by

$$P = \frac{e^2 \beta c}{2\pi\varepsilon_0\varepsilon_r R^2} \frac{1}{\varepsilon_r\beta^2 - 1} \sum_{s=1} \frac{1}{J_1^2(p_s)} . \tag{2.4.52}$$

Note that for ultra relativistic particle ($\beta \to 1$) the power is independent of the particle's energy. In order to have a measure of the radiation emitted consider a very narrow bunch of $N \sim 10^{11}$ electrons injected in a waveguide whose radius is 9.2 mm. The waveguide is filled with a material whose dielectric coefficient is $\varepsilon_r = 2.6$ and all electrons have the same energy 450 keV. If we were able to keep their velocity constant, then 23 MW of power at 11.4 GHz (first mode, $s = 1$) will trail the bunch. Further examining this expression we note that the average power is quadratic with the frequency i.e.,

$$P \equiv \sum_{s=1} P_s = \frac{(Ne)^2}{2\pi\varepsilon_0\varepsilon_r\beta c} \sum_{s=1} \frac{|\Omega_s|^2}{[p_s J_1(p_s)]^2} . \tag{2.4.53}$$

In addition, based on the definition of the Fourier transform of the current density in (2.4.10), we conclude that the current which this macro-particle excites in the sth mode is $I_s = eN\Omega_s/2\pi$. With this expression, the radiation impedance of the first mode ($s = 1$) is

$$R_{C,1} = \frac{P_1}{\frac{1}{2}|I_1|^2} = \eta_0 \frac{4\pi}{\varepsilon_r\beta\,[p_1 J_1(p_1)]^2} . \tag{2.4.54}$$

For a relativistic particle, $\beta \simeq 1$, a dielectric medium $\varepsilon_r = 2.6$ the radiation impedance corresponding to the first mode is $\simeq 1200\Omega$ which is one order of magnitude larger than that of a dipole in free space or between two plates. Note that this impedance is independent of the geometry of the waveguide and for an ultra-relativistic particle it is independent of the particle's energy.

2.4.3 Cerenkov Force

The radiation emitted when a particle moves with a velocity which exceeds the phase velocity of the electromagnetic wave in the medium, comes at the expense of its kinetic energy. In order to understand the source of this force we recall that attached to a moving particle there is a superposition of evanescent waves. As the particle moves in a vacuum channel of radius R surrounded by a dielectric medium ε_r, the evanescent waves hit the discontinuity at $r = R$ and they are partially reflected and partially transmitted. It is the reflected wave which acts back on the electron decelerating it. In this subsection we shall examine this process in a systematic way.

Consider a point charge (e) moving at a constant velocity v_0; the current density is described by (2.2.15) and its time Fourier transform by (2.4.10). It generates a magnetic vector potential, in the frequency domain, determined by

$$A_z(r<R,z,\omega) = 2\pi\mu_0 \int_{-\infty}^{\infty} \mathrm{d}z' \int_0^R \mathrm{d}r' r' G(r,z|r',z') J_z(r',z',\omega) + \int_{-\infty}^{\infty} \mathrm{d}k \varrho(k) \mathrm{e}^{-jkz} \mathrm{I}_0(\Gamma r) \,, \tag{2.4.55}$$

and

$$A_z(r>R,z,\omega) = \int_{-\infty}^{\infty} \mathrm{d}k \tau(k) \mathrm{e}^{-jkz} \mathrm{K}_0(\Lambda r) \,, \tag{2.4.56}$$

where $\Gamma^2 = k^2 - (\omega/c)^2$, $\Lambda^2 = k^2 - \varepsilon_\mathrm{r}(\omega/c)^2$, $G(r',z'|r,z)$ is the boundless Green's function as defined in (2.4.12,20) but for vacuum i.e.,

$$G(r',z'|r,z) = \frac{1}{(2\pi)^2} \int_{-\infty}^{\infty} \mathrm{d}k \mathrm{e}^{-jk(z-z')} \times \begin{cases} \mathrm{I}_0(\Gamma r)\,\mathrm{K}_0(\Gamma r') & \text{for } 0<r<r'<\infty \,, \\ \mathrm{K}_0(\Gamma r)\,\mathrm{I}_0(\Gamma r') & \text{for } 0<r'<r<\infty \,. \end{cases} \tag{2.4.57}$$

The amplitudes ϱ and τ represent the reflected and transmitted waves correspondingly. In order to determine these amplitudes we have to impose the boundary conditions at $r=R$. For this purpose it is convenient to write the solution of the magnetic vector potential off-axis as

$$A_z(0<r<R,z,\omega) = \int_{-\infty}^{\infty} \mathrm{d}k \mathrm{e}^{-jkz} \left[\varrho(k)\mathrm{I}_0(\Gamma r) + \alpha(k)\mathrm{K}_0(\Gamma r)\right] \,, \tag{2.4.58}$$

where

$$\alpha(k) = -\frac{e\mu_0}{(2\pi)^2} \delta\left(k - \frac{\omega}{v_0}\right) . \tag{2.4.59}$$

From the continuity of the longitudinal electric field (E_z) we conclude that

$$\frac{c^2}{j\omega}\left[\frac{\omega^2}{c^2} - k^2\right]\left[\varrho(k)\mathrm{I}_0(\Gamma R) + \alpha(k)\mathrm{K}_0(\Gamma R)\right] = \frac{c^2}{j\omega\varepsilon_\mathrm{r}}\left[\varepsilon_\mathrm{r}\frac{\omega^2}{c^2} - k^2\right]\tau(k)\mathrm{K}_0(\Lambda R) \,. \tag{2.4.60}$$

In a similar way the continuity of the azimuthal magnetic field implies

$$\Gamma\left[\varrho(k)\mathrm{I}_1(\Gamma R) - \alpha(k)\mathrm{K}_1(\Gamma R)\right] = -\Lambda\tau(k)\mathrm{K}_1(\Gamma R) \,. \tag{2.4.61}$$

At this stage we introduce the (normalized) impedances ratio

$$\zeta \equiv \frac{1}{\varepsilon_r} \frac{\Lambda}{\Gamma} \frac{\mathrm{K}_0(\Lambda R)}{\mathrm{K}_1(\Lambda R)}, \tag{2.4.62}$$

by whose means the amplitudes of the reflected waves are given by

$$\varrho = \alpha \frac{\zeta \mathrm{K}_1(\Gamma R) - \mathrm{K}_0(\Gamma R)}{\zeta \mathrm{I}_1(\Gamma R) + \mathrm{I}_0(\Gamma R)}. \tag{2.4.63}$$

The only non-zero field on axis is the longitudinal electric field and only the waves "reflected" from the radial discontinuity contribute to the force which acts on the particle, therefore

$$E_z(r=0, z=v_0 t, t) = \int_{-\infty}^{\infty} \mathrm{d}\omega \mathrm{d}k \frac{c^2}{j\omega} \left[\frac{\omega^2}{c^2} - k^2 \right] \varrho(\omega, k) \mathrm{e}^{j(\omega - k v_0)t}. \tag{2.4.64}$$

Substituting the explicit expression for ϱ and using the integral over the Dirac delta function [see (2.4.59)] and defining $x = \omega R / c\beta\gamma$, we obtain

$$\begin{aligned} &E_z(r=0, z=v_0 t, t) \\ &= \frac{-je}{(2\pi)^2 \varepsilon_0 R^2} \int_{-\infty}^{\infty} \mathrm{d}x\, x \frac{\zeta(x)\mathrm{K}_1(|x|) - \mathrm{K}_0(|x|)}{\zeta(x)\mathrm{I}_1(|x|) + \mathrm{I}_0(|x|)}. \end{aligned} \tag{2.4.65}$$

At this point it is convenient to define the normalized field which acts on the particle as

$$\begin{aligned} \mathcal{E} &\equiv E_z(r=0, z=v_0 t, t) \left(\frac{e}{4\pi\varepsilon_0 R^2} \right)^{-1} \\ &= \frac{2}{\pi} \int_0^{\infty} \mathrm{d}x\, x \,\mathrm{Re} \left[\frac{1}{j} \frac{\zeta(x)\mathrm{K}_1(|x|) - \mathrm{K}_0(|x|)}{\zeta(x)\mathrm{I}_1(|x|) + \mathrm{I}_0(|x|)} \right]. \end{aligned} \tag{2.4.66}$$

Clearly from this representation we observe that, for a non-zero force to act on the particle, the impedance ratio ζ has to be complex since the argument of the modified Bessel functions is real.

We can make one step further and simplify this expression by defining

$$\zeta(x) \equiv |\zeta(x)| \mathrm{e}^{j\psi(x)}, \tag{2.4.67}$$

and using $\mathrm{K}_0(x)\mathrm{I}_1(x) + \mathrm{K}_1(x)\mathrm{I}_0(x) = 1/x$, we obtain

$$\mathcal{E} = \frac{2}{\pi} \int_0^{\infty} \mathrm{d}x \frac{|\zeta(x)| \sin\psi(x)}{\mathrm{I}_0^2(x) + |\zeta(x)|^2 \mathrm{I}_1^2(x) + 2|\zeta(x)| \mathrm{I}_0(x)\mathrm{I}_1(x)\cos\psi(x)}. \tag{2.4.68}$$

In order to evaluate this integral for a dielectric medium and a particle whose velocity βc is larger than $c/\sqrt{\varepsilon_r}$, we go back to (2.4.62) which now reads

$$\zeta(x) = j \frac{\gamma}{\varepsilon_r} \sqrt{\varepsilon_r \beta^2 - 1} \frac{\mathrm{K}_0\left(jx\gamma\sqrt{\varepsilon_r \beta^2 - 1} \right)}{\mathrm{K}_1\left(jx\gamma\sqrt{\varepsilon_r \beta^2 - 1} \right)}, \tag{2.4.69}$$

and it can be further simplified if we assume that the main contribution occurs for large arguments of the Bessel function (i.e., $\gamma \gg 1$) thus

$$\zeta(x) \simeq j\frac{\gamma}{\varepsilon_r}\sqrt{\varepsilon_r\beta^2 - 1}\,. \qquad (2.4.70)$$

Since subject to this approximation $\psi = \pi/2$ and $|\zeta|$ is constant we can evaluate $\mathcal{E}$,

$$\mathcal{E} = \frac{2}{\pi}\int_0^\infty dx \frac{|\zeta|}{I_0^2(x) + |\zeta|^2 I_1^2(x)}\,, \qquad (2.4.71)$$

for two regimes: firstly when $|\zeta| \gg 1$ i.e., $\gamma \gg 1$, the contribution to the integral is primarily from small values of x thus

$$\begin{aligned}\mathcal{E} &\simeq \frac{2}{\pi}\int_0^\infty dx \frac{|\zeta|}{1 + |\zeta|^2 x^2/4}\,,\\ &\simeq \frac{4}{\pi}\int_0^\infty du \frac{1}{1+u^2}\,,\\ &\simeq 2\,. \end{aligned} \qquad (2.4.72)$$

At the other extreme ($|\zeta| \ll 1$) the normalized impedance has to be recalculated and the result is

$$\mathcal{E} \simeq \frac{\gamma^2(\varepsilon_r\beta^2 - 1)}{\varepsilon_r}\int_0^\infty dx \frac{x}{I_0^2(x)} \simeq 1.263\frac{\gamma^2(\varepsilon_r\beta^2 - 1)}{\varepsilon_r}\,, \qquad (2.4.73)$$

and we can summarize

$$\mathcal{E} \simeq \begin{cases} 0 & \text{for} \quad \beta < 1/\sqrt{\varepsilon_r}\,,\\ 1.263\gamma^2\sqrt{\varepsilon_r\beta^2 - 1}/\varepsilon_r & \text{for} \quad \gamma \ll \varepsilon_r/\sqrt{\varepsilon_r\beta^2 - 1}\,,\\ 2 & \text{for} \quad \gamma \gg \varepsilon_r/\sqrt{\varepsilon_r\beta^2 - 1}\,. \end{cases} \qquad (2.4.74)$$

It is interesting to note that for ultra-relativistic electrons the decelerating Cerenkov force reaches an asymptotic value which is independent of γ and the dielectric coefficient; it is given by $E = -e/2\pi\varepsilon_0 R^2$. In addition, we observe that the normalized impedance (ζ) determines the force.

2.4.4 Ohm Force

If in the Cerenkov case the electron has to exceed a certain velocity in order to generate radiation and therefore to experience a decelerating force, in the case of a lossy medium, the moving electron experiences a decelerating force starting from a vanishingly low speed. This is because it excites currents in the surrounding walls and as a result power is dissipated – which is equivalent to the emitted power in the Cerenkov case. The source of this power is the $\mathbf{J}\cdot\mathbf{E}$ [see (2.1.17)] term which infers the existence of a decelerating force acting on the electron. In order to evaluate this force we use the same formulation as in the previous subsection only that in this case the dielectric coefficient is complex and it is given by

$$\varepsilon_\mathrm{r} = 1 - j\frac{\sigma}{\varepsilon_0\omega}\,, \tag{2.4.75}$$

where σ is the (finite) conductivity of the surrounding medium. It is convenient to use the same notation as above, therefore the normalized impedance ζ from (2.4.70) is replaced by

$$\zeta \simeq \frac{1}{1 - j\bar{\sigma}/x}\sqrt{1 + j(\gamma\beta)^2\frac{\bar{\sigma}}{x}}\,. \tag{2.4.76}$$

In this expression $\bar{\sigma} \equiv \sigma\eta_0 R/\gamma\beta$ which for typical metals and $R \sim 1$ cm is of the order of $10^8/\gamma\beta$ thus for any practical purpose $\bar{\sigma} \gg 1$ hence

$$\zeta \simeq \gamma\beta \mathrm{e}^{j3\pi/4}\sqrt{\frac{x}{\bar{\sigma}}}\,. \tag{2.4.77}$$

Note that the phase of the normalized impedance is $\psi = 3\pi/4$. Substituting this expression in (2.4.68) we obtain

$$\begin{aligned}\mathcal{E} =& \frac{2}{\pi}\frac{\sqrt{2}}{2}\sqrt{\frac{(\gamma\beta)^3}{\sigma\eta_0 R}} \\ &\times \int_0^\infty \mathrm{d}x \frac{\sqrt{x}}{\mathrm{I}_0^2(x) + x\mathrm{I}_1^2(x)(\gamma\beta)^3/\sigma\eta_0 R - \sqrt{2x}\mathrm{I}_0(x)\mathrm{I}_1(x)\sqrt{(\gamma\beta)^3/\sigma\eta_0 R}}\,,\end{aligned} \tag{2.4.78}$$

which can be evaluated analytically for two extreme regimes: in the first case the (normalized) momentum of the particle is much smaller than the normalized conductivity term i.e., $(\gamma\beta)^3 \ll \sigma\eta_0 R$ in which case

$$\begin{aligned}\mathcal{E} &\simeq \frac{2}{\pi}\frac{\sqrt{2}}{2}\sqrt{\frac{(\gamma\beta)^3}{\sigma\eta_0 R}}\int_0^\infty \mathrm{d}x\frac{\sqrt{x}}{\mathrm{I}_0^2(x)} \\ &\simeq 0.54\sqrt{\frac{(\gamma\beta)^3}{\sigma\eta_0 R}}\,.\end{aligned} \tag{2.4.79}$$

The second case corresponds to a highly relativistic particle i.e., $(\gamma\beta)^3 \gg \sigma\eta_0 R$ implying that the main contribution to the integral is from the small values of x which justifies the expansion of the modified Bessel functions in Taylor series. Redefining $y^2 \equiv (x\gamma\beta)^3/4\sigma\eta_0 R$ we have

$$\begin{aligned}\mathcal{E} &\simeq \frac{4\sqrt{2}}{3\pi}\int_0^\infty \mathrm{d}y\frac{1}{1 + y^2 - y\sqrt{2}} \\ &\simeq 2\,.\end{aligned} \tag{2.4.80}$$

Therefore,

$$\mathcal{E} \simeq \begin{cases} 0.54\sqrt{(\gamma\beta)^3/\sigma\eta_0 R} & \text{for} \quad (\gamma\beta)^3 \ll \sigma\eta_0 R\,, \\ 2 & \text{for} \quad (\gamma\beta)^3 \gg \sigma\eta_0 R\,, \end{cases} \tag{2.4.81}$$

which, as in the Cerenkov case, indicates that for ultra-relativistic particles the decelerating force is independent of γ and of the material's characteristics. However, the threshold for γ in this regime is much higher. In both cases the particle is decelerated by a force which corresponds to a (positive) image charge located at a distance $R/\sqrt{2}$ behind the electron.

In the low energy regime the decelerating force increases rapidly with the momentum of the particle. The square root of the conductivity indicates that this is a skin-depth effect. A clearer interpretation of this statement is obtained if we define the characteristic frequency $\omega_0 \equiv 2c/R(\gamma\beta)^3$ by whose means

$$\delta = \sqrt{\frac{2}{\sigma\mu_0\omega_0}}\,, \tag{2.4.82}$$

and

$$\mathcal{E} \simeq \begin{cases} 0.54\,\delta/R & \text{for} \quad \delta \ll R\,, \\ 2 & \text{for} \quad \delta \gg R\,. \end{cases} \tag{2.4.83}$$

The characteristic frequency is low for very relativistic electrons and consequently the skin-depth is much larger than the radius and all the bulk material "participates" in the deceleration process. On the other hand, if the frequency is high, then the skin-depth is small (comparing to the radius) and only a thin layer dissipates power, therefore the loss is proportional to δ.

Finally, if the conductivity of the material is negative, that is to say that we have an active medium, then the phase in (2.4.77) is $\psi = 5\pi/4$ and the force is accelerating which means that energy can be transferred from the medium to the electron. This topic will be further discussed in Chap. 8 in the context of acceleration concepts.

2.5 Finite Length Effects

In all the effects discussed so far we assumed an infinite system with no reflected waves. In this section we consider three finite length systems and phenomena associated with reflected waves. When both forward and backward propagating waves coexist, there is a frequency selection associated with the interference of the two. Another byproduct of reflections is tunneling of the field in a region where the wave is below cutoff. We also examine more complex phenomena such as radiation generated by a single particle as it traverses a geometric discontinuity in a waveguide. We conclude with the evaluation of a wake field generated by a particle in a cavity.

2.5.1 Impedance Discontinuities

In most cases of interest the waveguide is not uniform and as a result more than one wave occurs. In order to illustrate the effect of discontinuities we shall consider the following problem: a cylindrical waveguide of radius R but, instead of being uniformly filled with one dielectric material, there are three different dielectrics in three different regions

$$\varepsilon_r(z) = \begin{cases} \varepsilon_1 & \text{for } -\infty < z < 0 \,, \\ \varepsilon_2 & \text{for } 0 \leq z \leq d \,, \\ \varepsilon_3 & \text{for } d < z < \infty \,, \end{cases} \tag{2.5.1}$$

as illustrated in Fig. 2.3.

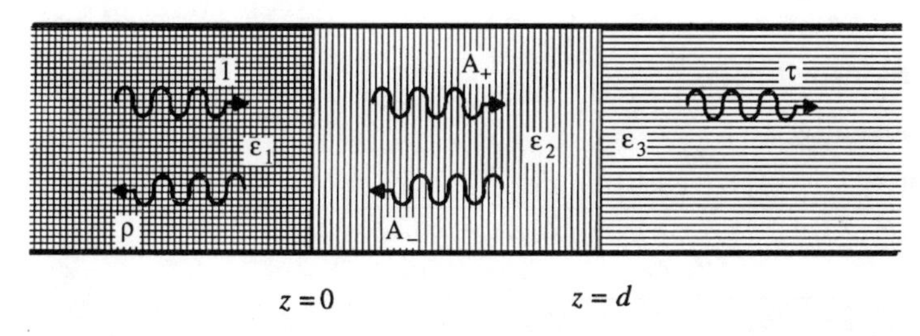

Fig. 2.3. Schematics of the system used to examine the reflected waves resulting from characteristic impedance discontinuities

A wave is launched from $z \to -\infty$ towards the discontinuity at $z = 0$. For simplicity we assume that this wave is composed of a single mode (TM_{01} i.e., $s = 1$). The z component of the magnetic vector in the first region ($-\infty < z < 0$) is given by

$$A_z(r, -\infty < z < 0, \omega) = A_{\text{in}} \left[e^{-\Gamma_1^{(1)} z} + \varrho e^{\Gamma_1^{(1)} z} \right] J_0 \left(p_1 \frac{r}{R} \right) , \tag{2.5.2}$$

where A_{in} is the amplitude of the incoming wave and ϱ is the reflection coefficient; $\Gamma_1^{(1)} = \sqrt{(p_1/R)^2 - \varepsilon_1(\omega/c)^2}$. Between the two discontinuities at $0 < z < d$ the solution has a similar form

$$A_z(r, 0 \leq z \leq d, \omega) = A_{\text{in}} \left[A_+ e^{-\Gamma_1^{(2)} z} + A_- e^{\Gamma_1^{(2)} z} \right] J_0 \left(p_1 \frac{r}{R} \right) , \tag{2.5.3}$$

where $\Gamma_1^{(2)} = \sqrt{(p_1/R)^2 - \varepsilon_2(\omega/c)^2}$. In the third region there is no reflected wave therefore

$$A_z(r, d < z < \infty, \omega) = A_{\text{in}} \tau e^{-\Gamma_1^{(3)} (z-d)} J_0 \left(p_1 \frac{r}{R} \right) , \tag{2.5.4}$$

and as above $\Gamma_1^{(3)} = \sqrt{(p_1/R)^2 - \varepsilon_3(\omega/c)^2}$; τ is the transmission coefficient. The four as yet unknown amplitudes ϱ, τ, A_+ and A_- are determined by imposing the boundary conditions at $z = 0, d$. Continuity of E_r at $z = 0$ implies

$$Z_1[1 - \varrho] = Z_2[A_+ - A_-]; \tag{2.5.5}$$

Z_1 and Z_2 are the characteristic impedances (2.3.26) in the first and second regions respectively. In a similar way the continuity of H_ϕ implies

$$1 + \varrho = A_+ + A_- . \tag{2.5.6}$$

An additional set of equations is found imposing the continuity of the same components at $z = d$:

$$Z_2[A_+ \, e^{-\psi} - A_- \, e^{\psi}] = Z_3 \, \tau , \tag{2.5.7}$$

and

$$A_+ \, e^{-\psi} + A_- \, e^{\psi} = \tau , \tag{2.5.8}$$

where $\psi \equiv \Gamma_1^{(2)} d$. From (2.5.5–8) the reflection (ϱ) and transmission (τ) coefficients are given by

$$\begin{aligned} \varrho &= \frac{\sinh(\psi)(Z_1 Z_3 - Z_2^2) + \cosh(\psi)(Z_1 Z_2 - Z_2 Z_3)}{\sinh(\psi)(Z_1 Z_3 + Z_2^2) + \cosh(\psi)(Z_1 Z_2 + Z_2 Z_3)} , \\ \tau &= \frac{2 Z_1 Z_2}{\sinh(\psi)(Z_1 Z_3 + Z_2^2) + \cosh(\psi)(Z_1 Z_2 + Z_2 Z_3)} . \end{aligned} \tag{2.5.9}$$

After we have established the amplitudes of the magnetic vector potential it is possible to determine the electromagnetic field in each one of the regions thus we can investigate the power flow in the system. Using Poynting's theorem the power conservation implies that

$$\mathrm{Re}(Z_1) \left[1 - |\varrho|^2\right] = \mathrm{Re}(Z_3) |\tau|^2 . \tag{2.5.10}$$

This expression relates the power in the first region to that in the third. It does not depend explicitly on the second region; if, for example, in the third region the wave is below cutoff, the characteristic impedance is imaginary and the right-hand side is zero. Consequently, the absolute value of the reflection coefficient is unity, regardless of what happens in the second region. On the other hand, if in regions 1 and 3 the wave is above cutoff, and in region 2 the wave is below cutoff, we still expect power to be transferred. However, the transmission coefficient decays exponentially with $\psi = \Gamma_1^{(2)} d$

$$\tau \sim \frac{4 Z_1 Z_2}{(Z_1 Z_3 + Z_2^2) + Z_2(Z_1 + Z_3)} e^{-\psi} . \tag{2.5.11}$$

In spite of the discontinuities there can be frequencies at which the reflection coefficient (ϱ) is zero if we design the structure such that

$$Z_1 Z_3 = Z_2^2 \quad \text{and} \quad \psi = j\pi/2\,, \tag{2.5.12}$$

as one can conclude by examining the numerator of ϱ. The expression in (2.5.12) defines the conditions for the so-called quarter-λ transformer. A typical picture of the transmission coefficient is illustrated in Fig. 2.4. Note that the peaks in the transmission correspond to constructive interference of the two waves in the central section; the valleys correspond to destructive interference of the same waves. Zero reflections also occur when

$$Z_1 = Z_3 \quad \text{and} \quad \psi = j\pi\,; \tag{2.5.13}$$

this is typically the choice of parameters for (high power) vacuum windows.

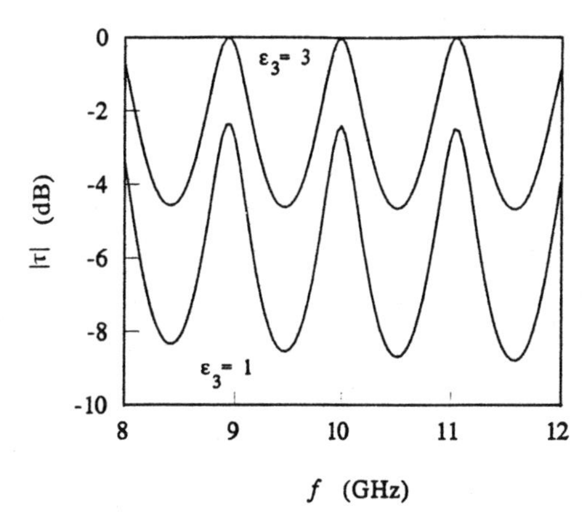

Fig. 2.4. Transmission coefficient as a function of the frequency for two cases: the upper trace represents a situation in which the dielectric coefficient in the third region equals that in the first, therefore at certain frequencies all the power is transferred – see (2.5.13). In the lower trace the two are different and the relation in (2.5.12) is not satisfied, therefore always a fraction of the energy is reflected. In both cases the constructive and destructive interference pattern is clearly revealed

If in the first and third region the wave is below cutoff but in the middle region a wave can propagate, then the system will determine a set of discrete frequencies at which the wave can bounce between the two sections. These eigen-frequencies are determined by the geometric parameters and the dielectric coefficients. We can calculate these frequencies from the poles of the transmission or reflection coefficient, namely from the condition that its denominator is zero:

$$\sinh(\psi)(Z_1 Z_3 + Z_2^2) + Z_2 \cosh(\psi)(Z_1 + Z_3) = 0\,. \tag{2.5.14}$$

Equivalently, one can write equations (2.5.5–8) in a matrix form, set the input term to zero ($A_{\text{in}} = 0$) and look for the non-trivial solution by requiring that the determinant of the matrix is zero – the result is identical with (2.5.14).

2.5.2 Geometric Discontinuity

Another source of reflected waves are geometric discontinuities. In a sense these can be conceived as impedance discontinuities but of a more complex character since geometric variations *couple* between the different modes in the waveguide. The simplest configuration which can be considered quasi-analytically consists of a waveguide of radius R_1 and another of radius $R_2 < R_1$; the discontinuity occurs at $z = 0$ as illustrated in Fig. 2.5. A detailed analysis when a single mode impinges upon a discontinuity can be found in the literature e.g., Mittra and Lee (1971) or Lewin (1975). We examine first the case when the source term is in the left-hand side ($z < 0$), therefore Green's function in the left-hand side has two components

$$G(z < 0, r|z' < 0, r') = \sum_{s=1}^{\infty} \frac{\mathrm{J}_0(p_s r/R_1)\, \mathrm{J}_0(p_s r'/R_1)}{\left[\frac{1}{2} R_1^2 \mathrm{J}_1^2(p_s)\right] \left[4\pi \Gamma_s^{(1)}\right]} \, \mathrm{e}^{-\Gamma_s^{(1)}|z-z'|} + \sum_{s=1}^{\infty} \varrho_s(r', z' < 0) \mathrm{J}_0\left(p_s \frac{r}{R_1}\right) \mathrm{e}^{\Gamma_s^{(1)} z} , \qquad (2.5.15)$$

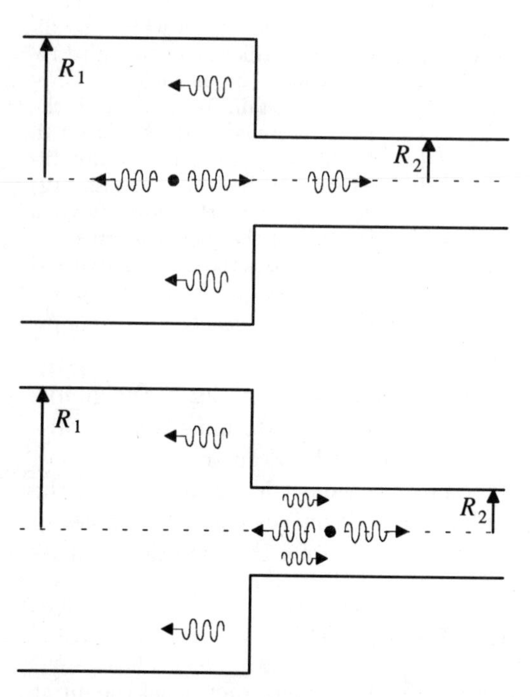

Fig. 2.5. Green's function calculation for one discontinuity in the geometry of a waveguide. In the upper figure the source is in the left and in the lower it is in the right

the non-homogeneous solution, which corresponds to an infinite waveguide and the homogeneous solution which is due to the discontinuity; $\Gamma_s^{(1)} = \sqrt{(p_s/R_1)^2 - (\omega/c)^2}$. In the right-hand side ($z > 0$),

$$G(z > 0, r|z' < 0, r') = \sum_{s=1}^{\infty} \tau_s(r', z' < 0) \mathrm{J}_0 \left(p_s \frac{r}{R_2} \right) \mathrm{e}^{-\Gamma_s^{(2)} z} , \tag{2.5.16}$$

where $\Gamma_s^{(2)} = \sqrt{(p_s/R_2)^2 - (\omega/c)^2}$. Continuity of the radial electric field at $z = 0$ entails

$$\frac{\partial^2}{\partial z \, \partial r} G(r, z = 0^- | r', z' < 0)$$

$$= \begin{cases} \frac{\partial^2}{\partial z \, \partial r} G(r, z = 0^+ | r', z' < 0) & \text{for} \quad 0 \leq r < R_2 , \\ 0 & \text{for} \quad R_1 \geq r \geq R_2 . \end{cases} \tag{2.5.17}$$

In order to determine the amplitudes ϱ_s and τ_s the last equation is multiplied by $\mathrm{J}_1(p_s r/R_1)$, the product is integrated from 0 to R_1 and using the orthogonality of the Bessel function [similar to (2.3.17) but for first order Bessel function] we obtain

$$g_s^{(1)}(r', z') - \varrho_s(r', z') = \sum_{\sigma=1}^{\infty} Z_{s,\sigma} \tau_\sigma (r', z') , \tag{2.5.18}$$

where

$$g_s^{(1)}(r', z') = \frac{\mathrm{J}_0(p_s r'/R_1)}{\frac{1}{2} R_1^2 \mathrm{J}_1^2(p_s)} \frac{1}{4\pi \Gamma_s^{(1)}} \mathrm{e}^{\Gamma_s^{(1)} z'} , \tag{2.5.19}$$

and

$$Z_{s,\sigma} \equiv \frac{\Gamma_\sigma^{(2)}}{\Gamma_s^{(1)}} \frac{p_\sigma}{p_s} \frac{R_1}{R_2} \frac{1}{\mathrm{J}_1^2(p_s)} \frac{2}{R_1^2} \int_0^{R_2} \mathrm{d}r \, r \, \mathrm{J}_1 \left(p_s \frac{r}{R_1} \right) \mathrm{J}_1 \left(p_\sigma \frac{r}{R_2} \right) . \tag{2.5.20}$$

Continuity of the azimuthal magnetic field in the domain $0 < r < R_2$ implies

$$\frac{\partial}{\partial r} G(r, z = 0^+ | r', z' < 0) = \frac{\partial}{\partial r} G(r, z = 0^- | r', z' < 0) . \tag{2.5.21}$$

As above, we use the fact that in the domain of interest, $\mathrm{J}_1(p_s r/R_2)$ form a complete orthogonal set of functions hence

$$\tau_\sigma(r', z') = \sum_{s=1}^{\infty} Y_{\sigma,s} \left[g_s^{(1)}(r', z') + \varrho_s(r', z') \right] , \tag{2.5.22}$$

where

$$Y_{\sigma,s} \equiv \frac{2}{R_1^2} \int_0^{R_2} \mathrm{d}r \, r \, \mathrm{J}_1 \left(p_s \frac{r}{R_1} \right) \mathrm{J}_1 \left(p_\sigma \frac{r}{R_2} \right) . \tag{2.5.23}$$

The integral in both expressions for Z and Y can be calculated analytically [Abramowitz and Stegun (1968) p. 484] and it is given by

$$\int_0^1 \mathrm{d}\xi \xi \mathrm{J}_1(p_n \xi) \mathrm{J}_1(p_m u \xi)$$

$$= \begin{cases} \frac{1}{2}\mathrm{J}_1^2(p_n) & \text{for } p_n = p_m u\,, \\ p_m u \left[p_n^2 - p_m^2 u^2\right]^{-1} \mathrm{J}_1(p_n)\mathrm{J}_0(p_m u) & \text{otherwise}\,. \end{cases} \tag{2.5.24}$$

From (2.5.18,22) one can determine the amplitudes of the reflected and transmitted waves. Adopting a vector notation, i.e., $\varrho_s(r', z' < 0) \rightarrow \mathbf{R}^{(-)}$, $\tau_s(r', z' < 0) \rightarrow \mathbf{T}^{(-)}$ and $g_s^{(1)}(r', z' < 0) \rightarrow \mathbf{g}^{(1)}$, these amplitudes can be formally written as

$$\mathbf{R}^{(-)} = (I + ZY)^{-1} (I - ZY)\, \mathbf{g}^{(1)} \tag{2.5.25}$$

and

$$\mathbf{T}^{(-)} = Y \left[I + (I + ZY)^{-1} (I - ZY)\right] \mathbf{g}^{(1)}\,. \tag{2.5.26}$$

In a similar way, if the source is in the right-hand side ($z' > 0$) then Green's function in the left-hand side can be written as

$$G(z < 0, r|z' > 0, r') = \sum_{s=1}^{\infty} \varrho_s(r', z' > 0)\, \mathrm{J}_0\left(p_s \frac{r}{R_1}\right) \mathrm{e}^{\Gamma_s^{(1)} z}\,, \tag{2.5.27}$$

and

$$G(z > 0, r|z' > 0, r') = \sum_{s=1}^{\infty} \frac{\mathrm{J}_0(p_s r/R_2)\, \mathrm{J}_0(p_s r'/R_2)}{\left[\frac{1}{2}R_2^2 \mathrm{J}_1^2(p_s)\right]\left[4\pi \Gamma_s^{(2)}\right]} \mathrm{e}^{-\Gamma_s^{(2)}|z-z'|}$$
$$+ \sum_{s=1}^{\infty} \tau_s(r', z' > 0)\, \mathrm{J}_0\left(p_s \frac{r}{R_2}\right) \mathrm{e}^{-\Gamma_s^{(2)} z}\,. \tag{2.5.28}$$

Continuity of E_r at $z = 0$ implies

$$\varrho_s(r', z') = \sum_{\sigma=1}^{\infty} Z_{s,\sigma} \left[g_\sigma^{(2)}(r', z') - \tau_\sigma(r', z')\right]\,, \tag{2.5.29}$$

where

$$g_s^{(2)}(r', z') = \frac{\mathrm{J}_0(p_s r'/R_2)}{\frac{1}{2}R_2^2 \mathrm{J}_1^2(p_s)} \frac{1}{4\pi \Gamma_s^{(2)}} \mathrm{e}^{-\Gamma_s^{(2)} z'}\,, \tag{2.5.30}$$

and the continuity of H_ϕ can be simplified to read

$$\tau_\sigma(r', z') + g_\sigma^{(2)}(r', z') = \sum_{s=1}^{\infty} Y_{\sigma,s}\, \varrho_s(r', z')\,. \tag{2.5.31}$$

Again, adopting a vector notation $\tau_s(r', z' > 0) \to \mathbf{T}^{(+)}$, $g_s^{(2)}(r', z' < 0) \to \mathbf{g}^{(2)}$ and $\varrho_s(r', z' > 0) \to \mathbf{R}^{(+)}$ we can write for the reflected and transmitted waves the following expressions

$$\mathbf{T}^{(+)} = -(I + YZ)^{-1}(I - YZ)\,\mathbf{g}^{(2)}, \tag{2.5.32}$$

and

$$\mathbf{R}^{(+)} = Z\left[I + (I + YZ)^{-1}(I - YZ)\right]\mathbf{g}^{(2)}. \tag{2.5.33}$$

With Green's function established we calculate now the energy emitted by a particle with a charge e as it traverses the discontinuity. The velocity v_0 of the charge is assumed to be constant, therefore the current distribution is given by (2.4.10) and the electric field which acts on the particle due to the discontinuity is given by

$$E_z(r, z, \omega) = \frac{ev_0}{j\omega\varepsilon_0}\left[\frac{\omega^2}{c^2} + \frac{\partial^2}{\partial z^2}\right]\int_{-\infty}^{\infty} \mathrm{d}z' G(r, z|0, z')\,\mathrm{e}^{-j(\omega/v_0)z'}. \tag{2.5.34}$$

With this field component we can examine the total power transferred by the particle i.e.,

$$P(t) = -2\pi \int_{-\infty}^{\infty} \mathrm{d}z \int_0^{R(z)} \mathrm{d}r\, r J_z(r, z, t) E_z(r, z, t), \tag{2.5.35}$$

and also the total energy defined by

$$W = \int_{-\infty}^{\infty} dt P(t), \tag{2.5.36}$$

which explicitly reads

$$\begin{aligned} W = &-\frac{e^2 v_0}{2\pi\varepsilon_0 R_1^2}\int_{-\infty}^{\infty} \mathrm{d}\omega \frac{1}{j\omega}\sum_{s=1}^{\infty} p_s^2 \int_{-\infty}^{0} \mathrm{d}z' \mathrm{e}^{-j(\omega/v_0)z'} \varrho_s(0, z') \\ &\times \int_{-\infty}^{0} dt \mathrm{e}^{j\omega t}\mathrm{e}^{\Gamma_s^{(1)} v_0 t} \\ &- \frac{e^2 v_0}{2\pi\varepsilon_0 R_2^2}\int_{-\infty}^{\infty} \mathrm{d}\omega \frac{1}{j\omega}\sum_{s=1}^{\infty} p_s^2 \int_0^{\infty} \mathrm{d}z' \mathrm{e}^{-j(\omega/v_0)z'} \tau_s(0, z') \\ &\times \int_0^{\infty} dt\, \mathrm{e}^{j\omega t}\mathrm{e}^{-\Gamma_s^{(2)} v_0 t}. \end{aligned} \tag{2.5.37}$$

According to (2.5.25,32) and the definitions of $g^{(1)}$ and $g^{(2)}$, we can write

$$\varrho_s(0, z' < 0) \equiv \sum_{s'=1} \alpha_{s,s'}\,\mathrm{e}^{\Gamma_{s'}^{(1)} z'}, \tag{2.5.38}$$

and

$$\tau_s(0, z' > 0) \equiv \sum_{s'=1} \chi_{s,s'} \, \mathrm{e}^{-\Gamma^{(2)}_{s'} z'} . \tag{2.5.39}$$

Consequently the expression for the total energy reads

$$\begin{aligned} W = & \frac{e^2 v_0^2}{2\pi\varepsilon_0 R_1^2} \int_{-\infty}^{\infty} \mathrm{d}\omega \sum_{s=1,s'=1}^{\infty} \frac{(p_s^2 \alpha_{s,s'})/j\omega}{j\omega - v_0 \Gamma^{(1)}_{s'}} \frac{1}{j\omega + v_0 \Gamma^{(1)}_{s}} \\ & - \frac{e^2 v_0^2}{2\pi\varepsilon_0 R_2^2} \int_{-\infty}^{\infty} \mathrm{d}\omega \sum_{s=1,s'=1}^{\infty} \frac{(p_s^2 \chi_{s,s'})/j\omega}{j\omega + v_0 \Gamma^{(2)}_{s'}} \frac{1}{j\omega - v_0 \Gamma^{(2)}_{s}} . \end{aligned} \tag{2.5.40}$$

The matrices α and χ are frequency dependent, therefore numerical methods have to be invoked in order to have a quantitative answer regarding the energy transfer. Nevertheless, the spectrum can be readily derived from these two expressions. The first term represents the energy emitted when the particle moves in the left-hand side and the second corresponds to the energy emitted when it moves in the right one. It should be pointed out that each one of the terms has two contributions: a fraction of the energy propagates to the left and the remainder to the right. In the next subsection we present a simpler configuration which allows one to trace analytically the way the electromagnetic field develops in time in the case of reflections.

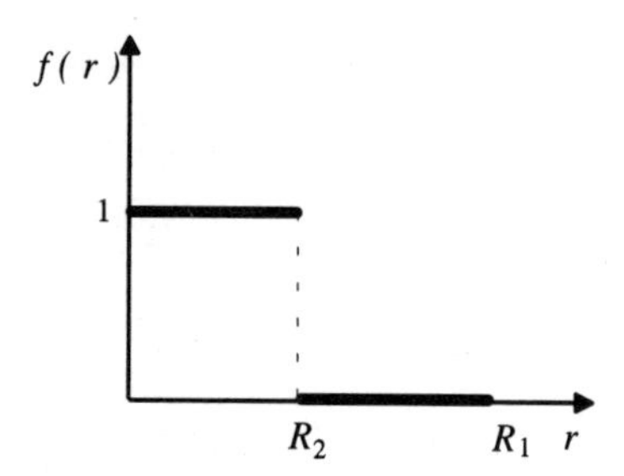

Fig. 2.6. Step function used to model the effect of truncation in a Bessel series representation

Before we conclude this subsection, there is one question we have to address. In principle the number of modes required to represent the field exactly is infinite, but practically only a finite number of terms is taken into consideration because of the need to invert the matrices numerically. The question is what should be the number of Bessel harmonics necessary for the representation of a discontinuity as the one presented above and what is the error associated with the truncation. In order to answer this question, let us consider a simple function

$$f(r) = \begin{cases} 1 & \text{for } 0 \le r < R_2 , \\ 0 & \text{for } R_2 < r \le R_1 , \end{cases} \tag{2.5.41}$$

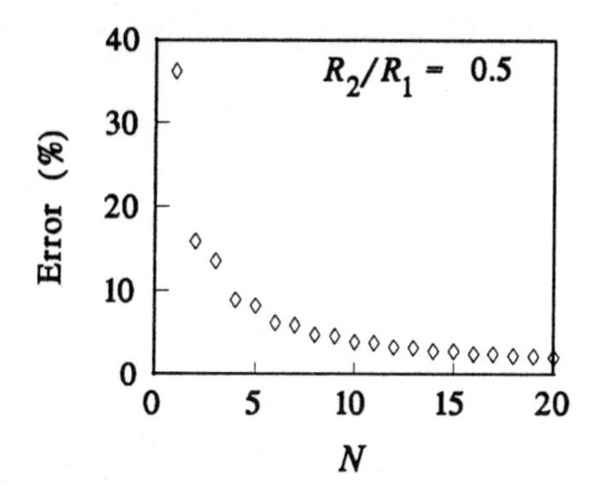

Fig. 2.7. Truncation error as a function of the number of terms

as illustrated in Fig. 2.6.

This function can also be represented by a superposition of Bessel functions:

$$f(r) = \sum_{s=1}^{\infty} f_s \mathrm{J}_0 \left(p_s \frac{r}{R_1} \right) , \tag{2.5.42}$$

where

$$f_s = 2 \frac{R_2}{R_1} \frac{\mathrm{J}_1(p_s R_2/R_1)}{p_s \mathrm{J}_1^2(p_s)} ; \tag{2.5.43}$$

here we used the fact that the integral

$$\int_0^x \mathrm{d}\xi \, \xi \mathrm{J}_0(\xi) = x \mathrm{J}_1(x) , \tag{2.5.44}$$

can be evaluated analytically [Abramowitz and Stegun (1968) p. 484]. We now define the relative error made when representing the function only with a finite number of Bessel harmonics as the

$$\mathrm{Error}(N) \equiv \frac{\int_0^{R_1} \mathrm{d}r \, r \left[f(r) - \sum_{s=1}^{N} f_s \mathrm{J}_0(p_s r/R_1) \right]^2}{\int_0^{R_1} \mathrm{d}r \, r \, f^2(r)} . \tag{2.5.45}$$

Using (2.5.43,44), the last relation can be simplified to read

$$\mathrm{Error}(N) = 1 - 4 \sum_{s=1}^{N} \left[\frac{\mathrm{J}_1(p_s R_2/R_1)}{p_s \mathrm{J}_1(p_s)} \right]^2 . \tag{2.5.46}$$

Figure 2.7 illustrates this error. Taking a single mode the normalized error is 36% for $R_2/R_1 = 0.5$ and it drops to 2% for 20 modes. Even with 20 modes the error can be significantly higher if the radii ratio is small and it is more than 15% for $N = 20$ and $R_2/R_1 \simeq 0.1$ – see Fig. 2.8. These facts become crucial when an accurate solution with multiple discontinuities is necessary.

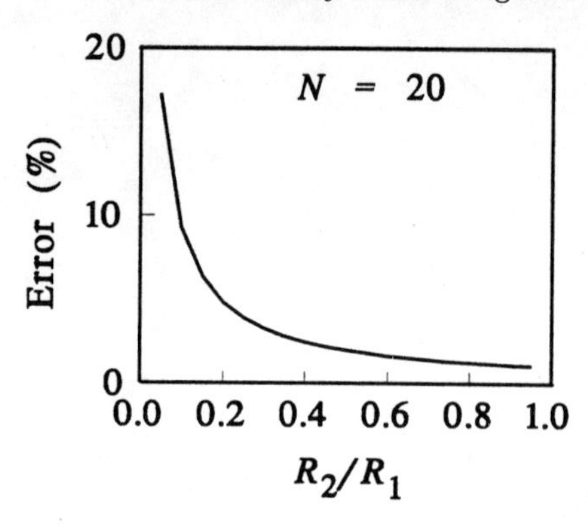

Fig. 2.8. Truncation error as a function of the radius ratio for a constant number of Bessel harmonics

2.5.3 Wake-Field in a Cavity

In order to examine transient phenomena associated with reflected waves we shall calculate the electromagnetic energy in a cavity as a single (point) charge traverses the structure. Consider a lossless cylindrical cavity of radius R and length d. A charged particle (e) moves along the axis at a constant velocity v_0. Consequently, the longitudinal component of the current density is the only non-zero term, thus

$$J_z(\mathbf{r},t) = -ev_0\frac{1}{2\pi r}\delta(r)\delta(z-v_0t)\,. \tag{2.5.47}$$

It excites the longitudinal magnetic vector potential $A_z(\mathbf{r},t)$ which for an azimuthally symmetric system satisfies

$$\left[\frac{1}{r}\frac{\partial}{\partial r}r\frac{1}{\partial r}+\frac{\partial^2}{\partial z^2}-\frac{1}{c^2}\frac{\partial^2}{\partial t^2}\right]A_z(r,z,t) = -\mu_0 J_z(r,z,t)\,. \tag{2.5.48}$$

In this section we shall consider only the internal problem, ignoring the electromagnetic phenomena outside the cavity. The boundary conditions on the internal walls of the cavity impose that $E_z(r=R,z,t)=0$, $E_r(r,z=0,t)=0$ and $E_r(r,z=d,t)=0$ therefore the magnetic vector potential reads

$$A_z(r,z,t) = \sum_{s=1,n=0}^{\infty} A_{s,n}(t)\,\mathrm{J}_0\left(p_s\frac{r}{R}\right)\cos\left(\frac{\pi n}{d}z\right)\,. \tag{2.5.49}$$

Using the orthogonality of the trigonometric and Bessel functions we find that the amplitude $A_{s,n}(t)$ satisfies

$$\begin{aligned}\left[\frac{\mathrm{d}^2}{\mathrm{d}t^2}+\Omega_{s,n}^2\right]A_{s,n}(t) &= -\frac{ev_0}{2\pi\varepsilon_0}\frac{1}{\frac{1}{2}R^2\mathrm{J}_1^2(p_s)}\frac{1}{g_n d}\\ &\quad\times\cos\left(\frac{\pi n}{d}v_0t\right)\left[h(t)-h\left(t-\frac{v_0}{d}\right)\right]\,,\end{aligned} \tag{2.5.50}$$

where

$$g_n = \begin{cases} 1 & \text{for } n = 0\,, \\ 0.5 & \text{otherwise}\,, \end{cases} \tag{2.5.51}$$

and

$$\Omega_{s,n} = c\sqrt{\left(\frac{p_s}{R}\right)^2 + \left(\frac{\pi n}{d}\right)^2}\,, \tag{2.5.52}$$

are the eigen-frequencies of the cavity. Before the particle enters the cavity ($t < 0$), no field exists, therefore

$$A_{s,n}(t < 0) = 0\,. \tag{2.5.53}$$

For the time the particle is in the cavity namely, $0 < t < d/v_0$, the solution of (2.5.50) consists of the homogeneous and the excitation term:

$$A_{s,n}\left(0 < t < \frac{d}{v_0}\right) = B_1 \cos(\Omega_{s,n} t) + B_2 \sin(\Omega_{s,n} t) + \alpha_{s,n} \cos(\omega_n t)\,, \tag{2.5.54}$$

where

$$\alpha_{s,n} = -\frac{e v_0}{2\pi\varepsilon_0} \frac{1}{\frac{1}{2}R^2 \mathrm{J}_1^2(p_s)} \frac{1}{g_n d} \frac{1}{\Omega_{s,n}^2 - \omega_n^2}\,, \tag{2.5.55}$$

and

$$\omega_n = \frac{\pi n}{d} v_0\,. \tag{2.5.56}$$

Since both the magnetic and the electric field are zero at $t = 0$, the function $A_{s,n}(t)$ and its first derivative are zero at $t = 0$ hence

$$B_1 + \alpha_{s,n} = 0\,, \tag{2.5.57}$$

and

$$B_2 = 0\,. \tag{2.5.58}$$

Consequently, the amplitude of the magnetic vector potential $[A_{s,n}(t)]$ reads

$$A_{s,n}(t) = \alpha_{s,n}\left[\cos(\omega_n t) - \cos(\Omega_{s,n} t)\right]\,. \tag{2.5.59}$$

Beyond $t = d/v_0$, the particle is out of the structure thus the source term in (2.5.50) is zero and the solution reads

$$A_{s,n}\left(t > \frac{d}{v_0}\right) = C_1 \cos\left[\Omega_{s,n}\left(t - \frac{d}{v_0}\right)\right] + C_2 \sin\left[\Omega_{s,n}\left(t - \frac{d}{v_0}\right)\right]\,. \tag{2.5.60}$$

As in the previous case, at $t = d/v_0$ both $A_{s,n}(t > d/v_0)$ and its derivative, have to be continuous:

$$\alpha_{s,n}\left[(-1)^n - \cos\left(\Omega_{s,n}\frac{d}{v_0}\right)\right] = C_1\,, \tag{2.5.61}$$

$$\alpha_{s,n}\Omega_{s,n}\sin\left(\Omega_{s,n}\frac{d}{v_0}\right) = C_2\,\Omega_{s,n}\,. \tag{2.5.62}$$

For this time period, the explicit expression for the magnetic vector potential is

$$\begin{aligned} A_{s,n}\left(t > \frac{d}{v_0}\right) &= \alpha_{s,n}\left[(-1)^n - \cos\left(\Omega_{s,n}\frac{d}{v_0}\right)\right]\cos\left[\Omega_{s,n}\left(t - \frac{d}{v_0}\right)\right] \\ &\quad + \alpha_{s,n}\sin\left(\Omega_{s,n}\frac{d}{v_0}\right)\sin\left[\Omega_{s,n}\left(t - \frac{d}{v_0}\right)\right], \end{aligned} \tag{2.5.63}$$

The expressions in (2.5.53,59,63) describe the magnetic vector potential in the cavity at all times. Figure 2.9 illustrates schematically this solution.

During the period the electron spends in the cavity, there are two frequencies which are excited: the eigen-frequency of the cavity $\Omega_{s,n}$ and the "resonances" associated with the motion of the particle, ω_n. The latter set corresponds to the case when the phase velocity, $v_{\rm ph} = \omega/k$, equals the velocity v_0. Since the boundary conditions impose $k = \pi n/d$ and the resonance implies

$$v_0 = v_{\rm ph} = c\left(\frac{\omega}{c}\frac{d}{\pi n}\right), \tag{2.5.64}$$

thus we can immediately deduce the resonance frequencies ω_n as given in (2.5.56).

Now that the magnetic vector potential has been determined, we consider the effect of the field generated in the cavity on the moving particle. The relevant component is

$$\begin{aligned} A_z\left(r, z, 0 < t < \frac{d}{v_0}\right) &= \sum_{s=1,n=0} \alpha_{s,n}\,\mathrm{J}_0\left(p_s\frac{r}{R}\right) \\ &\quad \times \cos\left(\frac{\pi n}{d}z\right)\left[\cos\left(\omega_n t\right) - \cos(\Omega_{s,n}t)\right]. \end{aligned} \tag{2.5.65}$$

Note that the upper limit in the double summation was omitted since in practice this limit is determined by the actual dimensions of the particle, which so far was considered infinitesimally small. In order to quantify this statement we realize that the summation is over all eigenmodes which have a wavenumber much longer than the particle's dimension i.e., $\Omega_{s,n}R_{\rm b}/c < 1$.

According to Maxwell's equations, the longitudinal electric field is

$$\varepsilon_0\frac{\partial}{\partial t}E_z(\mathbf{r}, t) = -J_z(\mathbf{r}, t) + \frac{1}{r}\frac{\partial}{\partial r}rH_\phi(\mathbf{r}, t)\,. \tag{2.5.66}$$

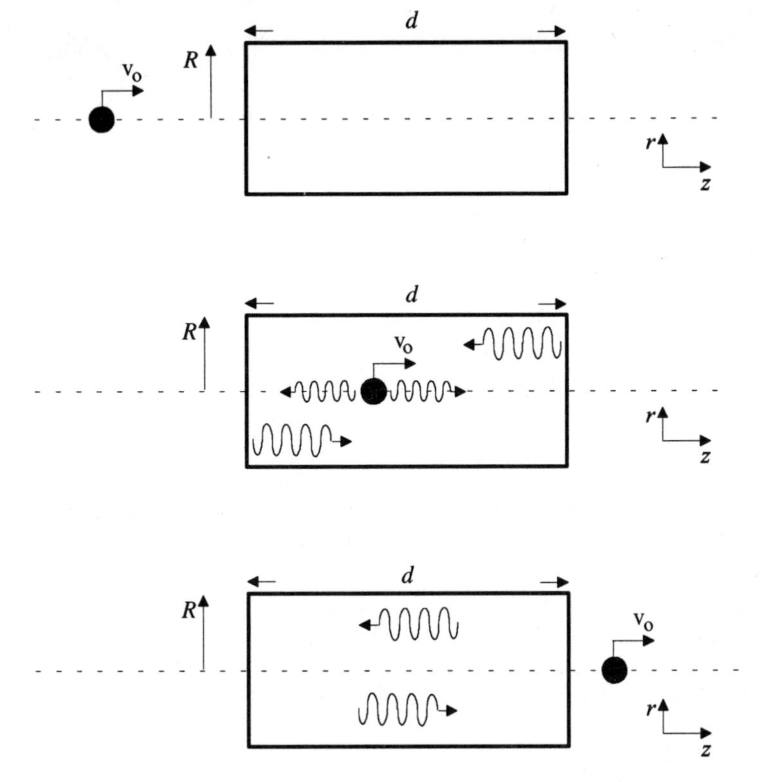

Fig. 2.9. Schematics of the field distribution generated by a particle as it traverses a cavity. Prior to its entrance no field exists in the cavity. When in the cavity the field has two contributions: directly from the source (non-homogeneous) and reflections from the walls (homogeneous). After the particle leaves the cavity only the homogeneous contribution remains

Furthermore, the field which acts on the particle does not include the self field, therefore we omit the current density term. Using the expression for the magnetic vector potential [(2.1.36)], we have

$$E_z(\mathbf{r},t) = -c^2 \int dt \, \frac{1}{r}\frac{\partial}{\partial r} r \frac{\partial}{\partial r} A_z(\mathbf{r},t) \,, \tag{2.5.67}$$

or explicitly,

$$\begin{aligned} E_z\left(r, z, 0 < t < \frac{d}{v_0}\right) = & \sum_{s=1,n=0} \alpha_{s,n} \left(\frac{cp_s}{R}\right)^2 \mathrm{J}_0\left(p_s \frac{r}{R}\right) \\ & \times \cos(\frac{\pi n}{d} z) \left[\frac{\sin(\omega_n t)}{\omega_n} - \frac{\sin(\Omega_{s,n} t)}{\Omega_{s,n}}\right] . \end{aligned} \tag{2.5.68}$$

In a lossless and closed cavity the total power flow is zero, therefore Poynting's theorem in its integral form reads

$$\frac{dW}{dt} = -2\pi \int_0^R dr\, r \int_0^d dz\, E_z(r,z,t)\, J_z(r,z,t)\,. \tag{2.5.69}$$

Thus substituting the current density [(2.5.47)] we obtain

$$W = ev_0 \int_0^{d/v_0} dt\, E_z(r, z = v_0 t, t)\,, \tag{2.5.70}$$

which has the following explicit form

$$\begin{aligned} W =& ev_0 \sum_{s=1,n=0} \alpha_{s,n} \left(\frac{cp_s}{R}\right)^2 \int_0^{d/v_0} dt\ \cos(\omega_n t) \\ &\times \left[\frac{\sin(\omega_n t)}{\omega_n} - \frac{\sin(\Omega_{s,n} t)}{\Omega_{s,n}}\right]. \end{aligned} \tag{2.5.71}$$

The time integral in this expression can be evaluated analytically. As can be readily deduced, the first term represents the non-homogeneous part of the solution and its contribution is identically zero whereas the second's reads

$$W = -ev_0 \sum_{s=1,n=0} \alpha_{s,n} \left(\frac{cp_s}{R}\right)^2 \frac{1-(-1)^n \cos(\Omega_{s,n} d/v_0)}{\Omega_{s,n}^2 - \omega_n^2}\,. \tag{2.5.72}$$

Substituting the explicit expression for $\alpha_{s,n}$ we have

$$\begin{aligned} \bar{W} \equiv& W \left(\frac{e^2}{4\pi\varepsilon_0 d}\right)^{-1} \\ =& \sum_{s=1,n=0} \left(\frac{2p_s}{J_1(p_s)}\right)^2 \frac{1}{g_n} \\ &\times \frac{1}{[p_s^2 + (\pi n R/d\gamma)^2]^2} \left[1-(-1)^n \cos\left(\frac{\Omega_{s,n}}{v_0} d\right)\right]. \end{aligned} \tag{2.5.73}$$

In Fig. 2.10 we illustrate two typical terms from the expression above as a function of the particle's momentum. The (normalized) energy stored at 10.7 GHz (corresponding to $s = 1, n = 1$) is shown in the left frame and we observe that for $\gamma\beta = 2.5$ the energy reaches its asymptotic value $[\bar{W}(s = 1, n = 1) \simeq 4]$. This is in contrast to the energy stored in the 35.5 GHz ($s = 3$, $n = 3$) wave which at the same momentum reaches virtually zero level; the asymptotic value $[\bar{W}(s = 3, n = 3) \simeq 0.5]$ is reached for a much higher momentum ($\gamma\beta = 15$). Figure 2.11 illustrates the normalized energy of all the frequencies below 300 GHz.

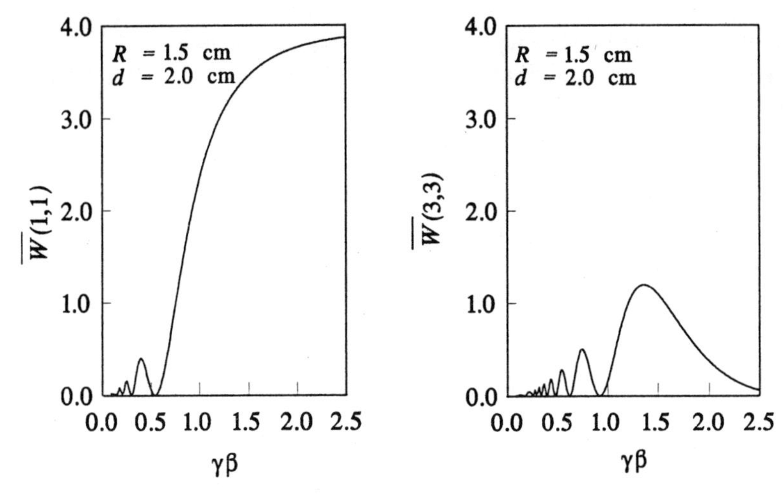

Fig. 2.10. Normalized energy generated at 10.7 GHz (left frame) as a function of the particle's momentum. In the right frame the normalized energy stored at 35.5 GHz, is presented

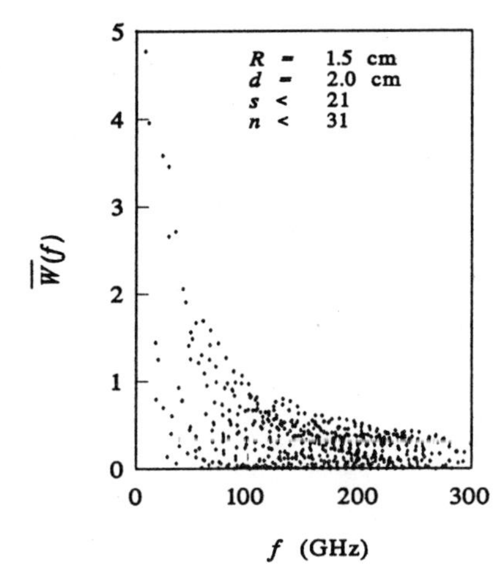

Fig. 2.11. Normalized energy as a function of the frequency as generated by a particle whose normalized momentum is $\gamma\beta = 15$

Exercises

2.1 Determine the boundary condition associated with charge conservation. How it relates to (2.1.13–16)?

2.2 In the context of Sect. 2.2.4, calculate the electromagnetic field associated with the moving charge (2.2.22-23). Calculate the Poynting vector associated with this field. With this result calculate the total power. Is there a force acting on the moving particle?

2.3 Show that the power radiated in free space by the current distribution in (2.3.2) is given by $P = \eta_0 I^2 \left(\omega d/c\right)^2 /12\pi$.

2.4 Calculate the energy velocity (Sect. 2.3.3) assuming two modes TM_{01} and TM_{02} above cut-off. Plot the energy velocity as a function of the ratio of the two modes $0.3 < \varrho = |A_{01}/A_{02}| < 3.0$.

2.5 Calculate the radiation impedance of the TM_{01} in a circular waveguide of radius R. Assume a current distribution

$$J_r(r,z,\omega) = I\Delta_z \, \mathrm{J}_1(p_1 r/R)\delta(z)p_1/2\pi r R \, .$$

2.6 Calculate Green's function associated with the system described in Sect. 2.5.1.

3. Elementary Electron Dynamics

There are numerous topics which can and probably should be discussed as background to the investigation of distributed electron-wave interaction. Among these, a fraction will be presented here with emphasis on basic concepts which are of relevance to the principles to be elaborated in the next chapters. All topics considered throughout the text rely on classical mechanics (Sect. 3.1) and without exception they are consistent with the special theory of relativity (Sect. 3.2), therefore the fundamentals of these two theories are summarized.

Beyond reviewing the fundamental concepts of relativistic classical dynamics, we consider in Sect. 3.3 some of the methods of electron generation and discuss the Child-Langmuir law which draws a limit on the maximum current achievable when applying a voltage on a cathode from which electrons are emitted. After electrons are generated, they are typically guided by magnetic fields and waveguides to the interaction region. In Sect. 3.4 we present some basics of beam propagation in free space with uniform or periodic magnetic field and in a uniform and disk loaded waveguide. The section concludes with the basic measures of beam quality: emittance and brightness.

Last section is dedicated to space-charge waves. After introducing the basic concepts of fast and slow space-charge waves, we consider two instabilities which can develop when these waves are present. One is the resistive wall instability and the other is the two-beam instability. The section and the chapter conclude with a discussion on the interference of two space-charge waves and their interaction with a cavity as it happens in relativistic klystrons.

3.1 Classical Dynamics

In a substantial fraction of the interaction schemes it is sufficient to describe the electron-wave interaction in the framework of classical mechanics and for this reason we shall not discuss here quantum mechanic effects. The classical approach includes either the Newtonian equation of motion, or Lagrangian or Hamiltonian formalism. In all cases the relativistic framework is considered. Furthermore, in all cases of interest many particles are involved and consequently, statistical methods are invoked and for this purpose, we present the kinetic and the fluid approximations which are used throughout the text.

3.1.1 Newtonian Equations of Motion

The elementary equations which describe the dynamics of a particle at the classical level are given by

$$\frac{\mathrm{d}}{\mathrm{d}t}\left[m\gamma(t)\mathbf{v}(t)\right] = \mathbf{F}(t)\,, \tag{3.1.1}$$

where $\mathbf{F}(t)$ is the force acting on the particle and if an electromagnetic field is present then the force is given by the so-called Lorentz force which reads

$$\mathbf{F}(t) = -e\left\{\mathbf{E}[\mathbf{r}(t),t] + \mathbf{v}(t)\times\mathbf{B}[\mathbf{r}(t),t]\right\}\,; \tag{3.1.2}$$

e and m represent the charge and the rest mass of the electron respectively, $\mathbf{v}(t)$ is its velocity vector at any point in time and

$$\gamma(t) = \frac{1}{\sqrt{1-\mathbf{v}(t)\cdot\mathbf{v}(t)/c^2}}\,. \tag{3.1.3}$$

The electromagnetic field, $\mathbf{E}(\mathbf{r}(t),t)$ and $\mathbf{B}(\mathbf{r}(t),t)$ is the field at the particle's location.

A full description of the particle's dynamics requires to determine also the location of the particle at each point in time; this is given by

$$\frac{\mathrm{d}}{\mathrm{d}t}\mathbf{r}(t) = \mathbf{v}(t)\,. \tag{3.1.4}$$

The state-vector of such a particle is a 6D vector and it consists of the relative location of the particle $\mathbf{r}(t)$ and its momentum i.e. $[\mathbf{r}(t),\, m\gamma(t)\mathbf{v}(t)]$.

As in the case of Maxwell's equations, the energy conservation can be deduced from these equations. For this purpose (3.1.1) is multiplied scalarly by $\mathbf{v}(t)$. After substituting (3.1.2) in the right-hand side we can find that the second term contribution is identically zero since the product $\mathbf{v}\times\mathbf{B}$ is orthogonal to both the velocity vector and the magnetic induction. In the left-hand side we have

$$\begin{aligned} &m\mathbf{v}(t)\cdot\frac{\mathrm{d}}{\mathrm{d}t}[\gamma(t)\mathbf{v}(t)] \\ &\quad = m\frac{1}{2}\gamma(t)\frac{\mathrm{d}}{\mathrm{d}t}\left[\mathbf{v}(t)\cdot\mathbf{v}(t)\right] + m\left[\mathbf{v}(t)\cdot\mathbf{v}(t)\right]\frac{\mathrm{d}}{\mathrm{d}t}\gamma(t)\,, \end{aligned} \tag{3.1.5}$$

which can be simplified if we now use the definition of $\gamma(t)$ in (3.1.3) to express $\mathbf{v}\cdot\mathbf{v}$ as $c^2[1-\gamma^{-2}(t)]$ which yields

$$mc^2\frac{\mathrm{d}}{\mathrm{d}t}\gamma(t) = -e\mathbf{v}(t)\cdot\mathbf{E}[\mathbf{r}(t),t]\,. \tag{3.1.6}$$

This is the expression for single particle energy conservation. Note that there are two different representations of the equations of motion: *(i)* it is possible to substitute (3.1.3) in the equation of motion and then we have to solve for the three components of $\mathbf{v}$ i.e. solve only (3.1.1–2). *(ii)* Another possibility is to postulate (3.1.1–2,6) [which anyhow yields (3.1.3)] and solve for the three components of $\mathbf{v}$ and for γ.

3.1.2 Lagrangian Formalism

It is convenient in many cases to use a different approach when formulating the dynamics of a single particle. The basic concept is to introduce a *scalar* function L, called Lagrangian, from which the *vector* equation of motion can be derived. This function depends on the velocity and location of the particle and in general it may also depend on time. Without loss of generality we can define the action as

$$I = \int_{t_1}^{t_2} \mathrm{d}t L(\mathbf{v}, \mathbf{r}; t) , \tag{3.1.7}$$

and require that the motion of the particle from time t_1 to time t_2 is such that the line integral is an extremum for the *path of motion*. To formulate this statement mathematically it implies to require that this action is at an extremum with respect to a virtual change $\delta\mathbf{r}$, hence

$$\begin{aligned}
\delta I &= \delta \int_{t_1}^{t_2} \mathrm{d}t L(\mathbf{v}, \mathbf{r}; t) = 0 , \\
&= \int_{t_1}^{t_2} \mathrm{d}t \delta L(\mathbf{v}, \mathbf{r}; t) = 0 , \\
&= \int_{t_1}^{t_2} \mathrm{d}t \left[\frac{\partial L}{\partial \mathbf{r}} \delta\mathbf{r} + \frac{\partial L}{\partial \mathbf{v}} \frac{\mathrm{d}}{\mathrm{d}t} \delta\mathbf{r} \right] = 0 , \\
&= \int_{t_1}^{t_2} \mathrm{d}t \delta\mathbf{r} \left[\frac{\partial L}{\partial \mathbf{r}} - \frac{\mathrm{d}}{\mathrm{d}t} \frac{\partial L}{\partial \mathbf{v}} \right] = 0 ,
\end{aligned} \tag{3.1.8}$$

In this context by "virtual" we mean an infinitesimal change in the configuration space due to an infinitesimal change of the coordinates system, $\delta\mathbf{r}$, consistent with the forces imposed on the particle at the given time. In the last line of the equation above we used the fact that after the integration by parts, the variation at t_1 and t_2 is identically zero. Thus, in order to satisfy $\delta I = 0$, the Lagrangian has to be a solution of the following differential equation

$$\frac{\mathrm{d}}{\mathrm{d}t} \left(\frac{\partial L}{\partial \mathbf{v}} \right) - \frac{\partial L}{\partial \mathbf{r}} = 0 . \tag{3.1.9}$$

This is called Lagrange's equation and it is identical with the (relativistic) equations of motion, provided that L is chosen to be

$$L = -mc^2\sqrt{1-\mathbf{v}\cdot\mathbf{v}/c^2} + e(\Phi - \mathbf{v}\cdot\mathbf{A}), \tag{3.1.10}$$

where Φ is the scalar electric potential and $\mathbf{A}$ is the magnetic vector potential.

At this point we are in position to define in a systematic way the momentum of a particle in the presence of an electromagnetic field. This will be referred to as the *canonical momentum* associated with the coordinate $\mathbf{r}$ and it is defined by

$$\begin{aligned}\mathbf{p} &= \frac{\partial L}{\partial \mathbf{v}} \\ &= m\frac{\mathbf{v}}{\sqrt{1-\mathbf{v}\cdot\mathbf{v}/c^2}} - e\mathbf{A} \\ &= m\gamma\mathbf{v} - e\mathbf{A}.\end{aligned} \tag{3.1.11}$$

With this definition in mind and Lagrange equation we can already point out one of the advantages of the Lagrangian formalism. If L is not an explicit function of one of the coordinates (say x) then the second term in (3.1.9) vanishes. This, in conjunction with the last definition, implies that the corresponding component of the canonical momentum (in this case p_x) is a constant. Therefore the constants of motion can be deduced from the symmetry of the system.

3.1.3 Hamiltonian Formalism

In particular if L does not depend explicitly on time then by its differentiation by parts and using Lagrange's equation we obtain

$$\begin{aligned}\frac{\mathrm{d}L}{\mathrm{d}t} &= \frac{\partial L}{\partial \mathbf{r}}\cdot\frac{\mathrm{d}\mathbf{r}}{\mathrm{d}t} + \frac{\partial L}{\partial \mathbf{v}}\cdot\frac{\mathrm{d}\mathbf{v}}{\mathrm{d}t} = \frac{\mathrm{d}}{\mathrm{d}t}\left(\frac{\partial L}{\partial \mathbf{v}}\right)\cdot\mathbf{v} + \frac{\partial L}{\partial \mathbf{v}}\cdot\frac{\mathrm{d}\mathbf{v}}{\mathrm{d}t} \\ \frac{\mathrm{d}L}{\mathrm{d}t} &= \frac{\mathrm{d}}{\mathrm{d}t}\left(\mathbf{v}\cdot\frac{\partial L}{\partial \mathbf{v}}\right) \\ \frac{\mathrm{d}}{\mathrm{d}t}&\left(L - \mathbf{v}\cdot\frac{\partial L}{\partial \mathbf{v}}\right) = 0.\end{aligned} \tag{3.1.12}$$

Subject to this condition, the expression in brackets is a constant and is proportional to the total energy in the system. Based on this last result it is convenient to define the so-called Hamiltonian of the system as

$$H = \mathbf{v}\cdot\mathbf{p} - L. \tag{3.1.13}$$

For a relativistic particle it reads

$$\begin{aligned}H &= \mathbf{v}\cdot(m\gamma\mathbf{v} - e\mathbf{A}) - \left[-mc^2\sqrt{1-\mathbf{v}\cdot\mathbf{v}/c^2} + e(\Phi - \mathbf{v}\cdot\mathbf{A})\right], \\ &= mc^2\gamma - e\Phi.\end{aligned} \tag{3.1.14}$$

According to the last expression and comparing it to the free particle case, the energy, in the presence of an electromagnetic field, is given by $E = H + e\Phi$.

Bearing in mind that, according to the special theory of relativity, the energy and the momentum are related by $E^2 = \mathbf{p}^2c^2 + m^2c^4$, we conclude that the Hamiltonian of a relativistic particle expressed in terms of momentum $\mathbf{p}$ [using (3.1.11)] is given by

$$H = \sqrt{(\mathbf{p} + e\mathbf{A})^2 c^2 + (mc^2)^2} - e\Phi . \tag{3.1.15}$$

This is also a scalar function and as in the case of the Lagrange's function, there are a set of equations which describe the motion of the system. These read

$$\frac{\mathrm{d}\mathbf{r}}{\mathrm{d}t} = \frac{\partial H}{\partial \mathbf{p}} , \tag{3.1.16}$$

and

$$\frac{\mathrm{d}\mathbf{p}}{\mathrm{d}t} = -\frac{\partial H}{\partial \mathbf{r}} . \tag{3.1.17}$$

Neither the Lagrangian nor the Hamiltonian formalisms include more information about the system than that provided by the Newtonian equations of motion; however as indicated in the case of the Lagrange function, the constants of motion can be determined in an easier and more systematic way. In addition, the formulation of the dynamics of more complex variables can be "naturally" formulated. Consider, for example, a dynamic variable $\varrho(\mathbf{p}, \mathbf{r}, t)$ and suppose it is required to determine its equation of motion. At first glance the vector equations of motion [(3.1.1–2)] give us a limited hint as to how to proceed whereas the Hamiltonian formalism is very helpful since firstly we can write

$$\begin{aligned} \frac{\mathrm{d}\varrho}{\mathrm{d}t} &= \frac{\partial \varrho}{\partial t} + \frac{\partial \varrho}{\partial \mathbf{r}}\frac{\partial \mathbf{r}}{\partial t} + \frac{\partial \varrho}{\partial \mathbf{p}}\frac{\partial \mathbf{p}}{\partial t} , \\ &= \frac{\partial \varrho}{\partial t} + \frac{\partial \varrho}{\partial \mathbf{r}}\frac{\mathrm{d}\mathbf{r}}{\mathrm{d}t} + \frac{\partial \varrho}{\partial \mathbf{p}}\frac{\mathrm{d}\mathbf{p}}{\mathrm{d}t} , \end{aligned} \tag{3.1.18}$$

and secondly substitute Hamilton's equations. The result is

$$\begin{aligned} \frac{\mathrm{d}\varrho}{\mathrm{d}t} &= \frac{\partial \varrho}{\partial t} + \frac{\partial \varrho}{\partial \mathbf{r}}\frac{\partial H}{\partial \mathbf{p}} - \frac{\partial \varrho}{\partial \mathbf{p}}\frac{\partial H}{\partial \mathbf{r}} , \\ &\equiv \frac{\partial \varrho}{\partial t} + \{\varrho, H\} . \end{aligned} \tag{3.1.19}$$

The latter definition is also known as the Poisson brackets. Hamiltonian formulation and a generalization of Poisson brackets provides the basis for the quantum formulation of microscopic electron's dynamic. A more detailed discussion on classical mechanics can be found in books by Goldstein (1950) and Landau and Lifshitz (1962).

3.1.4 Kinetic Approximation: Liouville's Theorem

The formulation presented above relies on a single particle interacting with the electromagnetic field and no direct interaction is considered other than through this field. Even in the absence of an external electromagnetic field (or a very simple one) there are numerous electrons in any system and it is not possible to solve instantaneously the equations of motion for all electrons, therefore statistical methods are invoked. Instead of information regarding each particle, we consider the probability density, $f(\mathbf{r}, \mathbf{p}, t)$, to find a particle at a given time t in the 6-dimensional phase-space element $\mathbf{rp} \rightarrow (\mathbf{r}+\delta\mathbf{r})(\mathbf{p}+\delta\mathbf{p})$; this probability density satisfies

$$\int_{-\infty}^{\infty} \mathrm{d}\mathbf{r} \int_{-\infty}^{\infty} \mathrm{d}\mathbf{p} f(\mathbf{r}, \mathbf{p}, t) \equiv 1 . \tag{3.1.20}$$

Although the notation is the same, it is important to realize the difference between $(\mathbf{r}, \mathbf{p})$ in this sub-section and the previous one: previously, $(\mathbf{r}, \mathbf{p})$ were the coordinates of a given particle in a 6D phase-space whereas here, we do not know the location of any of the particles and $(\mathbf{r}, \mathbf{p})$ in (3.1.20) are the variables of the probability density.

Assuming that we know this probability density function, the charge density is

$$\begin{aligned} \varrho(\mathbf{r}, t) &\equiv -en(\mathbf{r}, t) \\ &= -en_0 \int_{-\infty}^{\infty} \mathrm{d}\mathbf{p} f(\mathbf{r}, \mathbf{p}, t) , \end{aligned} \tag{3.1.21}$$

where n_0 is the average particle density and the current density is

$$\mathbf{J}(\mathbf{r}, t) = -en_0 \int_{-\infty}^{\infty} \mathrm{d}\mathbf{p}\, \mathbf{v}\, f(\mathbf{r}, \mathbf{p}, t) . \tag{3.1.22}$$

These two expressions indicate that, in principle, if we know this function we should be able to calculate the electromagnetic field. Motivated by this fact, we proceed and determine next the dynamics of this probability density function. According to Liouville's theorem, *the density of particles in phase-space as measured along the trajectory of a particle, is invariant.* This is valid, for non-interacting particles and closed system; however the formulation can be generalized to include collisions and external effects. For a collisionless ensemble, the Liouville theorem can be formulated as

$$\frac{\mathrm{d}}{\mathrm{d}t} f(\mathbf{r}, \mathbf{p}, t) = 0 . \tag{3.1.23}$$

Using the Hamiltonian dynamics in terms of Poisson brackets as formulated in (3.1.19) we have

$$
\begin{aligned}
\frac{\mathrm{d}}{\mathrm{d}t} f(\mathbf{r},\mathbf{p},t) = 0 & \\
&= \frac{\partial f(\mathbf{r},\mathbf{p},t)}{\partial t} + \{f(\mathbf{r},\mathbf{p},t), H\} \\
&= \frac{\partial f(\mathbf{r},\mathbf{p},t)}{\partial t} + \frac{\mathrm{d}\mathbf{r}}{\mathrm{d}t}\cdot\frac{\partial f(\mathbf{r},\mathbf{p},t)}{\partial \mathbf{r}} + \frac{\mathrm{d}\mathbf{p}}{\mathrm{d}t}\cdot\frac{\partial f(\mathbf{r},\mathbf{p},t)}{\partial \mathbf{p}} \\
&= \frac{\partial f(\mathbf{r},\mathbf{p},t)}{\partial t} + \mathbf{v}\cdot\frac{\partial f(\mathbf{r},\mathbf{p},t)}{\partial \mathbf{r}} \\
&\quad - e\left[\mathbf{E}(\mathbf{r},t) + \mathbf{v}\times\mathbf{B}(\mathbf{r},t)\right]\cdot\frac{\partial f(\mathbf{r},\mathbf{p},t)}{\partial \mathbf{p}} = 0\,. \qquad (3.1.24)
\end{aligned}
$$

This is also referred to as Vlasov equation.

Let us now present a very simple solution of this equation which also reflects on the character of the interaction of charges in plasma. In contrast to the case of a gas where each individual atom interacts only with its nearest neighbor due to the short range character of a neutral atom, in the case of charged particles the range of the Coulomb force is long and consequently many particles, in its vicinity, might be affected. We consider a static electric field ($\mathbf{E} = -\nabla\Phi$) which develops in a neutral system due to a local perturbation in the neutrality of the system. The Hamiltonian in this case reads

$$
H = \frac{\mathbf{p}^2}{2m} - e\Phi\,, \qquad (3.1.25)
$$

and the solution of (3.1.24) can be checked to read

$$
f(\mathbf{r},\mathbf{p},t) = f_0 \exp(-H/k_\mathrm{B}T)\,, \qquad (3.1.26)
$$

where $k_\mathrm{B} = 1.38066 \times 10^{-23}\mathrm{JK}^{-1}$ is the Boltzman constant and T is the absolute temperature of the particles; f_0 is determined using (3.1.20). Integration over the momentum can be performed analytically thus the density of the particles, according to (3.1.21), reads

$$
n(\mathbf{r}) = n_0 \mathrm{e}^{e\Phi(\mathbf{r})/k_\mathrm{B}T}\,. \qquad (3.1.27)
$$

A potential which develops in the distribution, causes a change $n(\mathbf{r}) - n_0$ in the particle density; n_0 is the average density of the particles. The latter affects in turn the electric scalar potential Φ hence

$$
\nabla^2\Phi(\mathbf{r}) = \frac{e}{\varepsilon_0} n_0 \left[\mathrm{e}^{e\Phi/k_\mathrm{B}T} - 1\right]. \qquad (3.1.28)
$$

Assuming $e|\Phi|/k_\mathrm{B}T \ll 1$, we can expand the right-hand side in Taylor series of which we keep only the first term. If we further assume spherical symmetry, we can readily solve (3.1.28) and the electrostatic potential is given by

$$
\Phi(r) = -\frac{e}{4\pi\varepsilon_0 r}\,\mathrm{e}^{-r/\lambda_\mathrm{D}}\,, \qquad (3.1.29)
$$

where $\lambda_{\rm D}$ is the Debye length, defined by

$$\lambda_{\rm D}^2 = \frac{\varepsilon_0 k_{\rm B} T}{n_0 e^2} \, . \tag{3.1.30}$$

The solution in (3.1.29) indicates that the potential generated by this perturbation is screened on a scale of the Debye length and beyond this radius its effect is vanishingly small. In order to get a feeling of this effect let us consider a relativistic beam of 3 mm radius which carries 1 kA and whose temperature is 1000° K. Debye length in this case is 80 μm. With this characteristic length parameter we can define the typical (Debye) sphere whose volume is $4\pi\lambda_{\rm D}^3/3$. In this range the charge has a non-negligible effect on adjacent particles. The number of the particles affected by the perturbation mentioned above, is proportional to the product of the averaged particles' density, n_0, and the volume of the Debye sphere. In order to avoid effects of such fluctuations it will be reasonable to require that no particles (other than the source) will be in this sphere i.e.,

$$n_0 \left(4\pi \frac{1}{3} \lambda_{\rm D}^3 \right) < 1 \, , \tag{3.1.31}$$

which also means that the density has to be larger than a critical value n_c given by

$$n_0 > n_c \equiv \left(\frac{4\pi}{3} \right)^2 \left(\frac{\varepsilon_0 k_{\rm B} T}{e^2} \right)^3 \, . \tag{3.1.32}$$

Whenever the kinetic approximation will be used it will be assumed that this condition is locally satisfied.

3.1.5 Hydrodynamic Approximation

In the framework of the kinetic approximation presented above, at a given location there is a finite probability to find particles of different velocities. In many cases the number of particles is sufficiently high such that in a small volume we can attribute to all particles a certain velocity and density. In other words, in an infinitesimal volume the average velocity of all the particles, is taken to be identical to that of a single particle which satisfies

$$\frac{\rm d}{{\rm d}t} \left[m\gamma(\mathbf{r},t)\mathbf{v}(\mathbf{r},t) \right] = -e \left[\mathbf{E}(\mathbf{r},t) + \mathbf{v}(\mathbf{r},t) \times \mathbf{B}(\mathbf{r},t) \right] \, . \tag{3.1.33}$$

In contrast to the case presented in Sect. 3.1.1, the velocity $\mathbf{v}$ represents a field, m is the rest mass of the particle and γ satisfies

$$\frac{\rm d}{{\rm d}t} \left[mc^2\gamma(\mathbf{r},t) \right] = -e\mathbf{v}(\mathbf{r},t) \cdot \mathbf{E}(\mathbf{r},t) \, . \tag{3.1.34}$$

The particles' density in the infinitesimal volume at the given time, $n(\mathbf{r}, t)$, satisfies the continuity equation

$$\nabla\cdot [n(\mathbf{r},t)\mathbf{v}(\mathbf{r},t)] + \frac{\partial}{\partial t} n(\mathbf{r},t) = 0, \qquad (3.1.35)$$

which is equivalent to the charge conservation introduced in context of Maxwell's equation (see Sect. 2.1.1). These three equations, (3.1.33–35), represent the basic equations of the *hydrodynamic approximation.* In the context of the equations above, the derivative $\mathrm{d}/\mathrm{d}t$ is given by

$$\frac{\mathrm{d}}{\mathrm{d}t} = \frac{\partial}{\partial t} + \mathbf{v}(\mathbf{r},t)\cdot\nabla. \qquad (3.1.36)$$

Assuming that the velocity and density fields were established, the charge and current densities read

$$\varrho(\mathbf{r},t) = -en(\mathbf{r},t), \qquad (3.1.37)$$

and

$$\mathbf{J}(\mathbf{r},t) = -en(\mathbf{r},t)\mathbf{v}(\mathbf{r},t). \qquad (3.1.38)$$

In order to quantify this approximation we can state that any variations of the velocity (or density) field on the scale of an infinitesimal volume are negligible on the scale of the distance between any two particles in this volume. If the density in the mentioned volume is n then the characteristic distance between each two particles is $l \simeq n^{-1/3}$ thus

$$\frac{|\nabla\cdot\mathbf{v}|}{|\mathbf{v}|} \ll \frac{1}{l}. \qquad (3.1.39)$$

If, for example, the largest spatial variation is determined by the radiation field i.e. λ, then the condition above implies that $\lambda \times n^{1/3} \ll 1$; for relativistic beam of radius 3 mm carrying a current of 1 kA, the density is $7\times 10^{11}\mathrm{m}^{-3}$, the characteristic length l is $l \simeq 110$ μm therefore as long as the radiation wavelength is larger than 1000 μm the approximation will be completely justified. However for a strongly bunched beam there will be regions in space where the density is orders of magnitude smaller than the (initial) average density and consequently the validity of the fluid approximation has to be properly re-examined. An equivalent formulation relies on the temperature of the beam which is assumed to be zero namely, the momentum distribution is described by a Dirac delta function. This is obviously an idealization and in practice we deal with low temperatures and the criterion can be based on (3.1.32) and formulated as $n \gg n_{\mathrm{c}}$.

3.1.6 Global Energy Conservation

In Chap. 2 we developed the Poynting's theorem from Maxwell's equations and it was indicated that it is associated with the power and energy conservation of the radiation field. It was formulated as

$$\nabla \cdot \mathbf{S} + \frac{\partial}{\partial t}\left[\frac{1}{2}\varepsilon_0\varepsilon_r \mathbf{E}\cdot\mathbf{E} + \frac{1}{2}\mu_0\mu_r \mathbf{H}\cdot\mathbf{H}\right] = -\mathbf{J}\cdot\mathbf{E}, \tag{3.1.40}$$

where $\mathbf{S} = \mathbf{E}\times\mathbf{H}$ is the Poynting vector and $\mathbf{J}$ was assumed to be given. At this stage we can release this constraint since the current density was determined in (3.1.38) in terms of the density and velocity fields. Our goal now is to formulate the global energy conservation of the electromagnetic, velocity and density fields as a conservation law e.g., charge conservation in (2.1.12). For this purpose, we substitute the current density definition in (3.1.40). In addition, we use the definition of the total electromagnetic energy density W from (2.1.19) to write

$$\nabla \cdot \mathbf{S} + \frac{\partial}{\partial t}W = en\mathbf{v}\cdot\mathbf{E}. \tag{3.1.41}$$

The scalar product on the right-hand side is identical to that in (3.1.34), therefore the last equation yields

$$\nabla \cdot \mathbf{S} + \frac{\partial}{\partial t}W = -mc^2\, n\frac{\mathrm{d}}{\mathrm{d}t}\gamma; \tag{3.1.42}$$

using the definition in (3.1.36) we have

$$\begin{aligned}\nabla \cdot \mathbf{S} + \frac{\partial}{\partial t}W &= -mc^2\left[n\frac{\partial}{\partial t}\gamma + n\mathbf{v}\cdot\nabla\gamma\right], \qquad (3.1.43)\\ &= -mc^2\left[\frac{\partial}{\partial t}(n\gamma) - \gamma\frac{\partial}{\partial t}n + \nabla\cdot(n\mathbf{v}\gamma) - \gamma\nabla\cdot(n\mathbf{v})\right].\end{aligned}$$

The continuity equation in (3.1.35) further simplifies this expression since the sum of the second and the fourth terms on the right-hand side is zero, hence

$$\nabla\cdot\left[\mathbf{S} + mc^2 n\gamma\mathbf{v}\right] + \frac{\partial}{\partial t}\left[W + mc^2 n\gamma\right] = 0. \tag{3.1.44}$$

This expression is the global energy conservation of the electromagnetic, velocity and density fields. The total energy flux is given by the first term and it is the sum of the electromagnetic Poynting vector and the kinetic energy flux: $\mathbf{S} + mc^2 n(\gamma - 1)\mathbf{v}$. The total energy density stored in the system is a superposition of the electromagnetic energy density W and the kinetic energy density $W + mc^2 n(\gamma - 1)$. For these interpretations we have subtracted from (3.1.44) the continuity equation

$$\nabla\cdot\left[\mathbf{S} + mc^2 n(\gamma - 1)\mathbf{v}\right] + \frac{\partial}{\partial t}\left[W + mc^2 n(\gamma - 1)\right] = 0, \tag{3.1.45}$$

multiplied by the rest energy of the electron i.e., mc^2.

3.2 Special Theory of Relativity

Modern high power radiation sources and accelerators rely on the interaction of electromagnetic waves with electrons whose velocity is very close to c. In these conditions one has to invoke relativistic dynamics.

3.2.1 Principles

The dynamics of the electrons as formulated so far is consistent with what is known as the *Special Theory of Relativity* as formulated by Albert Einstein in 1905. For an adequate formulation of its principles, we have to introduce the concept of the system of reference also referred to as frame of reference. It consists of a set of rulers to measure the distance in space and allow us to determine the location of an event in space and in addition, a series of clocks which show the time. An inertial frame of reference can be conceived as being attached to a free particle i.e., a particle which no forces act on, thus it moves with a constant velocity. Another frame of reference moving at a different but constant velocity is also inertial. The first principle of the theory states that: *(i) The laws of nature are form-invariant with respect to the transformation from one frame of reference to another.* In other words, the laws of nature can be written in the same form in all inertial systems of reference. As an example let us consider a motionless frame of reference $R(x, y, z, ct)$. The law of nature in the present example will be Maxwell's equations which read

$$\begin{aligned}
\nabla \times \mathbf{E} + \frac{\partial}{\partial t}\mathbf{B} &= 0\,, \\
\nabla \times \mathbf{H} - \frac{\partial}{\partial t}\mathbf{D} &= \mathbf{J}\,, \\
\nabla \cdot \mathbf{D} &= \varrho\,, \\
\nabla \cdot \mathbf{B} &= 0\,.
\end{aligned} \tag{3.2.1}$$

Now, in another frame of reference, $R'(x', y', z', ct')$, Maxwell's equations have an identical form. "Primed" notation indicates that the numbers which indicate the location and the time of the event under consideration are different than these measured in laboratory frame. Similarly, primed field components or the source terms are those measured by an observer in the moving frame and according to *(i)* they satisfy

$$\begin{aligned}
\nabla' \times \mathbf{E}' + \frac{\partial}{\partial t'}\mathbf{B}' &= 0\,, \\
\nabla' \times \mathbf{H}' - \frac{\partial}{\partial t'}\mathbf{D}' &= \mathbf{J}'\,, \\
\nabla' \cdot \mathbf{D}' &= \varrho'\,, \\
\nabla' \cdot \mathbf{B}' &= 0\,.
\end{aligned} \tag{3.2.2}$$

The two observers, one in the laboratory and the other in the moving frame of reference, intend to "compare notes" regarding data each one has measured. In this process they have to take into consideration the finite time it takes information to traverse the distance between two points. This brings us to the other principle of the special theory of relativity which states that *(ii) The phase velocity of a plane electromagnetic wave in vacuum is the same in all inertial frames of reference.*

3.2.2 Lorentz Transformation

In contrast with Newtonian mechanics, where the spatial coordinates are variables and the time is a parameter, according to special theory of relativity the attitude to time in the description of an event should be identical to the spatial coordinates. Therefore in order to describe the motion of a wave in vacuum, we denote by $\mathrm{d}r$ the space interval it traverses in a time interval $\mathrm{d}t$ hence, $\mathrm{d}r = c\,\mathrm{d}t$. The last expression can also be written as

$$\mathrm{d}s^2 = \mathrm{d}r^2 - c^2\,\mathrm{d}t^2 = \mathrm{d}x^2 + \mathrm{d}y^2 + \mathrm{d}z^2 - c^2\mathrm{d}t^2 = 0\,, \tag{3.2.3}$$

which is a generalization of the concept of distance in a regular three dimensional space. As clearly indicated, in the particular case of a plane wave, the distance it traverses in the four dimensional space (space-time) is zero. By virtue of the invariance of the phase velocity the same statement can be written by the moving observer as

$$(\mathrm{d}s')^2 = \mathrm{d}x'^2 + \mathrm{d}y'^2 + \mathrm{d}z'^2 - c^2\mathrm{d}t'^2 = 0\,. \tag{3.2.4}$$

Without loss of generality, we can assume that the relative motion of the two frames of reference is along the z axis and that at a certain point in space-time the two frames overlap; thus we assume the following general transformation

$$\begin{aligned} \mathrm{d}x' &= \mathrm{d}x\,, \\ \mathrm{d}y' &= \mathrm{d}y\,, \\ \mathrm{d}z' &= a_{11}\mathrm{d}z - a_{12}c\mathrm{d}t\,, \\ c\mathrm{d}t' &= a_{22}c\mathrm{d}t - a_{21}\mathrm{d}z\,. \end{aligned} \tag{3.2.5}$$

Substituting these relations in (3.2.3–4) and comparing coefficients we find the following relations

$$\begin{aligned} a_{11}^2 - a_{21}^2 &= 1\,, \\ a_{22}^2 - a_{12}^2 &= 1\,, \\ a_{11}a_{12} - a_{22}a_{21} &= 0\,. \end{aligned} \tag{3.2.6}$$

At the origin ($z' = 0$), we must have $z = v_0 t$ where v_0 is the relative velocity between the two frames, therefore $a_{12}/a_{11} = \beta \equiv v_0/c$. With this observation we can now determine the coefficients of (3.2.5). Firstly, we substitute $a_{12} =$

βa_{11} in the second and third equation. Secondly, we substitute a_{22} from one of the resulting equations. The equation obtained for a_{21} can be solved and the result is $a_{21} = \gamma\beta$ and $\gamma = [1 - \beta^2]^{-1/2}$. The other two coefficients can be readily determined and they are given by $a_{12} = \gamma\beta$, $a_{22} = \gamma$ and $a_{11} = \gamma$. These coefficients define the so-called Lorentz transformation which for the 4-vector of the coordinates $(\mathbf{r}, ct)$ can be formulated as

$$\begin{aligned} x' &= x\,, \\ y' &= y\,, \\ z' &= \gamma(z - \beta ct)\,, \\ ct' &= \gamma(ct - \beta z)\,. \end{aligned} \tag{3.2.7}$$

The transformation from the laboratory frame of reference to the moving one is determined by reversing the sign of β and replacing the prime and unprimed variables namely,

$$\begin{aligned} x &= x'\,, \\ y &= y'\,, \\ z &= \gamma(z' + \beta ct')\,, \\ ct &= \gamma(ct' + \beta z')\,. \end{aligned} \tag{3.2.8}$$

The same transformation relates the components of any 4-vector in the moving frame and the laboratory.

3.2.3 Phase Invariance

The phase velocity of a plane wave was defined as the velocity an observer has to move in order to measure the same phase – e.g. to be on the crest of the wave. If according to the special theory of relativity this velocity is the same in all frames of reference, then we may expect that the *phase* itself is also invariant otherwise, the observer will measure a phase which varies. Hence,

$$\omega t - \mathbf{k}\cdot\mathbf{r} = \omega' t' - \mathbf{k}'\cdot\mathbf{r}' = \text{const.}\,, \tag{3.2.9}$$

where ω is the angular frequency and $\mathbf{k}$ is the wavenumber vector. Substituting Lorentz transformations we obtain the following transformation for the frequency and wavenumbers:

$$\begin{aligned} k'_x &= k_x\,, \\ k'_y &= k_y\,, \\ k'_z &= \gamma\left(k_z - \beta\frac{\omega}{c}\right)\,, \\ \frac{\omega'}{c} &= \gamma\left(\frac{\omega}{c} - \beta k_z\right)\,. \end{aligned} \tag{3.2.10}$$

As in the space-time transformation, the inverse is obtained by reversing the sign of β and replacing the prime and unprime variables i.e.,

$$\begin{aligned}
k_x &= k'_x \,, \\
k_y &= k'_y \,, \\
k_z &= \gamma\left(k'_z + \beta\frac{\omega'}{c}\right) \,, \\
\frac{\omega}{c} &= \gamma\left(\frac{\omega'}{c} + \beta k'_z\right) \,.
\end{aligned} \tag{3.2.11}$$

Similar to the 4-vector of the coordinates $(\mathbf{r}, ct)$ which describes an event in an inertial frame of reference, the set $(\mathbf{k}, \omega/c)$ also forms a 4-vector which describes the propagation of an electromagnetic wave in vacuum.

Another 4-vector is that of the source densities $(\mathbf{J}, c\varrho)$ as can be concluded after applying invariance principle on the charge conservation law as expressed in (2.1.12). The detailed proof is left to the reader as an exercise. In addition, the potentials $(\mathbf{A}, \Phi/c)$ form a 4-vector provided that the Lorentz gauge is imposed. In order to prove the last statement one can check, by analogy to (3.2.3), that the wave equation operator

$$\frac{\partial^2}{\partial x^2} + \frac{\partial^2}{\partial y^2} + \frac{\partial^2}{\partial z^2} - \frac{1}{c^2}\frac{\partial^2}{\partial t^2} \,, \tag{3.2.12}$$

is relativistically invariant. Subject to the Lorentz gauge, we can write the wave equations for the potentials

$$\left[\frac{\partial^2}{\partial x^2} + \frac{\partial^2}{\partial y^2} + \frac{\partial^2}{\partial z^2} - \frac{1}{c^2}\frac{\partial^2}{\partial t^2}\right]\left(\mathbf{A}, \frac{\Phi}{c}\right) = -\mu_0\left(\mathbf{J}, c\varrho\right) \,. \tag{3.2.13}$$

Since the right-hand side is relativistically invariant, as is the wave-equation operator, we conclude that the three components of $\mathbf{A}$ and Φ, i.e. $(\mathbf{A}, \Phi/c)$, form a relativistically invariant 4-vector.

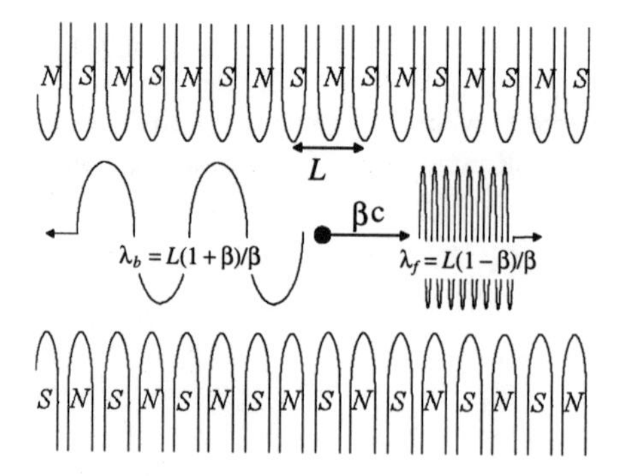

Fig. 3.1. Radiation emitted by an electron moving in a periodic magnetic field as measured in the laboratory frame of reference

It is instructive to examine a few of the principles presented above in a free electron laser - see Sect. 1.2.4. This consists of a static periodic magnetic field (wiggler) and an electron beam which is injected in this field with a velocity v_0 from $z \to -\infty$ to ∞ as illustrated in Fig. 3.1. If the period of the wiggler is L then the magnetostatic ($\omega = 0$) wavenumber associated with this field is $k_{\mathrm{w}} = 2\pi/L$ i.e., $B_x = B_0 \cos(k_{\mathrm{w}} z)$. This magnetostatic field is seen by the moving electron as a wave since if we substitute the third equation of (3.2.8) we obtain $\cos[k_{\mathrm{w}}\gamma(z' + \beta ct')]$. In the frame of reference attached to the electron this wave propagates from $z' \to \infty$ to $-\infty$. Based on the argument of the trigonometric function we can readily identify the characteristic frequency the electron oscillates as $\omega' = ck_{\mathrm{w}}\gamma\beta$ and the wave number as $k'_z = -k_{\mathrm{w}}\gamma$. The same result is achieved by employing the transformations in (3.2.10): firstly, we observe that the wiggler's wavenumber in the laboratory frame of reference is $k_z = -k_{\mathrm{w}}$ and as indicated previously, the field is static, therefore $\omega = 0$. Secondly, we substitute these two quantities in (3.2.10) and obtain:

$$
\begin{aligned}
k'_z &= \gamma[(-k_{\mathrm{w}}) - 0] = -\gamma k_{\mathrm{w}} \,, \\
\frac{\omega'}{c} &= \gamma[0 - \beta(-k_{\mathrm{w}})] = \gamma\beta k_{\mathrm{w}} \,.
\end{aligned}
\tag{3.2.14}
$$

Again, this corresponds to a backward propagating wave since

$$
k'_z = -\frac{1}{\beta}\frac{\omega'}{c} \,. \tag{3.2.15}
$$

Under the influence of this wiggler the electron oscillates around its initial location ($z' = 0$) thus, the force in its frame is expected to have the form $\cos(k_{\mathrm{w}}\gamma\beta ct')$. As it oscillates, it emits radiation in all directions; however we shall consider only the waves emitted along the z' axis. In the positive direction it emits a wave which oscillates at ω' and its wavenumber is $k'_z = \omega'/c$. Substituting in (3.2.11) we translate the parameters of this wave into the laboratory frame of reference, the result being

$$
\frac{\omega}{c} = \gamma\left(\frac{\omega'}{c} + \beta\frac{\omega'}{c}\right) = k_{\mathrm{w}}\gamma^2\beta(1+\beta) \,. \tag{3.2.16}
$$

For relativistic particles ($\beta \to 1$) the radiation wavelength is

$$
\lambda \simeq \frac{1}{2\gamma^2} L \,, \tag{3.2.17}
$$

and if $L = 4\,\mathrm{cm}$, the radiation wavelength for a $100\,\mathrm{MeV}$ electron is $0.5\,\mu\mathrm{m}$, thus this scheme has the potential of generating tunable radiation at frequencies which are not achievable with atomic or molecular lasers.

The wave emitted in the negative direction of the z' axis has a wave number $k'_z = -\omega'/c$ and in the laboratory frame of reference its frequency is

$$\frac{\omega}{c} = \gamma \left[\frac{\omega'}{c} + \beta \left(-\frac{\omega'}{c} \right) \right] = k_{\mathrm{w}} \frac{\beta}{1+\beta} . \tag{3.2.18}$$

Clearly in this direction (anti-parallel to the electron) the frequency is much lower and even for a relativistic particle it reaches the $\omega \simeq ck_{\mathrm{w}}/2$ level.

3.2.4 Field Transformation

Using the Lorentz transformations introduced above and the invariance of Maxwell's equations as formulated in (3.2.1–2) one can show that the relation between the electromagnetic field as measured in the laboratory and the one measured in the moving frame of reference is

$$\begin{aligned} E'_x &= \gamma(E_x - v_0 B_y) , \\ E'_y &= \gamma(E_y + v_0 B_x) , \\ E'_z &= E_z , \end{aligned} \tag{3.2.19}$$

and

$$\begin{aligned} H'_x &= \gamma(H_x + v_0 D_y) , \\ H'_y &= \gamma(H_y - v_0 D_x) , \\ H'_z &= H_z . \end{aligned} \tag{3.2.20}$$

With regard to (3.2.19) it is interesting to note that the equations of motion as written in (3.1.1) can be conceived as the translation of the equations of motion from the frame where the particle is momentarily at rest, to the laboratory frame of reference. In the former, only the electric field (**E**) affects the motion of the particle. When "translated" in terms of the variables as measured by an observer in the laboratory frame of reference, the term $\mathbf{v} \times \mathbf{B}$ appears.

The simplest manifestation of the field transformation to the topics to be encountered in this text is revealed by the different form the electromagnetic field of a moving electron has in the different frames of reference. In its rest frame of reference, the potentials it generates are given by

$$\begin{aligned} \mathbf{A}' &= 0 , \\ \Phi' &= \frac{-e}{4\pi\varepsilon_0} \frac{1}{r'} \equiv \frac{-e}{4\pi\varepsilon_0} \frac{1}{\sqrt{(x')^2 + (y')^2 + (z')^2}} , \end{aligned} \tag{3.2.21}$$

and correspondingly the electromagnetic field is

$$
\begin{aligned}
E'_x &= \frac{-e}{4\pi\varepsilon_0}\frac{x'}{(r')^3}\,,\\
E'_y &= \frac{-e}{4\pi\varepsilon_0}\frac{y'}{(r')^3}\,,\\
E'_z &= \frac{-e}{4\pi\varepsilon_0}\frac{z'}{(r')^3}\,,\\
H'_x &= 0\,,\\
H'_x &= 0\,,\\
H'_z &= 0\,.
\end{aligned}
\tag{3.2.22}
$$

In the laboratory frame of reference the 4-potential is given by

$$
\begin{aligned}
A_x &= 0\,,\\
A_y &= 0\,,\\
A_z &= \gamma\left(0+\beta\frac{\Phi'}{c}\right) = \gamma\beta\frac{1}{c}\frac{-e}{4\pi\varepsilon_0}\frac{1}{r'}\,,\\
\frac{\Phi}{c} &= \gamma\frac{1}{c}\frac{-e}{4\pi\varepsilon_0}\frac{1}{r'}\,,
\end{aligned}
\tag{3.2.23}
$$

which, in terms of the coordinates in this frame of reference, read

$$
\begin{aligned}
A_z &= \gamma\beta\frac{1}{c}\frac{-e}{4\pi\varepsilon_0}\frac{1}{\sqrt{x^2+y^2+\gamma^2(z-\beta ct)^2}}\,,\\
\frac{\Phi}{c} &= \gamma\frac{1}{c}\frac{-e}{4\pi\varepsilon_0}\frac{1}{\sqrt{x^2+y^2+\gamma^2(z-\beta ct)^2}}\,.
\end{aligned}
\tag{3.2.24}
$$

There are two ways now to calculate the electromagnetic field in the laboratory frame of reference: we can either use the 4-vector of the potentials and take the derivatives i.e.

$$
\begin{aligned}
E_x &= -\frac{\partial A_x}{\partial t} - \frac{\partial\Phi}{\partial x} = -\frac{\partial\Phi}{\partial x} = \gamma\frac{-e}{4\pi\varepsilon_0}\frac{x}{R^3}\,,\\
E_y &= -\frac{\partial A_y}{\partial t} - \frac{\partial\Phi}{\partial y} = -\frac{\partial\Phi}{\partial y} = \gamma\frac{-e}{4\pi\varepsilon_0}\frac{y}{R^3}\,,\\
E_z &= -\frac{\partial A_z}{\partial t} - \frac{\partial\Phi}{\partial z} = \gamma\frac{-e}{4\pi\varepsilon_0}\frac{z-\beta ct}{R^3}\,,\\
H_x &= \frac{1}{\mu_0}\frac{\partial A_z}{\partial y} = \gamma\beta\frac{1}{c\mu_0}\frac{-e}{4\pi\varepsilon_0}\frac{-y}{R^3}\,,\\
H_y &= \frac{-1}{\mu_0}\frac{\partial A_z}{\partial x} = \gamma\beta\frac{1}{c\mu_0}\frac{-e}{4\pi\varepsilon_0}\frac{x}{R^3}\,,\\
H_z &= 0\,,
\end{aligned}
\tag{3.2.25}
$$

or use transformations similar to (3.2.19–20) but from the moving frame to the laboratory namely,

$$
\begin{aligned}
E_x &= \gamma\left(E'_x + \beta c B'_y\right) = \gamma E'_x = \gamma \frac{-e}{4\pi\varepsilon_0}\frac{x}{R^3}\,,\\
E_y &= \gamma\left(E'_y - \beta c B'_x\right) = \gamma E'_y = \gamma \frac{-e}{4\pi\varepsilon_0}\frac{y}{R^3}\,,\\
E_z &= E'_z = \gamma \frac{-e}{4\pi\varepsilon_0}\frac{z-\beta ct}{R^3}\,,\\
H_x &= \gamma\left(H'_x - \beta c D'_y\right) = -\gamma\beta c D'_y = \gamma\beta c\varepsilon_0 \frac{-e}{4\pi\varepsilon_0}\frac{-y}{R^3}\,,\\
H_y &= \gamma\left(H'_y + \beta c D'_x\right) = \gamma\beta c D'_x = \gamma\beta c\varepsilon_0 \frac{-e}{4\pi\varepsilon_0}\frac{x}{R^3}\,,\\
H_z &= H'_z = 0\,;
\end{aligned}
\tag{3.2.26}
$$

in these expressions $R = \sqrt{x^2 + y^2 + \gamma^2(z-\beta ct)^2}$. This example emphasizes the fact that the electrostatic field in the particle's frame of reference is measured by an observer in the laboratory frame of reference as a time-dependent field. Nevertheless no net power is transmitted in the direction transverse to the motion but there is a non-zero Poynting flux flowing parallel to the particle and attached to it. Additional examples of the use of the Special Theory of Relativity can be found in books by Pauli (1958), Van Bladel (1984) and Schieber (1986).

3.3 Electron Generation

In all interaction mechanisms to be discussed in this text the electrons are free. But in nature electrons are attached to atoms or molecules, therefore they have to be extracted from the material in order to utilize them for conversion of energy. In this section we shall consider a few topics associated with electrons' generation. There are several ways to free electrons from the bulk material: they can be extracted from metals by applying an electric field perpendicular to the metal-vacuum interface – this is called field-emission. In recent years this emission gained a renewed interest due to the possibility of building small tips (on a microns level) using very large scale integration (VLSI) technology. With this technique each tip emits small currents but since many such tips can be made on one centimeter square the current density can be very high. The applied voltage is also low comparing to usual field-emitters. A second method to extract electrons is the thermionic emission. In this case, the emitting surface (cathode) is heated and a fraction of the electrons in the material can overcome the work function and they become free. A third mechanism relies on the photo-emission effect: a laser beam illuminates the cathode providing the electron with sufficient energy to overcome the work function and in this process it is being released from the metal. The fourth mechanism is based on the emission from ferro-electric ceramics. This is a material whose molecules are initially polarized in a preferred direction. Its internal polarization is screened by surface charge. As the polarization field is altered (by external means) the screening charge is no longer required and consequently, electrons may become available. A fifth possible mechanism relies on what is called secondary emission in which case one electron hits a surface and releases more electrons.

3.3.1 Child-Langmuir Limiting Current

The microscopic details of the field emission from metal have been considered in literature and a short review can be found in Miller's (1982) book. It is beyond the scope of our presentation to investigate the dynamics of electrons at the microscopic level in the metal, so we shall assume at the moment that whatever field is applied normal to the metallic cathode, electrons are being emitted. In fact, the discussion to follow is independent of the way electrons are extracted from the cathode. The question we shall address first is, what is the limiting current one can extract from a cathode by applying a voltage on the anode? The logic behind the search after this limit is simple: As a voltage is applied, electrons leave the cathode and move towards the anode. Since they traverse the anode-cathode gap in a finite time, the cathode is screened by these electrons and the field it sees is smaller – the situation is illustrated in Fig. 3.2. Maximum current limit is reached when the electric field on the anode is zero that is to say, the cathode is screened.

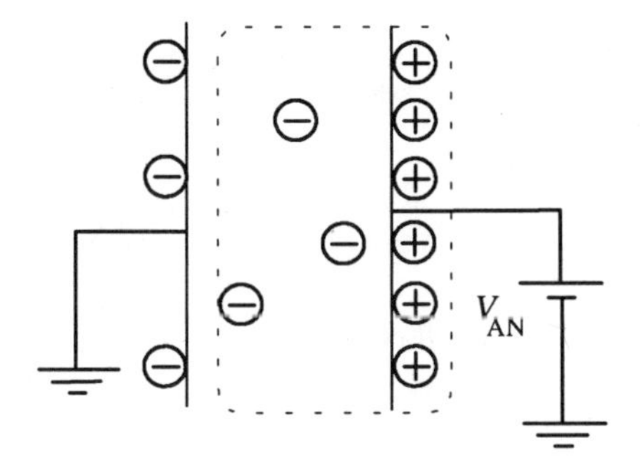

Fig. 3.2. Child-Langmuir limiting current: electrons in the gap screen the cathode

A typical voltage pulse on the anode is 100 nsec or longer and in comparison, the time an electron moving at $0.5c$ traverses a 3 cm gap is on the order of 0.2 nsec. The three orders of magnitude difference between these two time scales justifies the static approximation used next. In a one dimension system the electric scalar potential is a solution of the Poisson equation

$$\frac{d^2\Phi}{dz^2} = \frac{e}{\varepsilon_0}n(z)\,, \tag{3.3.1}$$

where $n(z)$ is the particle's density in the anode-cathode gap and it is yet to be determined; e is the charge of one electron. The dynamics of the particles' density is determined (within the framework of the hydro-dynamic approximation) by the continuity equation which in the case of a static problem reads

$$\frac{\mathrm{d}}{\mathrm{d}z}[n(z)v(z)] = 0\,. \tag{3.3.2}$$

The velocity $v(z)$, is governed by the equations of motion but in this particular case it is more convenient to use the single particle energy conservation (3.1.34) which in conjunction with (3.1.36) and the static case considered here ($\partial/\partial t = 0$) reads

$$\frac{\mathrm{d}}{\mathrm{d}z}\left[\gamma(z) - \frac{e\Phi(z)}{mc^2}\right] = 0\,. \tag{3.3.3}$$

At the cathode the potential is zero and the initial kinetic energy of the electrons is $mc^2(\gamma(0) - 1)$ implying,

$$\gamma(z) = \gamma(0) + \frac{e\Phi(z)}{mc^2}\,. \tag{3.3.4}$$

Expression (3.3.2) indicates that the current density J is constant in space and it reads

$$J = -en(z)v(z) = \text{const}\,.. \tag{3.3.5}$$

Equations (3.3.1–3) govern the dynamics of the electron in a static potential. In order to proceed to a solution of these equations it is convenient to substitute (3.3.4) in (3.3.1); in the resulting expression, we substitute the density from (3.3.5) and obtain

$$\frac{\mathrm{d}^2}{\mathrm{d}z^2}\gamma = \frac{eJ\eta_0}{mc^2}\frac{\gamma}{\sqrt{\gamma^2 - 1}}\,, \tag{3.3.6}$$

where we also used the fact that $\eta_0 \equiv 1/c\varepsilon_0$. The coefficient on the right-hand side has units of $1/\text{length}^2$, therefore we firstly define

$$K^2 = \frac{eJ\eta_0}{mc^2}\,, \tag{3.3.7}$$

and secondly define the normalized coordinate $\zeta = Kz$. Assuming that the electric field is not zero over the entire domain, the next step is to multiply (3.3.6) by $\mathrm{d}\gamma/\mathrm{d}\zeta$ and get

$$\frac{1}{2}\frac{\mathrm{d}}{\mathrm{d}\zeta}\left(\frac{\mathrm{d}\gamma}{\mathrm{d}\zeta}\right)^2 - \frac{\mathrm{d}}{\mathrm{d}\zeta}\sqrt{\gamma^2 - 1} = 0\,, \tag{3.3.8}$$

which implies that

$$\frac{1}{2}\left(\frac{\mathrm{d}\gamma}{\mathrm{d}\zeta}\right)^2 - \sqrt{\gamma^2 - 1} = \text{const}\,.. \tag{3.3.9}$$

As explained above, we consider the limit when the cathode is completely screened by the space-charge in the gap, therefore at $\zeta = 0$ the electric field vanishes and according to (3.3.3) $\mathrm{d}\gamma/\mathrm{d}\zeta = 0$. However for the generality let us assume that the electrons are injected in the gap with some initial energy i.e., $\gamma(\zeta = 0) = \gamma_0 > 1$, which implies

$$\frac{1}{\sqrt{2}}\frac{\mathrm{d}\gamma}{\mathrm{d}\zeta} = \sqrt{\sqrt{\gamma^2 - 1} - \sqrt{\gamma_0^2 - 1}}\,. \tag{3.3.10}$$

The last expression can be integrated and the formal result is

$$Kg\sqrt{2} = \int_{\gamma_0}^{\gamma_{\mathrm{AN}}} \frac{\mathrm{d}\gamma}{\sqrt{\sqrt{\gamma^2 - 1} - \sqrt{\gamma_0^2 - 1}}}\,, \tag{3.3.11}$$

where $\gamma_{\mathrm{AN}} = 1 + eV_{\mathrm{AN}}/mc^2$, V_{AN} is the voltage applied on the anode and g is the anode-cathode gap. The non-relativistic case can be readily calculated since $\gamma \simeq \gamma_0 + \delta\gamma$ and taking $\gamma_0 = 1$ we have

$$Kg = \left(\frac{\delta\gamma}{2}\right)^{3/4}\,, \tag{3.3.12}$$

which reads

$$J = \frac{16}{18\sqrt{2}}\frac{mc^2}{e\eta_0}\frac{1}{g^2}\left(\frac{eV_{\mathrm{AN}}}{mc^2}\right)^{3/2}\,,$$

$$J[\mathrm{kA/cm^2}] = 2.33\,\frac{V_{\mathrm{AN}}^{3/2}[\mathrm{MV}]}{g^2[\mathrm{cm}]}\,; \tag{3.3.13}$$

this is the Child-Langmuir limiting current. Although this expression was developed for a planar diode, the scaling of the current with the voltage remains the same in other geometries, therefore in analogy to the conductance in a metallic resistor, one can define the *perveance* as, $P \equiv I/V_{\mathrm{AN}}^{3/2}$, where I is the total current which flows in the diode. For reasons which will become clearer later in this chapter, the lower the perveance, the greater the possibility to achieve higher efficiency of energy conversion.

In order to have a feeling about this relation we can calculate the limiting current and for 1 MV and 3 cm gap we find it is 260 A/cm^2. It is important to realize that this is a supremum. For example, if the applied voltage is only 100 kV, then the limiting current is 8.2 A/cm^2 but a field $E \simeq V_{\mathrm{AN}}/g \simeq 3$ MV/m is not sufficient to extract virtually any current in this geometry, via field emission. In the case of thermionic emission, the current J emitted when a metallic cathode whose work function Φ_{w} is heated to an absolute temperature T is given by

$$J[\mathrm{kA/cm^2}] \simeq 0.12\,T^2\,\mathrm{e}^{-e\Phi_{\mathrm{w}}/k_{\mathrm{B}}T}\,; \tag{3.3.14}$$

for $T = 1400°$ K and Φ_w =2 V the current density is 15 A/cm^2. If we go back to the example above, we realize that for the first case (V_{AN} =1 MV) we can not reach the space-charge limiting current since the thermionic cathode can generate only one third of this limit. On the other hand applying V_{AN} =100 kV the system is space-charge limited and we do not utilize the entire potential of the thermionic emission. A similar situation occurs in the case of photo-emission.

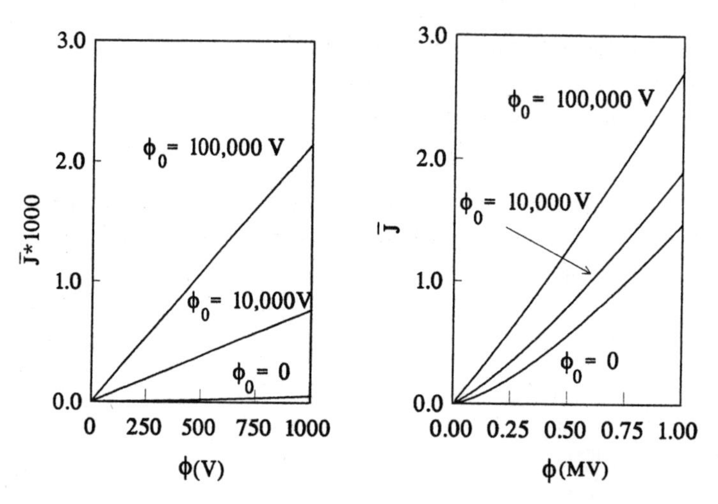

Fig. 3.3. Limiting current when the electrons enter the diode with an initial energy $e\Phi_0$

In the examples presented above the initial energy of the electrons at the surface of the cathode was zero i.e., ($\gamma_0 = 1$). If the electrons have an initial energy which corresponds to $\gamma_0 > 1$, then the current density at the anode is given by

$$J = \frac{I_e}{2g^2}\left(\int_{\gamma_0}^{\gamma_{AN}} \frac{d\gamma}{\sqrt{\sqrt{\gamma^2-1}-\sqrt{\gamma_0^2-1}}}\right)^2, \qquad (3.3.15)$$

where the characteristic electron current

$$I_e \equiv \frac{mc^2}{e\eta_0} \simeq 1.355\,\text{kA}\,; \qquad (3.3.16)$$

this current is related to the Alfven-Lawson current $I_A \equiv 4\pi mc^2/e\eta_0 = 4\pi I_e \simeq 17$ kA. In Fig. 3.3 the normalized current density $\bar{J} \equiv Jg^2/I_e$ is plotted as a function of the anode potential Φ; the energy of the electrons is a parameter, $\gamma_0 = 1 + e\Phi_0/mc^2$ and $\Phi_0 = 0, 10\,\text{kV}$ and $100\,\text{kV}$. The difference between the two frames is the applied anode voltage; in the left the maximum voltage applied is 1 kV whereas in the right, the maximum voltage is 1 MV. At low anode voltages the current density varies linearly with the voltage and the resistance of the anode-cathode gap is determined by the initial energy of the electrons; qualitatively it can be shown that the current density is proportional to the square root of the initial energy (provided that $\Phi \ll \Phi_0$). In order to emphasize the effect we observe that at $\Phi_0 = 100\,\mathrm{kV}$ the normalized current density is $\bar{J} \simeq 10^{-3}$ comparing to less than 2×10^{-5} for zero initial energy. This implies a factor of 50 larger limiting current. At $\Phi = 500\,\mathrm{kV}$ the difference is only a factor of 2 i.e., $\bar{J}(\Phi_0 = 0) \simeq 0.6$ comparing to $\bar{J}(\Phi_0 = 100\,\mathrm{kV}) \simeq 1.2$. Note that for the cases the electrons have an initial energy, the current presented above is the increment due to the applied voltage.

The Child-Langmuir limiting current is a byproduct of the energy conservation, so to exceed this limit, one has to provide the necessary energy. In the examples presented above the required energy was provided by the initial kinetic energy of the electrons. If (neutral) plasma is formed on the cathode surface this plasma can, in principle, traverse the diode gap and in the process the current increases. Assuming that the front of this plasma cloud advances with a velocity v_{pl} then the current density is given by

$$J(t) = \frac{2I_{\mathrm{e}}}{9(g - v_{\mathrm{pl}}t)^2} \left(\frac{2eV_{\mathrm{AN}}}{mc^2} \right)^{3/2} ; \tag{3.3.17}$$

this effect is called *gap closure*. The enhanced current associated with this process is accompanied by fluctuations which in turn may alter the beam quality.

3.3.2 Beyond Child-Langmuir Limit: Ferro-Electric Emission

One way to exceed the Child-Langmuir limit is to provide energy additional to that applied by the source which controls the anode voltage. And here there are several possibilities: it is possible, for example, to inject electrons with certain kinetic energy, into the diode gap by enforcing phase transition from ferro-electric to antiferro-electric – see Riege (1993). Another possibility is to provide energy by electrostatically coupling the diode gap to a capacitor located in the back of a gridded cathode. As in the previous case ferro-electric ceramic is involved but no phase transition occurs. We shall consider next the latter case.

The basic concept here is to utilize the electrons which screen the polarization field P of a ferro-electric ceramic. It is important to emphasize that we do not know, as for now, what is *microscopic* emission process. Charge is released once the internal polarization field is altered e.g., by applying an external voltage – see Fig. 3.4.

In order to envision the process consider two charges $\pm Q$ separated by a distance d; the positive charge is located at $z = -d$ and it represents

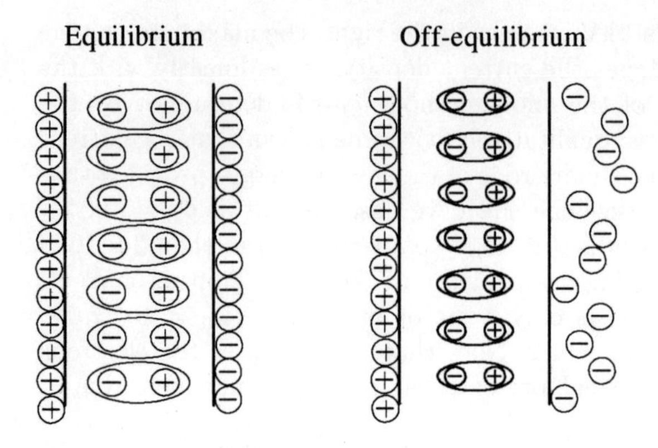

Fig. 3.4. Surface charge screens the polarization field in the material. As the latter changes the electrons on the gridded surface redistribute

the screening charge on the back electrode of the ferro-electric ceramic. The negative charge is located at $z = 0$ and represents the interface to the diode gap. As the polarization field is altered, the number of charges required for screening drops to $Q - q$. Charge located at the back electrode can not leave but it can do so in the front electrode and the charge $-q$ is redistributed in the gap. As we did for the screening charge, we shall consider here the charge distribution in the diode gap to be point-like. The potential which it experiences is given by

$$\Phi(z) = \frac{Q}{4\pi\varepsilon_0(z+d)} - \frac{Q-q}{4\pi\varepsilon_0|z|} - \frac{Q}{4\pi\varepsilon_0|2g+d-z|} + \frac{Q-q}{4\pi\varepsilon_0|2g-z|} \,. \quad (3.3.18)$$

The quantity $\bar{V} \equiv \Phi(4\pi\varepsilon_0 g)/qQ$ is plotted for $q/Q = 0.95$ and $d/g = 0.2$ in Fig. 3.5 where it is represented by the continuous line.

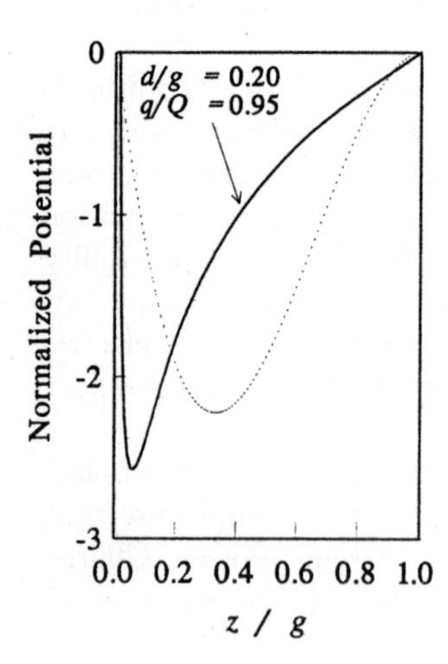

Fig. 3.5. The potential induced in the diode gap by the polarization field from the ferroelectric ceramic

We do not have to know the details of the exact distribution of the electrons in order to estimate their impact on the potential distribution in the gap. For such an evaluation we assume that *(i)* the plane of zero potential is actually on the cathode. If no current is measured on the anode for zero voltage ($V_{\rm AN} = 0$), we deduce that *(ii)* the electric field near the anode behaves as if no electrons were injected into the gap. Bearing in mind that *(iii)* the anode potential $V_{\rm AN}$ is known, we consider a solution of the 1D Poisson equation which satisfies the boundary conditions mentioned above i.e.,

$$\Phi(z) = V_{\rm AN}\frac{z}{g} + \Phi\frac{z}{g}\frac{(g-z)^2}{g^2} . \tag{3.3.19}$$

The amplitude Φ is determined by substituting (3.3.19) in the Poisson equation and integrating the resulting expression over the entire length of the diode. The source term in the Poisson equation is then proportional to the charge in the gap and therefore so is Φ; explicitly $\Phi = g|Q_{\rm gap}|/3\varepsilon_0 A$ where A is the diode area. According to this simplified potential distribution the minimum occurs at $z = g/3$ and its value there is $\Phi_{\rm min} \simeq \Phi/7$ – see dashed line in Fig. 3.5. If all the charge (say 10μC) which initially was on the surface of the ferro-electric is repelled into the gap, $\Phi_{\rm min}$ approaches the 1 MV level.

In equilibrium ($V_{\rm AN} = 0$), the potential in (3.3.19) supports stable oscillation of the electrons and consequently, the net current is zero. The average kinetic energy, $mc^2(\gamma_0 - 1)$, associated with the oscillation can be calculated by averaging the expression for the energy conservation over the gap spacing. Using (3.3.19), it can be shown that γ_0 reads:

$$\gamma_0 = 1 + \frac{1}{36}\bar{Q} , \tag{3.3.20}$$

where

$$\bar{Q} \equiv \frac{eQ_{\rm gap}g}{\varepsilon_0 Amc^2} \tag{3.3.21}$$

As we may have expected, the average kinetic energy of the electrons increases linearly with the total amount of charge in the gap ($Q_{\rm gap}$). To complete the description of the equilibrium state, we denote the average particle density in the cloud with n and a crude estimate of this quantity is the total number of particles divided by the effective gap volume i.e., $\tilde{n} \simeq Q_{\rm gap}/egA$.

When a positive anode voltage ($V_{\rm AN}$) is applied, the potential in the gap is described by the first term in (3.3.19). It will be assumed now that this potential is much smaller than the induced potential, therefore the charge density does not change. As in equilibrium, we can now calculate the average change in the velocity field of the two flows we mentioned above. Using again energy conservation we find

$$\delta\beta_{\pm} = \pm\frac{1}{2}\frac{1}{\beta_0\gamma_0^3}\frac{e}{mc^2}V_{\rm AN} , \tag{3.3.22}$$

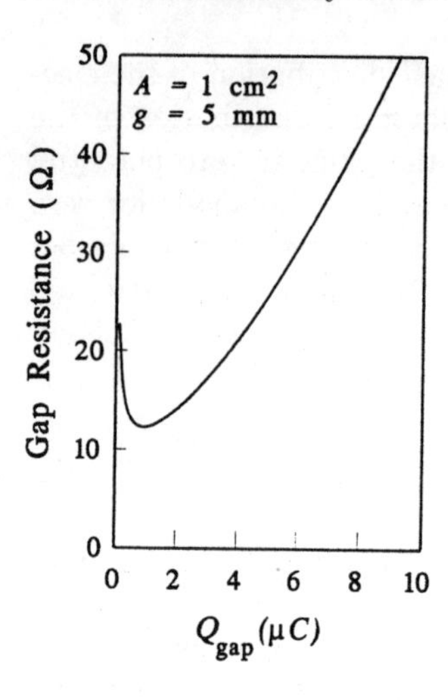

Fig. 3.6. Gap resistance as a function of the charge repelled in the gap

corresponding to the outgoing and backflowing electrons. The total current in the system is determined by these two quantities and the density of particles $I_{\rm AN} = eAnc(\delta\beta_+ - \delta\beta_-)$ which finally allows us to determine the gap resistance $R_{\rm gap}$:

$$R_{\rm gap} \equiv \frac{V_{\rm AN}}{I_{\rm AN}} = \eta_0 \frac{1}{36} \frac{g^2}{A} \gamma_0^2 \sqrt{\frac{\gamma_0 + 1}{\gamma_0 - 1}} . \tag{3.3.23}$$

As in the case when energetic electrons were injected, the $I - V$ curve is linear. Experimental results indicate that for anode voltages of $300 - 600V$ currents in excess of 100 times the Child-Langmuir limit were measured [Ivers (1993)]. Figure 3.6 illustrates the gap resistance as a function of the total charge released into the gap. As indicated previously, the Child-Langmuir limiting current relies on energy conservation. The energy associated with the excess of current is provided by the the ferro-electric ceramic by electrostatic coupling. The excess of current was measured for several hundreds of volts applied on the anode. A typical potential induced in the gap, is on the order of $\Phi_{\rm min} = 100\,{\rm kV}$ thus for an anode voltage much larger than this value (as in the case of energetic electrons) the Child-Langmuir limit is expected to be reached again.

3.4 Beam Propagation

Once the electron beam has been generated, it has to be confined to a small region in space and guided towards the interaction region. Naturally, most beams tend to diverge under the repelling effect of the electrostatic force, therefore external means have to be applied in order to preserve the beam shape. The most common way to guide electron beams is to apply a static magnetic field which can be either uniform or periodic. In this section we shall consider the propagation of a cylindrical beam i.e., assuming an azimuthally symmetric system.

3.4.1 Beam Propagation in Free Space: Uniform B-Field

In the absence of the guiding magnetic field a non-neutral beam will diverge under the influence of the repelling electrostatic force. If the applied magnetic field is too low, the beam will also diverge but after a longer duration. As the guiding field exceeds a certain value, which will be determined here, the trajectories are stable. In order to investigate the electron motion let us consider the two transverse components of the equations of motion. The radial component reads

$$\frac{\mathrm{d}}{\mathrm{d}t}v_r + \frac{1}{\gamma}\frac{\mathrm{d}\gamma}{\mathrm{d}t}v_r - \frac{1}{r}v_\phi^2 = -\frac{e}{m}\frac{1}{\gamma}\left(E_r + v_\phi B_z - v_z B_\phi\right), \tag{3.4.1}$$

whereas the azimuthal component is given by

$$\frac{1}{r}\frac{\mathrm{d}}{\mathrm{d}t}\left(\gamma r v_\phi\right) = -\frac{e}{m}\left(E_\phi - v_r B_z + v_z B_r\right). \tag{3.4.2}$$

Before we actually proceed to the stability analysis it will be instructive to examine the last equation more thoroughly. It will be shown next that this equation, for a static and an azimuthally symmetric magnetic field, contains the information regarding the conservation of the canonical angular momentum i.e., $\mathbf{r} \times \mathbf{p}$. For this purpose, we define the magnetic flux encompassed by an electron which is at a distance r from the axis by

$$\Psi(r,z) = \int_0^r \mathrm{d}r' 2\pi r' B_z(r',z). \tag{3.4.3}$$

The absolute time variation of this quantity is given by

$$\frac{\mathrm{d}\Psi}{\mathrm{d}t} = \frac{\partial\Psi}{\partial r}\frac{\mathrm{d}r}{\mathrm{d}t} + \frac{\partial\Psi}{\partial z}\frac{\mathrm{d}z}{\mathrm{d}t}, \tag{3.4.4}$$

and it can be simplified by using the definition of the magnetic flux and also the fact that the divergence of the magnetic inductance $\mathbf{B}$ vanishes i.e., $\nabla \cdot \mathbf{B} = 0$:

$$\frac{\mathrm{d}\Psi}{\mathrm{d}t} = 2\pi r B_z v_r - 2\pi r B_r v_z. \tag{3.4.5}$$

If we now compare the right-hand side of the last equation with the right-hand side of (3.4.2) we observe that in the absence of azimuthal electric field, they are virtually identical and therefore by substituting (3.4.2) in (3.4.5) we obtain

$$\frac{\mathrm{d}}{\mathrm{d}t}\left[\Psi - \frac{2\pi}{e} r(m\gamma v_\phi)\right] = 0\,. \tag{3.4.6}$$

This expression indicates that for a particle "born" with zero azimuthal motion in a magnetic field, the azimuthal motion is determined entirely by the local flux, its radial location and the energy. A different interpretation of the same result can be achieved by noting that we can represent the magnetic field in terms of the azimuthal magnetic vector potential as $B_z = [\partial(rA_\phi)/\partial r]/r$ therefore substituting in the definition of the flux we obtain

$$r(m\gamma v_\phi - eA_\phi) = rp_\phi = \text{const.}\,, \tag{3.4.7}$$

which is the longitudinal component of the *canonical angular momentum.*

With this result we can now investigate the radial component of the equation of motion. For this purpose, it is assumed that the guiding magnetic field is uniform (B_0), thus the azimuthal component of the magnetic vector potential is $A_\phi = \frac{1}{2} r B_0$. Consequently, the canonical angular momentum conservation implies

$$\begin{aligned} rp_\phi &= r\,(m\gamma v_\phi - eA_\phi) \\ &= r\left(m\gamma v_\phi - e\frac{1}{2} r B_0\right) \\ &= rm\left(\gamma v_\phi - \frac{1}{2}\Omega_\mathrm{c} r\right)\,. \end{aligned} \tag{3.4.8}$$

where Ω_c is the (non-relativistic) cyclotron angular frequency defined by

$$\Omega_\mathrm{c} \equiv \frac{eB_0}{m}\,. \tag{3.4.9}$$

We can use the last expression in (3.4.8) and substitute v_ϕ in (3.4.1); the result is

$$\frac{\mathrm{d}^2}{\mathrm{d}t^2} r - \frac{\Omega_\mathrm{c}^2}{4\gamma^2 r^3}(r^2 - r_0^2)^2 + \frac{1}{2}\frac{\Omega_\mathrm{c}^2}{r\gamma}(r^2 - r_0^2) = -\frac{e}{m\gamma}\left(E_r - v_z B_\phi\right)\,. \tag{3.4.10}$$

In this expression we neglected the energy variation and r_0 is the radius where the electron was "born" – it is tacitly assumed that at this location the azimuthal velocity vanishes ($v_\phi = 0$).

For further simplification of the equation which describes the radial motion we shall next evaluate the two field components in the right-hand side of (3.4.10). The radial component of the electric field represents the field generated by electrons which are located at radii smaller than that of the particle hence using (2.1.3) which in our case reads

$$\frac{1}{r}\frac{\mathrm{d}}{\mathrm{d}r} r E_r(r) = -\frac{e}{\varepsilon_0} n(r)\,, \tag{3.4.11}$$

and assuming uniform distribution of particles, we find that

$$E_r(r) = -\frac{en}{2\varepsilon_0} r \,. \tag{3.4.12}$$

In a similar way, we consider (2.1.2) to calculate the azimuthal component of the magnetic field; the relevant component reads

$$\frac{1}{r}\frac{\mathrm{d}}{\mathrm{d}r}\left[rH_\phi(r)\right] = -en(r)v_z(r) \,. \tag{3.4.13}$$

As in the previous case, we assume that both the density and the velocity are uniform in space hence

$$H_\phi(r) = -\frac{env_z}{2} r \,. \tag{3.4.14}$$

These two expressions [(3.4.12,14)] can be substituted in the right-hand side of the radial component of the equation of motion [(3.4.10)] which then reads

$$\frac{\mathrm{d}^2}{\mathrm{d}t^2} r - \frac{\Omega_\mathrm{c}^2}{4\gamma^2 r^3}(r^2 - r_0^2)^2 + \frac{1}{2}\frac{\Omega_\mathrm{c}^2}{r\gamma}(r^2 - r_0^2) = \frac{1}{2}\frac{e^2 n}{m\varepsilon_0 \gamma^3} r \,, \tag{3.4.15}$$

In equilibrium, there are no time variations, therefore the beam radius (R_b)is a solution of

$$-\frac{\Omega_\mathrm{c}^2}{4\gamma^2 R_\mathrm{b}^3}(R_\mathrm{b}^2 - r_0^2)^2 + \frac{1}{2}\frac{\Omega_\mathrm{c}^2}{R_\mathrm{b}\gamma}(R_\mathrm{b}^2 - r_0^2) = \frac{1}{2\gamma^3}\omega_\mathrm{p}^2 R_\mathrm{b} \,. \tag{3.4.16}$$

which is

$$R_\mathrm{b}^2 = \frac{r_0^2}{\sqrt{1 - 2\omega_\mathrm{p}^2/\Omega_\mathrm{c}^2\gamma}} \,; \tag{3.4.17}$$

in these expressions we used the non relativistic definition of the plasma frequency

$$\omega_\mathrm{p}^2 = \frac{e^2 n}{m\varepsilon_0} \,, \tag{3.4.18}$$

From the expression in (3.4.17) we conclude that a real (thus stable) solution exists only if

$$\Omega_\mathrm{c}^2 \geq 2\omega_\mathrm{p}^2 \frac{1}{\gamma} \,. \tag{3.4.19}$$

Note that from (3.4.15) we can have a clear measure of the way the beam diverges in the absence of a guiding magnetic field. When $\Omega_\mathrm{c} = 0$ it is convenient to define $\bar{r} \equiv r(t)/r_0$ and $\tau \equiv t\omega_\mathrm{p}/\gamma^{3/2}\sqrt{2}$. With this notation (3.4.15) reads

$$\left[\frac{\mathrm{d}^2}{\mathrm{d}\tau^2} - 1\right]\bar{r} = 0 \,. \tag{3.4.20}$$

A general solution of this equation is a superposition of $\sinh(\tau)$ or $\cosh(\tau)$. If, for simplicity, we assume that the radial velocity at $\tau = 0$ is zero then $\bar{r} = \cosh(\tau)$ which clearly indicates that the beam diverges.

Existence of equilibrium [(3.4.19)] does not ensure stability since a small deviation might evolve into an unstable state i.e., the beam might diverge. However in this particular case it can be shown (see also Exercise 3.4) that the condition in (3.4.19) also ensures the stability of the beam. Note that there are two characteristic length parameters which determine the beam radius: the first (r_0) is trivial and denotes the radius where the particle was "born". In order to establish the second, we use the definition of the total current $I = env_0\pi R_b^2$ and substitute it in the expression for the plasma frequency, to rewrite (3.4.17) as

$$R_b^2 = \frac{r_0^2}{\sqrt{1 - L_0^2/R_b^2}}, \tag{3.4.21}$$

where $L_0^2 \equiv (2/\pi)(1/\gamma\beta)(e\eta_0 I)mc^2/(ecB_0)^2$ and finally the beam radius is determined by

$$R_b^2 = \frac{1}{2}\left(L_0^2 + \sqrt{L_0^4 + 4r_0^4}\right). \tag{3.4.22}$$

From this expression, we conclude that the beam radius increases monotonically with L_0, therefore for a given r_0, the beam is compressed when the guiding field or the kinetic energy is increased. The radius increases when the current is raised. In addition, it is possible to deduce the equilibrium radius of the beam corresponding to electrons "born" in a zero magnetic field and on axis ($r_0 \simeq 0$), its value ($R_b \simeq L_0$) is determined entirely by the guiding field, current and kinetic energy.

3.4.2 Beam Propagation in Free Space: Periodic B-Field

Generation of a uniform magnetic field for guiding an intense relativistic beam may require a substantial amount of energy. This magnetic field is generated by discharging a large bank of capacitors in solenoids and it can become quite energy consuming when high repetition rate is required. In the latter case the main alternative is to consider the use of permanent periodic magnets (PPM) for guiding the beam. The magnetic field configuration is

$$B_z(r,z) = B_0\cos(k_w z)\mathrm{I}_0(k_w r),$$
$$B_r(r,z) = B_0\sin(k_w z)\mathrm{I}_1(k_w r), \tag{3.4.23}$$

where $k_w = 2\pi/L$ and L is the period of the permanent magnetic field. In order to present the stability condition, the analysis will be limited to a narrow pencil beam of maximum radius R_b which is much smaller than the periodicity of the field namely, $k_w R_b \ll 1$ hence

$$B_z(r,z) = B_0 \cos(k_\mathrm{w} z)\,,$$
$$B_r(r,z) = B_0 \sin(k_\mathrm{w} z)\frac{1}{2}k_\mathrm{w} r\,. \tag{3.4.24}$$

This field can be derived from the azimuthal component of the magnetic vector potential:

$$A_\phi = B_0 \cos(k_\mathrm{w} z)\frac{1}{2} r\,. \tag{3.4.25}$$

For simplicity we shall further assume that the electrons are "born" in a region of zero magnetic field and their initial azimuthal motion is zero; therefore according to the conservation of the canonical angular momentum, [(3.4.7)] we conclude that

$$v_\phi = \frac{1}{2\gamma} r \Omega_\mathrm{c} \cos(k_\mathrm{w} z)\,. \tag{3.4.26}$$

Neglecting energy variations and substituting the last expression in the radial component of the equation of motion, (3.4.1), we obtain

$$\frac{\mathrm{d}^2}{\mathrm{d}t^2} r - \frac{1}{r}\left[\frac{1}{2\gamma} r \Omega_\mathrm{c} \cos(k_\mathrm{w} z)\right]^2 + \frac{1}{\gamma}\Omega_\mathrm{c}\cos(k_\mathrm{w} z)\left[\frac{1}{2\gamma} r \Omega_\mathrm{c} \cos(k_\mathrm{w} z)\right]$$
$$= -\frac{e}{m\gamma}\left[E_r - v_z B_\phi\right]\,. \tag{3.4.27}$$

Using the expressions for the electric and magnetic field developed in the previous section [(3.4.12,14)] and assuming that the longitudinal motion is predominant i.e., $|v_z| \gg |v_r|, |v_\phi|$ we find

$$\frac{\mathrm{d}^2}{\mathrm{d}t^2} r + \left[\frac{1}{4\gamma^2}\Omega_\mathrm{c}^2 \cos^2(\Omega_\mathrm{w} t) - \frac{1}{2\gamma^3}\omega_\mathrm{p}^2\right] r = 0\,, \tag{3.4.28}$$

where $\Omega_\mathrm{w} = k_\mathrm{w} v_0$ and v_0 is the velocity of the electron. For a zero order stability estimate, one can average the square brackets on time, thus the resulting coefficient of $r(t)$ has to be positive for the electrons to follow confined trajectories i.e.,

$$\Omega_\mathrm{c}^2 \geq \frac{4}{\gamma}\omega_\mathrm{p}^2\,. \tag{3.4.29}$$

Comparing this expression with the condition for a uniform field [(3.4.19)] it is evident that in the periodic case, on axis, the amplitude has to be by a factor of $\sqrt{2}$ larger.

3.4.3 Beam Propagation in a Uniform Waveguide

In the previous sub-section we determined the necessary condition for the propagation of an electron beam in free space when guided by a static magnetic field. In this sub-section, we shall investigate the propagation of the beam in a waveguide assuming that an infinite magnetic field is applied such that only longitudinal motion is permitted – the basic configuration is illustrated schematically in Fig. 3.7. The region under investigation is far away from the entrance to the waveguide, therefore longitudinal variations are neglected and only radial variations are of interest. The presence of the metallic boundary at $r = R$ changes the potential experienced by the beam and consequently since the total energy is the sum of kinetic and potential energy, the former varies across the beam.

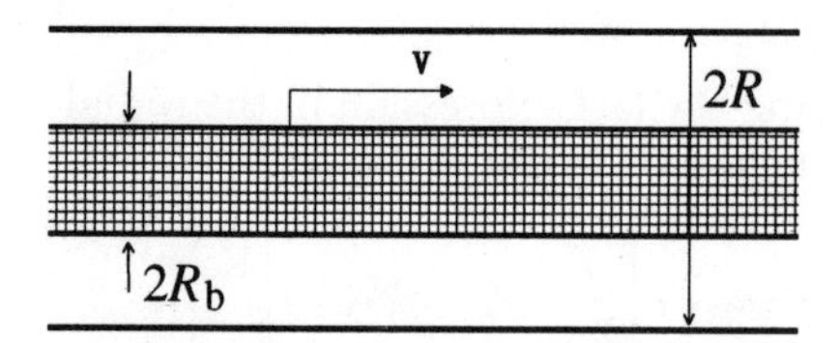

Fig. 3.7. Electron beam in a circular waveguide

In order to envision the effect, consider a beam which at the input (free space) has a uniform spatial distribution and, more important, all the electrons have the same kinetic energy $mc^2(\gamma_0 - 1)$. Since the potential varies in the waveguide's space, electrons which are at a different distance from the wall experience different potential and therefore their kinetic energy differs. The potential energy comes at the expense of the kinetic energy. That is to say that if electrons injected into a waveguide have at the input the same kinetic energy, then by increasing their number (N) we increase the potential and decrease their velocity (v). In terms of the injected current, which is proportional to the product Nv, it reaches a maximum value and beyond it, even if we increase the number of electrons injected, the current remains unchanged since electrons bounce back. Let us now investigate this phenomenon in a systematic way. In the waveguide the electrostatic potential satisfies the Poisson equation:

$$\frac{1}{r}\frac{\mathrm{d}}{\mathrm{d}r}r\frac{\mathrm{d}}{\mathrm{d}r}\Phi(r) = \frac{1}{\varepsilon_0}en(r)\,. \tag{3.4.30}$$

Green's function associated with this equation is a solution of

$$\frac{1}{r}\frac{\mathrm{d}}{\mathrm{d}r}r\frac{\mathrm{d}}{\mathrm{d}r}G(r|r') = -\frac{1}{2\pi r}\delta(r - r')\,, \tag{3.4.31}$$

and it reads

$$G(r|r') = -\frac{1}{2\pi}\begin{cases}\ln(r/R) & \text{for} \quad 0 \le r' \le r < R\,, \\ \ln(r'/R) & \text{for} \quad 0 < r \le r' < R\,. \end{cases} \tag{3.4.32}$$

Based on Green's scalar theorem the potential reads

$$\Phi(r) = \frac{-e}{\varepsilon_0}\left[\int_0^r \mathrm{d}r' r' \ln\left(\frac{r}{R}\right) n(r') + \int_r^R \mathrm{d}r' r' \ln\left(\frac{r'}{R}\right) n(r')\right], \tag{3.4.33}$$

and in particular at the edge of the beam ($r = R_\mathrm{b}$) it reads

$$\Phi(R_\mathrm{b}) = -\frac{1}{\varepsilon_0} e \ln\left(\frac{R_\mathrm{b}}{R}\right)\int_0^{R_\mathrm{b}} \mathrm{d}r' r' n(r')\,. \tag{3.4.34}$$

Using energy conservation

$$\begin{aligned}\gamma - \frac{e\Phi}{mc^2} &= \text{const.}\,, \\ &= \gamma_0\,,\end{aligned} \tag{3.4.35}$$

and charge conservation

$$J = -env = \text{const.}\,, \tag{3.4.36}$$

we can write

$$\Phi(R_\mathrm{b}) = -\frac{J}{c\varepsilon_0}\ln\left(\frac{R}{R_\mathrm{b}}\right)\int_0^{R_\mathrm{b}} \mathrm{d}r' r' \left[1 - (\gamma_0 + \frac{e\Phi(r')}{mc^2})^{-2}\right]^{-1/2}. \tag{3.4.37}$$

The expression $mc^2(\gamma_0 - 1)$ represents the kinetic energy before the particles entered the waveguide. For a narrow pencil beam, the potential is not expected to vary significantly on the beam cross-section, therefore we can replace $\Phi(r')$ with $\Phi(R_\mathrm{b})$ thus

$$\begin{aligned}\xi\left(\frac{e\Phi(R_\mathrm{b})}{mc^2}\right) &\equiv \left(\frac{e\Phi(R_\mathrm{b})}{mc^2}\right)\sqrt{1 - \left(\gamma_0 + \frac{e\Phi(R_\mathrm{b})}{mc^2}\right)^{-2}}\,, \\ &= -\frac{1}{2\pi}\frac{eI\eta_0}{mc^2}\ln\left(\frac{R}{R_\mathrm{b}}\right).\end{aligned} \tag{3.4.38}$$

The function $\xi(x)$ is plotted in Fig. 3.8 for $\gamma_0 = 3$. The extremum occurs for $x_0 = -\gamma_0 + \gamma_0^{1/3}$ and $\xi(x_0) = -\left(\gamma_0^{2/3} - 1\right)^{3/2}$ therefore the maximum current which can flow in the waveguide is given by

$$I_\mathrm{max} = \frac{mc^2}{e\eta_0}\frac{2\pi}{\ln(R/R_\mathrm{b})}\left(\gamma_0^{2/3} - 1\right)^{3/2}. \tag{3.4.39}$$

Any attempt to inject a higher current in the waveguide will cause electrons to bounce back to the diode and a virtual cathode may develop.

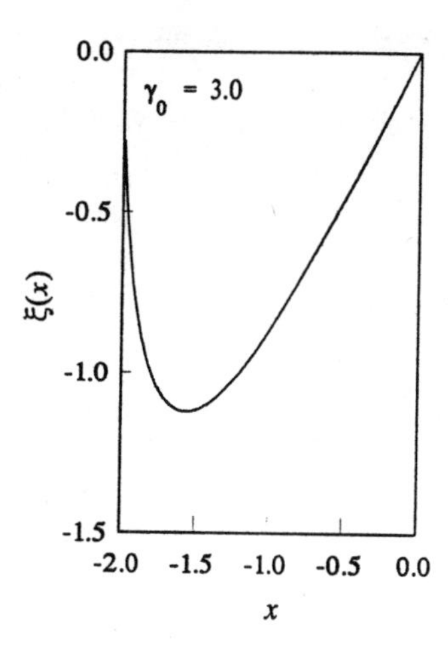

Fig. 3.8. ξ as defined in (3.4.38) is plotted as a function of the normalized potential ($x \equiv e\Phi(R_b)/mc^2$) on the beam envelope

Another implication of (3.4.38) regards the effective kinetic energy of the beam. For a given initial energy, the kinetic energy decreases with the increasing current due to the potential energy associated with the space-charge effect. Figure 3.9a illustrates the effective kinetic energy of the electrons as a function of the current, for R=2 cm, $R_b = 3$ mm and $\gamma_0 = 3$. For this particular set of parameters the effective kinetic energy drops to $\gamma = 2.0$ when 3.9 kA are injected; the limiting current in this case is 5 kA in which case $\gamma = 1.56$. In other words, in the diode the electrons are accelerated to the 1 MeV level; however in order to propagate in the waveguide, the electrons have to overcome the potential barrier associated with the collective effect of their presence in the waveguide - this potential barrier is almost 500 keV. This effect inflicts limitations on the maximum efficiency of a device since if, for example, the diode had generated a 1 MeV $\times$ 3.9 kA $\simeq$ 3.9 GeV beam, in the waveguide, the maximum power available for radiation generation is 0.5 MeV $\times$ 3.9 kA $\simeq$ 2 GeV and the maximum radiation conversion efficiency is 50% since what counts is the effective kinetic (available) energy in the waveguide. Figure 3.9b illustrates the maximum (theoretical) efficiency, defined as $(\gamma - 1)/(\gamma_0 - 1)$, as a function of the perveance.

The limiting current which can propagate in the waveguide is directly related to the Child-Langmuir limiting current in a diode as it becomes evident in particular at low voltages ($eV_{\rm AN}/mc^2 \ll 1$) since $\gamma_0 = 1 + eV_{\rm AN}/mc^2$ and

$$I_{\rm max} = \frac{mc^2}{e\eta_0} \frac{2\pi}{\ln(R/R_b)} \left(\frac{2}{3} V_{\rm AN}\right)^{3/2}, \qquad (3.4.40)$$

which has an identical form to (3.3.13). For this reason a high efficiency device has always to be designed based on a low perveance diode.

It was indicated that the space charge effect causes a transverse spatial variation of the electron's kinetic energy which clearly may alter the interaction with electromagnetic waves. However in the analysis above, it was

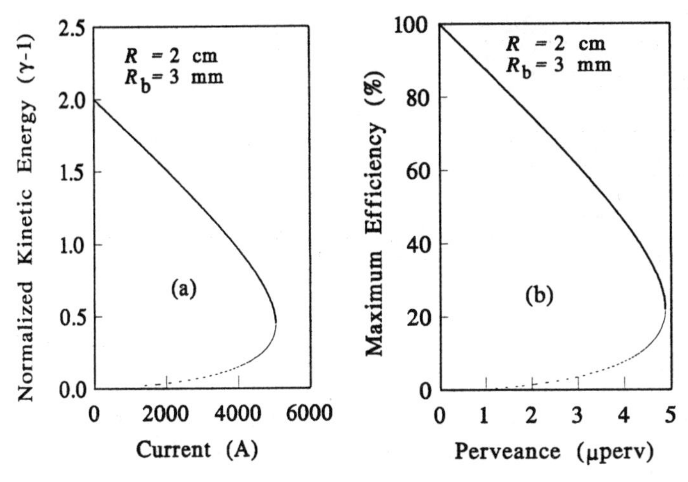

Fig. 3.9. (**a**) Normalized kinetic energy of the electrons in the waveguide as a function of the injected current. The initial energy is alway the same. (**b**) Maximum potential efficiency as a function of the perveance

assumed that the beam is sufficiently narrow such that variations across its section could be neglected. A parameter of importance is the location of the beam relative to the waveguide's wall as it becomes evident from (3.4.39) that by increasing the distance between the external wall and the beam, the limiting current becomes smaller. In order to examine this effect more accurately, one can examine the limiting current of an annular beam of radius R_b and δr thickness (much smaller than R_b). The approach is similar to the above and the result is similar except for the different meaning R_b has in this case. The closer the beam is to the waveguide's wall, the higher the limiting current and therefore the lower the potential depression. To emphasize this effect even further, consider two thin annular beams of two different radii – for details see Exercise 3.5. The kinetic energy of the electrons in the outer beam is larger than that of the electrons in the inner beam.

3.4.4 Beam Propagation in a Disk-Loaded Waveguide

If the geometry of the waveguide is not uniform the limiting current differs from one region to another and so is the kinetic energy of the electrons. We shall now investigate the limiting current in a disk loaded waveguide as drawn in Fig. 3.10. The radius of the beam R_b is assumed to be small such that the potential can be considered uniform on its cross-section. The periodicity of the waveguide is denoted by L, the disk width is d, its internal radius is R_{int} and the external radius of the waveguide is R_{ext}. The electrostatic field is derived from the electric scalar potential $\Phi(r, z)$ which satisfies

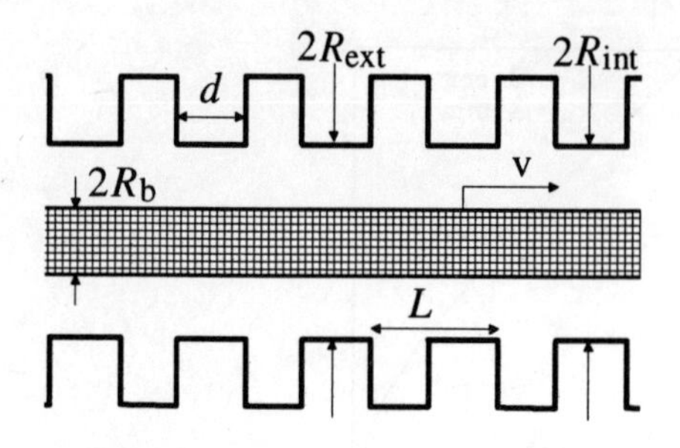

Fig. 3.10. Beam propagation in a corrugated waveguide

$$\left[\frac{1}{r}\frac{\partial}{\partial r}r\frac{\partial}{\partial r} + \frac{\partial^2}{\partial z^2}\right]\Phi(r,z) = \frac{e}{\varepsilon_0}n(r,z). \tag{3.4.41}$$

We shall assume that the beam is guided by a very strong magnetic field which confines the motion of the electrons only to the longitudinal direction. Consequently, for a static case the charge conservation reads

$$J = -en(r,z)v(r,z) = \text{const.}, \tag{3.4.42}$$

where $v(r,z)$ is the longitudinal velocity. From the energy conservation law, we can conclude that

$$\gamma(r,z) = \gamma_0 + \frac{e\Phi(r,z)}{mc^2}, \tag{3.4.43}$$

hence

$$\left[\frac{1}{r}\frac{\partial}{\partial r}r\frac{\partial}{\partial r} + \frac{\partial^2}{\partial z^2}\right]\bar{\Phi}(r,z) = -K^2\left[1-\left(\gamma_0+\bar{\Phi}(r,z)\right)^{-2}\right]^{-1/2}, \tag{3.4.44}$$

where $\bar{\Phi} = e\Phi/mc^2$ and $K^2 = eJ\eta_0/mc^2$. The solution can be formally written as

$$\bar{\Phi}(r,z) = 2\pi K^2\int_0^{R_{\rm int}}\mathrm{d}r'r'\int_0^L\mathrm{d}z'$$
$$\times\, G(r,z|r',z')\left[1-\left(\gamma_0+\bar{\Phi}(r',z')\right)^{-2}\right]^{-1/2}; \tag{3.4.45}$$

$G(r,z|r',z')$ being Green's function of the system. It satisfies

$$\left[\frac{1}{r}\frac{\partial}{\partial r}r\frac{\partial}{\partial r} + \frac{\partial^2}{\partial z^2}\right]G(r,z|r',z') = \frac{-1}{2\pi r}\delta(r-r')\sum_{n=-\infty}^{\infty}\delta(z-z'-nL) \tag{3.4.46}$$

subject to the boundary conditions on the walls of the waveguide. Let us now determine the explicit expression for this function. In the inner cylinder ($0\le r<R_{\rm int}$) we have

$$
\begin{aligned}
G_1(r,z|r',z') &\equiv G(r<R_{\rm int},z|r',z') \\
&= \sum_{n=-\infty}^{\infty} g_n(r|r')\mathrm{e}^{-jk_n(z-z')} \\
&\quad + \sum_{n=-\infty}^{\infty} h(r',z')\mathrm{I}_0(|k_n|r)\mathrm{e}^{-jk_n z}\,; \qquad (3.4.47)
\end{aligned}
$$

$k_n = 2\pi n/L$. The first term represents Green's function for the boundless case whereas the second is the effect of the boundaries. Consequently, $g_n(r|r')$ is given by

$$
g_0(r|r') = -\frac{1}{2\pi L}\begin{cases} \ln(r/R_0) & \text{for} \quad 0\le r'\le r\le R_0\,, \\ \ln(r'/R_0) & \text{for} \quad 0<r\le r'\le R_0 \end{cases} \qquad (3.4.48)
$$

and for $n\neq 0$

$$
g_n(r|r') = \frac{1}{2\pi L}\begin{cases} \mathrm{K}_0(|k_n|r)\mathrm{I}_0(|k_n|r') & \text{for} \quad 0\le r'\le r<\infty\,, \\ \mathrm{I}_0(|k_n|r)\mathrm{K}_0(|k_n|r') & \text{for} \quad 0\le r'\le r<\infty\,. \end{cases} \qquad (3.4.49)
$$

In the grooves ($R_{\rm int}<r<R_{\rm ext}$) Green's function reads

$$
\begin{aligned}
G_2(r,z|r',z') &\equiv G(R_{\rm int}<r<R_{\rm ext},z|r',z')\,, \\
&= \sum_{\nu=1}^{\infty} f_\nu T_\nu(r)\sin\left[q_\nu(z-d)\right]\,, \qquad (3.4.50)
\end{aligned}
$$

where $q_\nu = \pi\nu/(L-d)$ and

$$
T_\nu(r) = \mathrm{I}_0(q_\nu r)\mathrm{K}_0(q_\nu R_{\rm ext}) - \mathrm{K}_0(q_\nu r)\mathrm{I}_0(q_\nu R_{\rm ext})\,. \qquad (3.4.51)
$$

In addition, we shall use

$$
T'_\nu(r) = \mathrm{I}_1(q_\nu r)\mathrm{K}_0(q_\nu R_{\rm ext}) + \mathrm{K}_1(q_\nu r)\mathrm{I}_0(q_\nu R_{\rm ext})\,. \qquad (3.4.52)
$$

Next we impose the boundary conditions at $r=R_{\rm int}$; the continuity of the potential implies

$$
G_1(r=R_{\rm int},z|r',z') = \begin{cases} 0 & \text{for } 0\le z\le d\,, \\ G_2(r=R_{\rm int},z|r',z') & \text{for } d\le z\le L\,. \end{cases} \qquad (3.4.53)
$$

This can be simplified by multiplying by $\mathrm{e}^{jk_n z}$ and integrated over one period of the structure; the result reads

$$
\begin{aligned}
&g_n(R_{\rm int}|r')\mathrm{e}^{jk_n z'} + h_n(r',z')\mathrm{I}_0(|k_n|R_{\rm int}) \\
&\quad = \frac{L-d}{L}\sum_{\nu=1}^{\infty} f_\nu T_\nu(R_{\rm int})U_{n,\nu}\,, \qquad (3.4.54)
\end{aligned}
$$

where

$$U_{n,\nu} \equiv \frac{1}{L-d}\int_d^L \mathrm{d}z e^{jk_n z}\sin[q_\nu(z-d)]\,. \tag{3.4.55}$$

Continuity of the radial electric field in the aperture ($d < z < L$) implies

$$\left[\frac{\partial}{\partial r}G_2(r,z|r',z')\right]_{r=R_{\rm int}} = \left[\frac{\partial}{\partial r}G_1(r,z|r',z')\right]_{r=R_{\rm int}} \tag{3.4.56}$$

which after using the orthogonality of the sin function in the domain, implies

$$\begin{aligned}\bar{f}_\nu(r',z') &\equiv f_\nu(r',z')q_\nu T'_\nu(R_{\rm int})\\ &= 2\sum_n e^{jk_n z'}\frac{\partial}{\partial r}g_n(r|r')|_{r=R_{\rm int}}U^*_{n,\nu}\\ &\quad + 2\sum_n h_n(r',z')|k_n|\mathrm{I}_1(|k_n|R_{\rm int})U^*_{n,\nu}\,.\end{aligned} \tag{3.4.57}$$

Equations (3.4.54,57) determine Green's function. It is convenient to substitute (3.4.54) in the latter; the result reads

$$\sum_\mu \chi_{\nu,\mu}\bar{f}_\mu(r',z') = s_\nu(r',z')\,, \tag{3.4.58}$$

where

$$\chi_{\nu,\mu} = \delta_{\nu,\mu} - 2\frac{L-d}{L}\frac{T_\mu(R_{\rm int})}{T'_\mu(R_{\rm int})}\frac{1}{q_\mu}\sum_{n=-\infty}^{\infty}|k_n|\frac{\mathrm{I}_1(|k_n|R_{\rm int})}{\mathrm{I}_0(|k_n|R_{\rm int})}U^*_{n,\nu}U_{n,\mu} \tag{3.4.59}$$

and

$$\begin{aligned}s_\nu(r',z') &= \sum_n U^*_{n,\nu}e^{jk_n z'}\\ &\quad\times\left[\frac{\partial g_n(r|r')}{\partial r} - g_n(r|r')|k_n|\frac{\mathrm{I}_1(|k_n|r)}{\mathrm{I}_0(|k_n|R_{\rm int})}\right]_{r=R_{\rm int}}\,.\end{aligned} \tag{3.4.60}$$

With Green's function determined, we proceed and simplify (3.4.45) by using the fact that the potential does not vary substantially over the beam cross-section and average in the z direction over one period of the structure i.e., $\bar{\Phi}_b \equiv \int_0^L \bar{\Phi}(r=R_{\rm b},z)\mathrm{d}z/L$:

$$\begin{aligned}\bar{\Phi}_b &= 2\pi K^2\left[1-(\gamma_0+\bar{\Phi}_b)^{-2}\right]^{-1/2}\\ &\quad\times\int_0^{R_{\rm int}}\mathrm{d}r'r'\int_0^L\mathrm{d}z'\frac{1}{L}\int_0^L\mathrm{d}zG(r=R_{\rm b},z|r',z')\,.\end{aligned} \tag{3.4.61}$$

Substituting the explicit expression for Green's function established above, performing the integration over z and r coordinates and following the same procedure as in the previous section, we find the following expression for the limiting current

$$I_{\max} = \frac{mc^2}{e\eta_0} \frac{2\pi}{F_b \ln(R_{\text{int}}/R_{\text{b}})} \left(\gamma_0^{2/3} - 1\right)^{3/2} , \tag{3.4.62}$$

where F_b is the boundary factor and is given by

$$\begin{aligned} F_b \equiv & 1 - \left(\frac{2}{\pi}\right)^3 \frac{(L-d)^2}{LR_{\text{int}}} \frac{1}{\ln(R_{\text{int}}/R_{\text{b}})} \\ & \times \sum_{\nu,\mu} \frac{T_{2\nu+1}(R_{\text{int}})}{T'_{2\nu+1}(R_{\text{int}})} \frac{\left[\chi^{-1}\right]_{2\nu+1,2\mu+1}}{(2\nu+1)^2(2\mu+1)} . \end{aligned} \tag{3.4.63}$$

The dependence of this factor on the external radius R_{ext}, disk width d and periodicity of the structure is presented in Fig. 3.11a–c for $R_{\text{b}} = 3\,\text{mm}$ and $R_{\text{int}} = 9\,\text{mm}$. Figure 3.11a shows the boundary factor as a function of $R_{\text{ext}}/R_{\text{int}}$ and we observe that for $L = 7.7\,\text{mm}$ and $d = 1\,\text{mm}$, F_{b} reaches a maximum value of 1.1 when $R_{\text{ext}} = 1.5R_{\text{int}}$ and any further increase does not change the boundary factor. Figure 3.11b illustrates this factor as a function of the disk thickness and we observe that it decreases with the increasing d. Finally, Fig. 3.11c indicates that the boundary factor increases with the increasing periodicity ($d = L/2$).

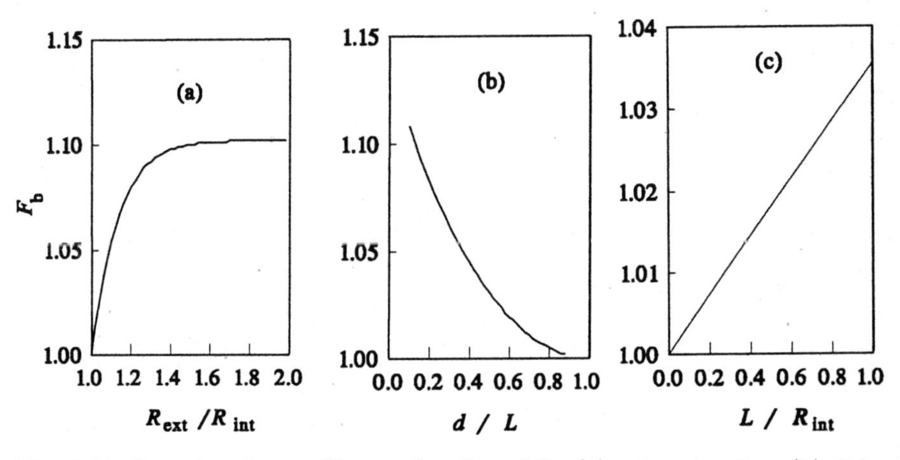

Fig. 3.11. Boundary factor, F_{b}, as a function of the (**a**) external radius, (**b**) disk thickness and (**c**) period

3.4.5 Beam Emittance and Brightness

A permanent periodic magnetic field can be conceived as a set of magnetic lenses which guide the electrons. In an ideal *periodic* structure without beam-wave interaction the shape of the beam is conserved. In order to examine the propagation of the beam we shall assume that the longitudinal motion is predominant such that $v_z \gg |v_x|, |v_y|$. Consequently, we can replace the time derivative $\partial/\partial t$ with $v_z \partial/\partial z$ thus the transverse motion in a longitudinally periodic structure can be modeled by

$$\frac{\mathrm{d}^2}{\mathrm{d}z^2}\delta x_i + K_0^2 \delta x_i = 0 . \tag{3.4.64}$$

At this stage we shall consider only one component of the transverse motion δx_i, which represents the distance of the ith particle from axis. An ideal system is represented by a constant periodicity, therefore K_0 is space independent. Consequently, the trajectory is described by

$$\delta x_i(z) = A_i \cos(K_0 z + \phi_i) , \tag{3.4.65}$$

and its first derivative, denoted by $\delta x_i'(z)$, is given by

$$\delta x_i'(z) = -A_i K_0 \sin(K_0 z + \phi_i) . \tag{3.4.66}$$

In the phase-space the locus of each trajectory is an ellipse

$$\left(\frac{\delta x_i}{A_i}\right)^2 + \left(\frac{\delta x_i'}{K_0 A_i}\right)^2 = 1 . \tag{3.4.67}$$

If we denote by $a \equiv \max(A_i)$ the largest amplitude among all particles, then the area of the ellipse is given by

$$\pi\varepsilon \equiv \pi a(aK_0) = \pi K_0 a^2 , \tag{3.4.68}$$

and it defines the emittance ε of an ideal beam. This corresponds to the area in the $\delta x, \delta x'$ space. It is natural to define the emittance in the "normal" phase space i.e., x, p_x. For this purpose the $\delta x'$ is multiplied by $\gamma\beta$ since $p_x = mc\gamma\beta\delta x'$. Consequently, the normalized emittance, ε_n, is defined as

$$\varepsilon_n = \gamma\beta\varepsilon , \tag{3.4.69}$$

omitting the (constant) mc term from the momentum definition. It should be pointed out that occasionally the emittance is also defined with the π included; in this text we prefer the definition of above.

An ideal periodic beam is only a convenient model which provides us with an intuitive interpretation of the emittance. In all practical cases the system is not periodic either due to imperfections in the guiding system or due to the amplification or acceleration process. In the model introduced above, these effects can be represented by a space dependent $K_0(z)$ i.e.,

$$\frac{d^2}{dz^2}\delta x_i + K_0^2(z)\delta x_i = 0 . \tag{3.4.70}$$

The solution in this case has the form

$$\delta x_i(z) = A_i(z)\cos[\psi(z) + \phi_i(z)] , \tag{3.4.71}$$

where for a constant K_0 we have $\psi(z) = K_0 z$. Substituting in (3.4.70) and using the orthogonality of the trigonometric functions sin and cos we obtain two equations

$$\cos[..]: \quad \frac{d^2 A_i}{dz^2} + K_0^2(z)A_i - A_i\left(\frac{d\psi}{dz}\right)^2 = 0 ,$$

$$\sin[..]: \quad 2\frac{dA_i}{dz}\frac{d\psi}{dz} + A_i\frac{d^2\psi}{dz^2} = 0 . \tag{3.4.72}$$

The second equation can be multiplied by A_i and written as

$$\frac{d}{dz}\left[A_i^2\frac{d\psi}{dz}\right] = 0 \rightarrow A_i^2\frac{d\psi}{dz} = C_i . \tag{3.4.73}$$

In particular, a particle which oscillates with a maximum amplitude a satisfies

$$a^2\frac{d\psi}{dz} = C . \tag{3.4.74}$$

For a constant K_0 and comparing with (3.4.68) we have

$$C = \varepsilon . \tag{3.4.75}$$

With this relation we can determine the equation for the envelope a by substituting (3.4.74) in the first equation of (3.4.72); the result is

$$\frac{d^2}{dz^2}a + K_0^2(z)a - \varepsilon^2\frac{1}{a^3} = 0 . \tag{3.4.76}$$

Next stage is to introduce a general definition of the emittance which is not dependent on the solutions presented above. Such a definition was used by Lapostolle (1971) and it is given by

$$\bar{\varepsilon} \equiv 4\left[\langle\delta x^2\rangle\langle(\delta x')^2\rangle - \langle\delta x\delta x'\rangle^2\right]^{1/2} . \tag{3.4.77}$$

This definition can be tested against the trivial solution in (3.4.65–66). Assuming that the phases ϕ_i and the amplitudes A_i are (statistically) independent, it can be readily seen that the second term is identically zero for a uniform distribution of phases and

$$\langle(\delta x)^2\rangle = \frac{1}{2}\langle A_i^2\rangle ,$$

$$\langle(\delta x')^2\rangle = \frac{1}{2}K_0^2\langle A_i^2\rangle , \tag{3.4.78}$$

therefore

$$\bar{\varepsilon} = 2K_0\langle A_i^2\rangle . \tag{3.4.79}$$

If the quantity $\langle A_i^2\rangle$ corresponds to the mean square value of the amplitudes $\langle A_i^2\rangle = a^2/2$ the emittance obtained is identical with that determined in (3.4.68). In order to have an idea as for the emittance value we can quote here the emittance measured at the Cornell Electron Storage Ring as reported by Rubin (1992): for a 5.289 GeV electron beam $\bar{\varepsilon} \simeq 0.25 \times 10^{-6}$m rad.

In the discussion so far we have considered only one out of the two transverse dimensions; in order to attribute the emittance to the adequate dimension we denote the emittance associated with the motion in the x direction by ε_x and in a similar way we define ε_y as the emittance associated with the motion in the y direction. With these two definitions it is convenient to introduce another quantity which provides a figure of merit regarding the beam quality. This is the brightness:

$$B \equiv \frac{I}{(\pi\varepsilon_x)(\pi\varepsilon_y)} F_{\mathrm{f}} . \tag{3.4.80}$$

F_{f} is a geometrical form factor on the order of unity. As the emittance the normalized brightness can be defined as $B_n \equiv B/\beta^2\gamma^2$. Both brightness and emittance provide information regarding the beam quality which is crucial in particle accelerators or sources of very high frequency radiation such as ultra violet or X-Rays. More detailed discussion on emittance can be found in the book by Lawson (1988).

3.5 Space-Charge Waves

All the effects considered so far were either static or quasi-static. In this section we shall introduce some elementary concepts of waves which propagate along electron beams. For this purpose consider a beam whose unperturbed beam density is n_0 and its zero order velocity is v_0 (the effect of potential depression is already included); the beam is guided by a very strong magnetic field. Consider an electric field E_z is excited in the system and its form is

$$E_z(r, z, \omega) = E(r)\mathrm{e}^{-jkz} . \tag{3.5.1}$$

The z component of the linearized equation of motion implies that the linear perturbation in the velocity field denoted by δv is given by

$$m\gamma^3 j(\omega - v_0 k)\delta v = -eE_z ; \tag{3.5.2}$$

here m and e are the mass and the charge of an electron respectively; $\gamma = 1/\sqrt{1-(v_0/c)^2}$. Next we can determine the perturbation in density (δn) using the continuity equation; the result is

$$\delta n = n_0 \frac{k}{\omega - v_0 k} \delta v \,. \tag{3.5.3}$$

The current density defined in (3.1.38) is linearized in the perturbation terms i.e., $J_z = -e(\delta n v_0 + n_0 \delta v)$ and it reads

$$J_z = -j\omega\varepsilon_0 \frac{\omega_{\rm p}^2}{\gamma^3(\omega - v_0 k)^2} E_z \,, \tag{3.5.4}$$

where $\omega_{\rm p}$ is the plasma frequency as defined in (3.4.18). Next we use (2.1.38,41) and substitute the result in the wave equation for the magnetic vector potential ((2.1.39)) which then reads

$$\left[\frac{1}{r}\frac{\rm d}{{\rm d}r} r \frac{\rm d}{{\rm d}r} + \left(\frac{\omega^2}{c^2} - k^2\right)\left(1 - \frac{\omega_{\rm p}^2}{\gamma^3(\omega - v_0 k)^2}\right)\right] A_z(r, k; \omega) = 0 \,. \tag{3.5.5}$$

In an infinite system $\partial/\partial r = 0$, there are two sets of solutions

$$\begin{aligned} k^2 - \frac{\omega^2}{c^2} &= 0 \,, \\ (k - \frac{\omega}{v_0})^2 - \frac{\omega_{\rm p}^2}{v_0^2 \gamma^3} &= 0 \,. \end{aligned} \tag{3.5.6}$$

The first set corresponds to a pure electromagnetic wave and the second represents the dispersion relation of the so-called space-charge waves. There are two such waves, both propagating parallel to the beam with a phase velocity close to the average velocity of the beam

$$\begin{aligned} k_+ &= \frac{\omega}{v_0} + K_{\rm p} \,, \\ k_- &= \frac{\omega}{v_0} - K_{\rm p} \,, \\ K_{\rm p}^2 &\equiv \frac{\omega_{\rm p}^2}{\gamma^3 v_0^2} \,. \end{aligned} \tag{3.5.7}$$

In contrast to regular electromagnetic waves, the space-charge wave has only an electric field and its magnetic component, in the absence of transverse variation, is identically zero even if the time variations are very rapid.

In a waveguide of radius R the boundary condition $[E_z(r = R) = 0]$ imposes the following dispersion relation

$$\left(\frac{\omega^2}{c^2} - k^2\right)\left[1 - \frac{\omega_{\rm p}^2}{\gamma^3(\omega - v_0 k)^2}\right] = \frac{p_s^2}{R^2} \tag{3.5.8}$$

and as above there are two groups of solutions: the electromagnetic modes group whose asymptotic behavior can be determined from the limit when no beam is present ($\omega_{\rm p}^2 = 0$) namely, $(\omega/c)^2 - k^2 = (p_s/R)^2$. We discussed this group in the context of electromagnetic TM modes in Chap. 2. The second group represents waves which propagate along the beam and they can be approximated by

$$\begin{aligned} k_s^{(\pm)} &\simeq \frac{\omega}{v_0} \pm \delta k_s \,, \\ \delta k_s^2 &= \frac{\omega_{\rm p}^2}{\gamma^3 v_0^2} \xi_s^2 \,, \\ \xi_s &= \left[1 + \left(\frac{\gamma \beta p_s c}{\omega R} \right)^2 \right]^{-1/2} \,, \end{aligned} \tag{3.5.9}$$

provided that the plasma wavenumber $K_{\rm p}$ is much smaller than ω / v_0. Note that the factor ξ_s is always smaller than unity such that each mode sees a *reduced plasma frequency.* An extensive discussion on space-charge wave can be found in a book by Beck (1958). In this section we shall review five instructive topics.

3.5.1 Slow and Fast Space-Charge Waves

The waves which correspond to $k_s^{(+)}$ have a phase velocity

$$\begin{aligned} v_{\rm slow} &= \frac{\omega}{k_s^{(+)}} \\ &= \frac{v_0}{1 + v_0 \delta k_s / \omega} < v_0 \,, \end{aligned} \tag{3.5.10}$$

which is slower than the beam average velocity (v_0) therefore these are called *slow space-charge waves.* The waves which correspond to $k_s^{(-)}$ have a phase velocity

$$\begin{aligned} v_{\rm fast} &= \frac{\omega}{k_s^{(-)}} \\ &= \frac{v_0}{1 - v_0 \delta k_s / \omega} > v_0 \,, \end{aligned} \tag{3.5.11}$$

which is greater than the average velocity of the beam and these are referred to as *fast space-charge waves.*

3.5.2 "Negative" and "Positive" Energy

The contribution of the space-charge wave to the average kinetic energy density is determined based on the global energy conservation in (3.1.45) which is given by

$$\delta E = mc^2 \frac{1}{4} \left[\delta n \delta \gamma^* + \delta n^* \delta \gamma \right] . \tag{3.5.12}$$

We can now express $\delta\gamma$ in terms of δv namely $\delta\gamma = \gamma^3\beta\delta v/c$ and then use the expression in (3.5.3) to write

$$\delta E = \frac{1}{2}mc^2|\delta n|^2\gamma^3\beta\frac{1}{n_0 c}\left(\frac{\omega}{k} - v_0\right) . \tag{3.5.13}$$

This result indicates that the slow space-charge waves have a total kinetic energy density which is smaller than the average kinetic energy of the beam i.e.,

$$\delta E_{\text{slow}} = \frac{1}{2}mc^2|\delta n|^2\gamma^3\beta\frac{1}{n_0 c}\left(v_{\text{slow}} - v_0\right) < 0 . \tag{3.5.14}$$

For this reason these waves are also referred to as "negative" energy waves. Fast space-charge waves have "positive" energy since

$$\delta E_{\text{fast}} = \frac{1}{2}mc^2|\delta n|^2\gamma^3\beta\frac{1}{n_0 c}\left(v_{\text{fast}} - v_0\right) > 0 . \tag{3.5.15}$$

When *distributed* interaction between electrons and electromagnetic waves is possible we will see that the slow space-charge waves play an important role in the process. Both fast and slow space-charge waves play a very important role in klystrons where the interaction is limited to the close vicinity of a cavity gap.

The real (averaged) power change in each one of the modes is *identically zero* as is readily seen by examining the current density term in (3.5.4) and

$$\text{Re}\left(\frac{1}{2}E_z J_z^*\right) = \text{Re}\left[j\omega\varepsilon_0\frac{\omega_{\text{p}}^2}{\gamma^3(\omega - v_0 k)^2}\frac{1}{2}\left|E_z\right|^2\right] . \tag{3.5.16}$$

Since all the k's are real then clearly the right hand side is zero. However in a superposition of two space-charge waves the real power may be non-zero. This is the basis for the operation of relativistic klystrons.

3.5.3 Resistive Wall Instability

When electromagnetic waves propagate in a waveguide with lossy wall, a fraction of the power is absorbed in the wall. Since the process is linear, namely the power absorbed per unit length is proportional to the local power flow, the wave decays exponentially in space. In the case of space-charge waves the situation is different – it will be shown that the slow space-charge wave can actually be amplified due to the complex impedance at the metallic surface. Before we examine a realistic case it is instructive to investigate a simplified model. For this purpose we assume that the beam propagates instead of vacuum in a lossy medium which is characterized by $\varepsilon_{\text{r}} = 1 + \sigma/j\omega\varepsilon_0$ – where σ is the conductivity of the medium. It can be shown that the dispersion relation of the space-charge waves as formulated in (3.5.6) should be updated by replacing $\varepsilon_0 \to \varepsilon_0\varepsilon_r$. This implies that $\omega_p^2 \to \omega_p^2/\varepsilon_r$. In the right-hand side the expression is complex and since the solution for k has the form $k_\pm = \omega/v_0 \pm K_p/\sqrt{\varepsilon_r}$ we clearly see that one of the solutions corresponds to a growing wave.

Let us now examine this process in a more realistic system. Consider a waveguide of radius R made of a material of finite conductivity $\sigma \gg \omega\varepsilon_0$. The beam which propagates is electromagnetically characterized by (3.5.4) and for simplicity, we shall assume that it fills the entire waveguide. The magnetic vector potential for, $r < R$, is a solution of (3.5.5) and its solution reads

$$A_z(r,k;\omega) = A\mathrm{I}_0(\Lambda r)\,, \tag{3.5.17}$$

where

$$\Lambda^2 \equiv \left(k^2 - \frac{\omega^2}{c^2}\right)\left[1 - \frac{\omega_p^2}{\gamma^3(\omega - v_0 k)^2}\right]. \tag{3.5.18}$$

The impedance at $r = R$ is the ratio

$$\begin{aligned} Z_{\text{beam}} &\equiv \frac{E_z(r=R)}{H_\phi(r=R)}\,, \\ &= j\eta_0 \frac{(\omega/c)^2 - k^2}{\Lambda\omega/c}\frac{\mathrm{I}_0(\Lambda R)}{\mathrm{I}_1(\Lambda R)}\,. \end{aligned} \tag{3.5.19}$$

In the metallic wall, the magnetic vector potential satisfies

$$\left[\frac{1}{r}\frac{\mathrm{d}}{\mathrm{d}r}r\frac{\mathrm{d}}{\mathrm{d}r} - k^2 - j\omega\mu_0\sigma\right]A_z(r,k;\omega) = 0\,, \tag{3.5.20}$$

and the solution for $r \geq R$ reads

$$A_z(r,k;\omega) = B\mathrm{K}_0(\Xi r)\,, \tag{3.5.21}$$

where $\Xi^2 = k^2 + j\omega\mu_0\sigma$. The impedance at the discontinuity is

$$\begin{aligned} Z_{\text{wall}} &\equiv \frac{E_z(r=R)}{H_\phi(r=R)}\,, \\ &= -j\eta_0 \frac{(\omega/c)^2 - k^2}{\Xi\omega/c}\frac{\mathrm{K}_0(\Xi R)}{\mathrm{K}_1(\Xi R)}\,. \end{aligned} \tag{3.5.22}$$

Imposing the boundary condition, as we have done before, is equivalent to the requirement that the two impedances are continuous at the metallic wall $(r = R)$ i.e.,

$$Z_{\text{wall}} = Z_{\text{beam}}\,, \tag{3.5.23}$$

and this determines the dispersion relation in a lossy waveguide. In order to illustrate the effect of the lossy material on the propagation of a space-charge wave, we consider a solution which has the form

$$k = \frac{\omega}{v_0} + q\,, \tag{3.5.24}$$

and $\omega/v_0 \gg |q|$. Furthermore, for sufficiently high conductivity we have $\sigma \gg \omega\varepsilon_0$ and for simplicity we consider the limit $\omega R/c \gg 1$. Consequently, the dispersion relation can be simplified to read

$$\sqrt{1 - \frac{\omega_{\rm p}^2}{\gamma^3 v_0^2 q^2}} = -\gamma\beta\frac{c}{\omega}\sqrt{j\omega\mu_0\sigma} = -\gamma\beta\sqrt{j\frac{\sigma}{\omega\varepsilon_0}}\,. \tag{3.5.25}$$

According to the last term on the right, for a relativistic beam the right-hand side is much larger than unity, therefore on the left-hand side the second term has to satisfy $|q|^2 \ll \omega_{\rm p}^2/v_0^2\gamma^3$, hence

$$\sqrt{\frac{\omega_{\rm p}^2}{\gamma^3 v_0^2 q^2}} = j\gamma\beta\frac{c}{\omega}\sqrt{j\omega\mu_0\sigma}\,. \tag{3.5.26}$$

The expression for the dispersion relation can be simplified by defining

$$q_0^2 \equiv \frac{eJ\eta_0}{mc^2}\frac{\omega\varepsilon_0}{\sigma}\frac{1}{(\gamma\beta)^5}\,, \tag{3.5.27}$$

and

$$\bar{q} = \frac{q}{q_0}\,; \tag{3.5.28}$$

with these definitions the dispersion relation reads

$$\sqrt{j\bar{q}^2} + j = 0\,. \tag{3.5.29}$$

We used above the definition of the plasma frequency in (3.4.18) and of the current density $|J| = en_0v_0$. This simplified dispersion equation has two solutions: $\bar{q} = (1+j)/\sqrt{2}$ representing a slow space-charge wave which grows exponentially in space

$$\mathrm{e}^{-jzk} = \mathrm{e}^{-jz[\omega/v_0+q_0(1+j)/\sqrt{2}]} = \mathrm{e}^{-jz(\omega/v_0+q_0/\sqrt{2})}\mathrm{e}^{zq_0/\sqrt{2}}\,, \tag{3.5.30}$$

whose imaginary part determines the spatial growth rate of what is called the *resistive wall instability*. Note that the "negative" energy wave is the one which grows. This is the case in all the schemes of collective beam-wave interaction. The second solution, $q = -(1+j)q_0/\sqrt{2}$, describes a fast space-charge wave which decays exponentially in space. The similarity to the simple model presented at the beginning of this sub-section is evident.

3.5.4 Two-Beam Instability

The phase velocity of space-charge waves is close to the average velocity of the beam. If two beams move close to each other at two different but close velocities the space-charge waves which may develop along these beams can convert energy from dc to rf. In order to investigate the effect of two different velocities we examine two annular beams which have two different velocities v_1 and v_2; their radius is $R_{b,1}$ and $R_{b,2}$. For simplicity it is assumed that the thickness of both beams is much smaller than the wavelength and it is equal in both cases ($\Delta \ll \lambda$). The inner beam (subscript 1) carries a current I_1, and I_2 is the current carried by the outer beam – subscript 2. Both beams propagate along a very strong magnetic field, in a metallic cylindrical (lossless) waveguide of radius R – see Fig. 3.12. The static problem, of determining the kinetic energy of each one of the beams given the energy at the input of the waveguide, is left as an exercise to the reader (Exercise 3.5). In this section we examine the dynamic problem of propagation of wave in this system. It will be shown that if the velocity difference is larger than some threshold value, the space-charge wave can grow exponentially in space similar to the case of the resistive wall instability.

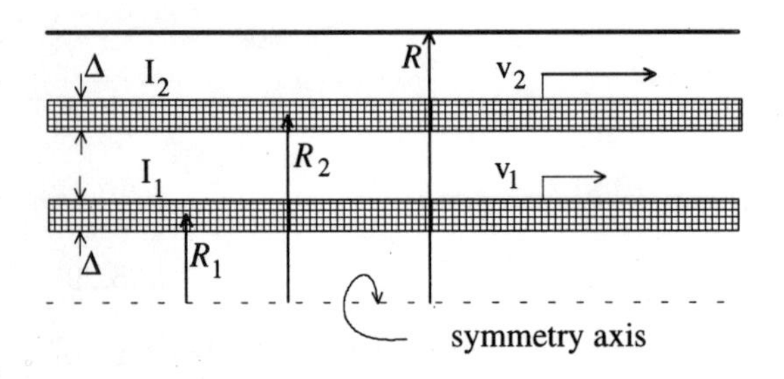

Fig. 3.12. Two-beam instability: two beams of different currents and different velocities move in a cylindrical waveguide. Under certain circumstances dc energy is converted into rf

The electromagnetic field is derived from a magnetic vector potential which is a solution of

$$\left[\nabla^2 + \frac{\omega^2}{c^2}\right] A_z(r, z, \omega) = -\mu_0 J_z(r, z, \omega) . \tag{3.5.31}$$

We are looking for a solution which, as a function of z, has the form e^{-jkz} therefore the current density in each one of the beams is

$$J_z(r = R_\nu, k, \omega) = -j\omega\varepsilon_0 \frac{\omega_{p,\nu}^2}{\gamma_\nu^3(\omega - kv_\nu)^2} E_z(r = R_\nu, k, \omega) , \tag{3.5.32}$$

where $\nu = 1, 2$. The solution is split now in three different vacuum regions: between the waveguide wall and the outer beam ($R > r > R_{b,2} + \Delta/2$) the solution is

$$A_z(r, k, \omega) = AT_0(r) , \tag{3.5.33}$$

where $T_0(r) = \mathrm{I}_0(\Gamma r)\mathrm{K}_0(\Gamma R) - \mathrm{K}_0(\Gamma r)\mathrm{I}_0(\Gamma R)$ and $\Gamma^2 = k^2 - (\omega/c)^2$. In what follows we shall also use $T_1'(r) = \mathrm{I}_1(\Gamma r)\mathrm{K}_0(\Gamma R) + \mathrm{K}_1(\Gamma r)\mathrm{I}_0(\Gamma R)$. Between the two beams ($R_{\mathrm{b},2} - \Delta/2 > r > R_{\mathrm{b},1} + \Delta/2$) the solution is

$$A_z(r,k,\omega) = B\mathrm{K}_0(\Gamma r) + C\mathrm{I}_0(\Gamma r), \tag{3.5.34}$$

and in the last region, $R_{\mathrm{b},1} - \Delta/2 > r > 0$, the solution is

$$A_z(r,k,\omega) = D\mathrm{I}_0(\Gamma r). \tag{3.5.35}$$

In order to determine the dispersion relation we impose the boundary conditions at the discontinuities imposed by the current density sheets. Integrating (3.5.31) in the vicinity of each one of the discontinuities results in the following expression

$$\left[r\frac{\partial A_z(r,k,\omega)}{\partial r}\right]_{r=R_1}^{R_2} = -\mu_0 \int_{R_1}^{R_2} \mathrm{d}r\, r\, J_z(r,k,\omega). \tag{3.5.36}$$

At $r = R_{\mathrm{b},2}$ this implies

$$\begin{aligned} &\Gamma R_{\mathrm{b},2} A T_0'(R_{\mathrm{b},2}) - \Gamma R_{\mathrm{b},2}\left[-B\mathrm{K}_1(\Gamma R_{\mathrm{b},2}) + C\mathrm{I}_1(\Gamma R_{\mathrm{b},2})\right] \\ &\quad = \left[\frac{\omega_{p,2}^2}{\gamma_2^3(\omega - kv_2)^2}\left(\frac{\omega^2}{c^2} - k^2\right)\right] A T_0(R_{\mathrm{b},2}) R_{\mathrm{b},2}\Delta. \end{aligned} \tag{3.5.37}$$

Continuity of the longitudinal electric field implies

$$A T_0(R_{\mathrm{b},2}) = B\mathrm{K}_0(\Gamma R_{\mathrm{b},2}) - C\mathrm{I}_0(\Gamma R_{\mathrm{b},2}). \tag{3.5.38}$$

In a similar way at $r = R_{\mathrm{b},1}$ we have

$$\begin{aligned} &\Gamma R_{\mathrm{b},1}\left[-B\mathrm{K}_1(\Gamma R_{\mathrm{b},1}) + C\mathrm{I}_1(\Gamma R_{\mathrm{b},1})\right] - \Gamma R_{\mathrm{b},1} D\mathrm{I}_1(\Gamma R_{\mathrm{b},1}) \\ &\quad = \left[\frac{\omega_{p,1}^2}{\gamma_1^3(\omega - kv_1)^2}\left(\frac{\omega^2}{c^2} - k^2\right)\right] D\mathrm{I}_0(\Gamma R_{\mathrm{b},1}) R_{\mathrm{b},1}\Delta, \end{aligned} \tag{3.5.39}$$

and

$$D\mathrm{I}_0(\Gamma R_{\mathrm{b},1}) = B\mathrm{K}_0(\Gamma R_{\mathrm{b},1}) - C\mathrm{I}_0(\Gamma R_{\mathrm{b},1}). \tag{3.5.40}$$

It is convenient to introduce the following notation for the description of the dispersion relation: $\alpha_\nu = \Gamma R_{\mathrm{b},\nu}$, $t_\nu = T_0(R_{\mathrm{b},\nu})$ and $t'_\nu = T'_0(R_{\mathrm{b},\nu})$. In addition, we define $\tau_1 = \mathrm{I}_1(\alpha_1)/\mathrm{I}_0(\alpha_1) - \omega_{\mathrm{p},1}^2/(\omega - kv_1)^2$ and $\tau_2 = t'_2/t_2 + \omega_{\mathrm{p},2}^2/(\omega - kv_2)^2$; with these definitions the dispersion relation reads

$$\begin{aligned} &\mathrm{K}_1(\alpha_2)\mathrm{I}_1(\alpha_1) - \mathrm{K}_1(\alpha_1)\mathrm{I}_1(\alpha_2) \\ &\qquad + \tau_2\left[\mathrm{K}_0(\alpha_2)\mathrm{I}_1(\alpha_1) + \mathrm{K}_1(\alpha_1)\mathrm{I}_0(\alpha_2)\right] \\ &\qquad - \tau_1\left[\mathrm{K}_1(\alpha_2)\mathrm{I}_0(\alpha_1) + \mathrm{K}_0(\alpha_1)\mathrm{I}_1(\alpha_2)\right] \\ &\qquad + \tau_1\tau_2\left[\mathrm{I}_0(\alpha_2)\mathrm{K}_0(\alpha_1) - \mathrm{K}_0(\alpha_2)\mathrm{I}_0(\alpha_1)\right] = 0. \end{aligned} \tag{3.5.41}$$

We shall now simplify this expression on the basis of several realistic assumptions: *(i)* the beam radius is large on the scale of the wavelength i.e., $|\alpha_\nu| \gg 1$ hence

$$\tau_1 = \tau_2 + (\tau_1\tau_2 - 1)\tanh(\alpha_2 - \alpha_1)\,, \tag{3.5.42}$$

where $\tau_1 \simeq 1 - \omega_{\mathrm{p},1}^2/(\omega - kv_1)^2$ and $\tau_2 = -c\tanh(\alpha_1 - \alpha_2) + \omega_{\mathrm{p},2}^2/(\omega - kv_2)^2$. *(ii)* Close enough to resonance the plasma terms in the last two expressions are assumed to dominate thus

$$\tau_1 \simeq -\omega_{\mathrm{p},1}^2/(\omega - kv_1)^2\,,$$
$$\tau_2 \simeq \omega_{\mathrm{p},2}^2/(\omega - kv_2)^2\,. \tag{3.5.43}$$

Next we assume that *(iii)* the velocities of the two beams are close such that we have $\delta v = (v_2 - v_1)/2$, $|\delta v| \ll v_0$ where $v_0 = (v_1 + v_2)/2$. We consider a solution of the form

$$k = \frac{\omega}{v_0} + \delta k\,, \tag{3.5.44}$$

subject to the constraint that *(iv)* the change of the amplitude and phase is small in the scale of one wavelength i.e. that $|\delta k| \ll \omega/v_0$. Consequently,

$$\alpha_\nu = \left(\frac{\omega}{c} R_{\mathrm{b},\nu}\right)\frac{1}{\gamma\beta}\,, \tag{3.5.45}$$

with $\beta = v_0/c$ and $\gamma = [1 - \beta^2]^{-1/2}$. With these simplifying assumptions the dispersion relation (in terms of δk) is a simple quadratic equation which reads

$$\delta k^2 + 2\delta k\,\delta k_0 \frac{K_1^2 - K_2^2}{K_1^2 + K_2^2} + \delta k_0^2 - \frac{K_1^2 K_2^2}{K_1^2 + K_2^2}\tanh(\alpha_2 - \alpha_1) = 0\,, \tag{3.5.46}$$

where

$$\delta k_0 = \frac{\omega}{v_0}\frac{\delta v}{v_0}\,,$$
$$K_\nu^2 = \frac{I_\nu}{I_\mathrm{e}}\frac{1}{2\pi R_{\mathrm{b},\nu}}\frac{\omega}{c}\frac{1}{\gamma^4\beta^3}\,, \tag{3.5.47}$$

and it has two solutions. If the velocity difference δv satisfies

$$\left(\frac{\omega}{v_0}\frac{\delta v}{v_0}\right)^2 > \frac{1}{4}\left(K_1^2 + K_2^2\right)\tanh\left[\frac{\omega}{c}(R_{\mathrm{b},2} - R_{\mathrm{b},1})\frac{1}{\gamma\beta}\right], \tag{3.5.48}$$
$$> \frac{1}{8\pi I_\mathrm{e}\gamma^4\beta^3}\frac{\omega}{c}\left(\frac{I_1}{R_{\mathrm{b},1}} + \frac{I_2}{R_{\mathrm{b},2}}\right)\tanh\left[\frac{\omega}{c}(R_{\mathrm{b},2} - R_{\mathrm{b},1})\frac{1}{\gamma\beta}\right],$$

then the two solutions are complex. One growing in space and the other one decaying. This is a simplified version of the condition for the occurrence of what is called the *two-beam instability.* Assuming that this condition is satisfied then the instability is represented by the imaginary part of the wave number

$$\mathrm{Im}(\delta k) = \frac{2K_1K_2}{K_1^2+K_2^2}\sqrt{\delta k_0^2 - \frac{1}{4}(K_1^2+K_2^2)\tanh(\alpha_2-\alpha_1)}\,. \tag{3.5.49}$$

Note that the spatial growth rate decreases when the separation between the two beams is increased and it is maximum in case of overlap.

3.5.5 Interference of Space-Charge Waves

We shall examine next the effect of local perturbations on space-charge waves. For a description of this kind of effect it is convenient to adopt a transmission-line notation. The radial electric field is associated with the voltage

$$V(z) = \mathrm{e}^{-jk_e z}\left[V_+\mathrm{e}^{-jK_\mathrm{p}z} + V_-\mathrm{e}^{jK_\mathrm{p}z}\right], \tag{3.5.50}$$

and the azimuthal magnetic field is associated with the current

$$I(z) = \mathrm{e}^{-jk_e z}\left[\frac{V_+}{Z_+}\mathrm{e}^{-jK_\mathrm{p}z} + \frac{V_-}{Z_-}\mathrm{e}^{jK_\mathrm{p}z}\right]. \tag{3.5.51}$$

In these expressions $k_e = \omega/v_0$, $K_\mathrm{p} = \omega_\mathrm{p}/v_0\gamma^{3/2}$,

$$Z_\pm = \eta_0\left(\frac{1}{\beta} \pm \frac{cK_\mathrm{p}}{\omega}\right), \tag{3.5.52}$$

$\beta = v_0/c$ and $\gamma = [1-\beta^2]^{-1/2}$. Note that, in contrast to transmission lines where the two possible waves propagate in opposite directions, here both waves propagate in the same direction but with two different phase velocities. Let us assume now that both voltage and current are known at $z=0$ hence

$$V(0) = U_0\,,$$
$$I(0) = I_0\,. \tag{3.5.53}$$

Subject to these two conditions the amplitudes $V_\pm$ can be calculated and the voltage and current modulation on the beam read

$$V(z) = \mathrm{e}^{-jk_e z}\Bigg[U_0\cos(K_\mathrm{p}z) + j\sin(K_\mathrm{p}z)\frac{1}{Z_+ - Z_-} \tag{3.5.54}$$
$$\times\left(Z_+(I_0Z_- - U_0) + Z_-(I_0Z_+ - U_0)\right)\Bigg],$$

$$I(z) = \mathrm{e}^{-jk_e z}\left[I_0\cos(K_\mathrm{p}z) + j\sin(K_\mathrm{p}z)\frac{1}{Z_+ - Z_-}\left(I_0(Z_+ + Z_-) - 2U_0\right)\right].$$

When the initial current modulation is zero ($I_0 = 0$) these equations read

$$V(z) = U_0 e^{-jk_e z}\left[\cos(K_p z) - j\sin(K_p z)\frac{Z_+ + Z_-}{Z_+ - Z_-}\right],$$

$$I(z) = U_0 e^{-jk_e z}\left[j\sin(K_p z)\frac{-2}{Z_+ - Z_-}\right], \tag{3.5.55}$$

and associated with this modulation the total (real) power which develops along the beam oscillates in space and it is given by

$$\begin{aligned} P(z) &= \frac{1}{2}\mathrm{Re}\left[V(z)I^*(z)\right] \\ &= \frac{U_0^2}{2\eta_0\beta}\left(\frac{\omega}{cK_p}\right)^2 \sin^2(K_p z)\,. \end{aligned} \tag{3.5.56}$$

Note that this power is zero at $z = 0$ and it grows monotonically until it peaks at $z = \lambda_p/4(\lambda_p \equiv 2\pi/K_p)$ as is the current modulation. Beyond this point both the real power and the current decrease monotonically until they reach zero at $z = \lambda_p/2$. This fact is utilized in relativistic klystrons for further amplification of the modulation by placing a cavity at $z = \lambda_p/4$. Let us now examine this effect.

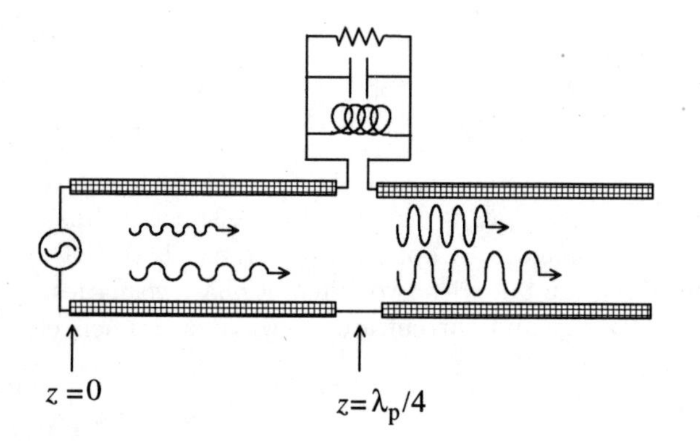

Fig. 3.13. Schematic of beam-cavity interaction in the framework of the transmission line formulation. The amplitude of the space-charge waves is significantly affected by the cavity

In the framework of the current model the cavity will be represented by an RLC circuit whose impedance is

$$Z_{\mathrm{cav}} = Z_0\frac{j\omega\omega_0}{\omega_0^2 - \omega^2 + 2j\omega\omega_0/Q}, \tag{3.5.57}$$

where Q is known as the quality factor of the cavity and together with Z_0 they determine the impedance at resonance i.e., $Z_{\mathrm{cav}} = \frac{1}{2}Z_0 Q$ as illustrated in Fig. 3.13. The incident waves are given by (3.5.50–51) and the transmitted ones by

$$V_{\rm tr}(z) = {\rm e}^{-jk_e(z-\lambda_{\rm p}/4)} \left[V'_+ {\rm e}^{-jK_{\rm p}(z-\lambda_{\rm p}/4)} + V'_- {\rm e}^{jK_{\rm p}(z-\lambda_{\rm p}/4)} \right] ,$$

$$I_{\rm tr}(z) = {\rm e}^{-jk_e(z-\lambda_{\rm p}/4)} \left[\frac{V'_+}{Z_+} {\rm e}^{-jK_{\rm p}(z-\lambda_{\rm p}/4)} + \frac{V'_-}{Z_-} {\rm e}^{jK_{\rm p}(z-\lambda_{\rm p}/4)} \right] . \quad (3.5.58)$$

The boundary condition at $z = \lambda_{\rm p}/4$ can be determined from the fact that the current associated with the incident and transmitted waves has to be continuous i.e.

$$I\left(\frac{\lambda_{\rm p}}{4}\right) = I_{\rm tr}\left(\frac{\lambda_{\rm p}}{4}\right) , \quad (3.5.59)$$

and the voltage associated with the incident waves is the sum of the transmitted and the voltage on the cavity aperture:

$$V\left(\frac{\lambda_{\rm p}}{4}\right) = V_{\rm tr}\left(\frac{\lambda_{\rm p}}{4}\right) + V_{\rm cav} . \quad (3.5.60)$$

Subject to these boundary conditions the transmitted waves are given by

$$V_{\rm tr}(z) = V(z) + Z_{\rm cav} I\left(\frac{\lambda_{\rm p}}{4}\right) {\rm e}^{-jk_e(z-\lambda_{\rm p}/4)}$$
$$\times \left[-\cos\left(K_{\rm p}\left(z - \frac{\lambda_{\rm p}}{4}\right)\right) + j\frac{Z_+ + Z_-}{Z_+ - Z_-} \sin\left(K_{\rm p}\left(z - \frac{\lambda_{\rm p}}{4}\right)\right)\right]$$

$$I_{\rm tr}(z) = I(z) + 2jI\left(\frac{\lambda_{\rm p}}{4}\right) {\rm e}^{-jk_e(z-\lambda_{\rm p}/4)}$$
$$\times \sin\left(K_{\rm p}\left(z - \frac{\lambda_{\rm p}}{4}\right)\right) \frac{Z_{\rm cav}}{Z_+ - Z_-} . \quad (3.5.61)$$

In Fig. 3.14a we illustrate the normalized current density as it develops in space; the normalization is with the average current injected. The parameters in this case are: $\gamma = 3$, $R_{\rm b} = 3\,{\rm mm}$, $I = 1\,{\rm kA}$, $U_0 = 1\,{\rm V}$ and $Z_{\rm cav}$=5000 Ω. Prior to the cavity the modulation is negligible (0.001%); however the cavity causes a dramatic (local) change in the amplitudes of both fast and slow space-charge waves. Beyond the cavity the modulation increases monotonically by almost 5 orders of magnitude to almost 90%. This figure has to be interpreted as the general trend (or the potential) of the method rather than literally since within the framework of the linearization we performed, only relatively small perturbation of the average value of the current is permitted.

The interference between these two waves is responsible to the propagation of real power. Assuming that the terms associated with the cavity are dominant the power conversion beyond the cavity is

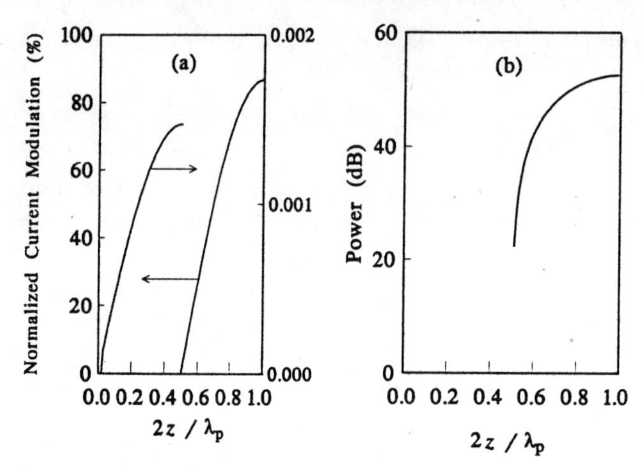

Fig. 3.14. (**a**) Current modulation as it varies in space. (**b**) Normalized power (in dB) as it develops after the cavity

$$
\begin{aligned}
P_{\rm tr}(z) &\equiv {\rm Re}\left[\frac{1}{2}V_{\rm tr}(z)I^*_{\rm tr}(z)\right] \\
&\simeq \frac{U_0^2}{2}\left(\frac{2|Z_{\rm cav}|}{(Z_+ - Z_-)^2}\right)^2 \sin^2\left(K_{\rm p}\left(z-\frac{\lambda_{\rm p}}{4}\right)\right) \\
&\simeq P\left(\frac{\lambda_{\rm p}}{4}\right)\left(\frac{|Z_{\rm cav}|}{\eta_0}\frac{\omega}{cK_{\rm p}}\right)^2 \sin^2\left(K_{\rm p}\left(z-\frac{\lambda_{\rm p}}{4}\right)\right)\,,
\end{aligned}
\tag{3.5.62}
$$

which is plotted in Fig. 3.14b normalized to $P_0 = U_0^2/2\beta\eta_0$ and presented in dB. For this particular example the cavity causes an increase of more than 50 dB. It is important to emphasize here that in contrast to the case of the two instabilities introduced in the last two sections, the amplitudes of each one of the space-charge waves are *constant* in space and only the interference between the two generates the average power mentioned above. The maximum power occurs at the location of maximum current which is after an additional $z = \lambda_{\rm p}/4$. Hence,

$$
\begin{aligned}
P_{\rm tr}(\frac{\lambda_{\rm p}}{2}) &\equiv {\rm Re}\left[\frac{1}{2}V_{\rm tr}\left(\frac{\lambda_{\rm p}}{4}\right) I^*_{\rm tr}\left(\frac{\lambda_{\rm p}}{4}\right)\right] \\
&= \frac{U_0^2}{2\eta_0\beta}\left(\frac{\omega}{cK_{\rm p}}\right)^4 \\
&= P\left(\frac{\lambda_{\rm p}}{4}\right)\left[\frac{|Z_{\rm cav}|}{\eta_0}\frac{\omega}{cK_{\rm p}}\right]^2 .
\end{aligned}
\tag{3.5.63}
$$

The last term in the square brackets is the amplification term due to the presence of the cavity.

Exercises

3.1 Show by substituting (3.1.10) in Lagrange's equation of motion that they are identical to the Newtonian equations of motion. Repeat the exercise with Hamilton's equations of motion.

3.2 Show in a systematic way that if the charge of an electron is the same in all frames of reference then the set $(\mathbf{J}, c\varrho)$ forms a 4-vector.

3.3 Show that the condition in (3.4.19) is sufficient to ensure stability of the beam. Hint: linearize the equation of motion and assume that in spite oscillation the total current is preserved.

3.4 Apply Lorentz transformation to Maxwell's equations [(3.2.2)] and get the transformation for the electromagnetic field in (3.2.19-20).

3.5 Calculate the limiting currents of two thin annular beams of radii R_1 and R_2 moving in a waveguide of radius R $(R > R_2 > R_1)$; both beams are generated by the same diode, thus the initial kinetic energy $[mc^2(\gamma_0 - 1)]$ is the same. Determine the kinetic energy of the electrons in each one of the beams. Repeat the exercise for two beams of different initial energies and currents. In both cases the beams propagate in cylindrical waveguide of radius R.

3.6 Show that in dielectric loaded waveguide the limiting current of a pencil beam is given by

$$I_{\mathrm{max}} = 2\pi \frac{mc^2}{e\eta_0} \left(\gamma_0^{2/3} - 1\right)^{3/2} \ln^{-1} \left[\frac{R_{\mathrm{b}}}{R_d} \left(\frac{R_d}{R} \right)^{1/\varepsilon_{\mathrm{r}}} \right] .$$

The dielectric $(\varepsilon_{\mathrm{r}} > 1)$ fills the region $R \geq r \geq R_d$; the beam radius is R_{b} and it is smaller than the inner radius of the dielectric, R_d. Note that with the dielectric, for the same waveguide radius R, the limiting current is larger which means that the potential depression is smaller and consequently, the kinetic energy is larger.

3.7 Calculate the limiting current for an annular beam of radius R_{b} which carries a current I and its thickness is Δ and it moves *outside* a metallic waveguide of radius R.

3.8 Calculate the two-beam instability for the case of a motionless plasma and a pencil beam moving through the plasma in a waveguide of radius R. The radius of the beam is R_{b}, its plasma density is ω_{p} and the density of the background plasma is denoted by n_b.

3.9 Compare the Debye length with the plasma wavelength. For a temperature of 1000° K, which one is larger? What is your conclusion regarding the use of the hydrodynamic and kinetic approximations.

3.10 Develop the Child-Langmuir limiting current using the Lagrangian formulation. Hint: consult Chodorow and Susskind (1964) p. 125.

3.11 Use the integral $\oint \mathbf{E} \cdot d\mathbf{l} = 0$ in order to calculate the potential depression of a pencil beam in a waveguide. Hints: (a) include the diode in your contour, (b) assume perfect conductors and (c) in the beam region the contour closes on axis.

4. Models of Beam-Wave Interaction in Slow-Wave Structures

In this chapter we investigate the fundamentals of distributed beam-wave interaction in a slow-wave structure. A dielectric loaded waveguide was chosen as the basic model in the first sections because it enables us to illustrate the essence of the interaction without the complications associated with complex boundary conditions. Throughout this chapter the electron beam is assumed to be guided by a very strong magnetic field such that the electrons' motion is confined to the longitudinal direction. Furthermore, the kinetic energy of the electrons is assumed to take into consideration the potential depression associated with the injection of a beam into a metallic waveguide.

In the first section we present part of Pierce's theory for the traveling-wave amplifier applied to dielectric loaded structure and extended to the relativistic regime. The interaction for a semi-infinitely long system is formulated in terms of the interaction impedance introduced in Chap. 2. Finite length effects are considered in the second section where we first examine the other extreme of the beam-wave interaction namely, the oscillator. In the context of an amplifier it is shown that reflections affect the bandwidth and in addition, the beam shifts the frequency where maximum transmission occurs.

The macro-particle approach is described in Sect. 4.3 where the beam dynamics instead of being considered in the framework of the hydrodynamic approximation i.e. as a single fluid flow, is represented by a large number of clusters of electrons. Each one of the clusters is free to move at a different velocity according to the local field it experiences but the electrons which constitute the cluster are "glued" together. This formalism enables us to examine the interaction in phase-space either in the linear regime of operation or close to saturation. It also permits investigation of tapered structures and analysis of the interaction of pre-bunched beams in tapered structures.

In Sect. 4.4 we extend the concept of the interaction impedance to a complex quantity and formulate the equations accordingly. This impedance is allowed to vary in space and no prior assumption on the form of the electromagnetic field is made but it is assumed that the effect of the interaction is local. In other words, the power at a given location can be determined by the electric field which acts on the electrons at the same location. The chapter concludes with a further extension of the macro-particle approach formalism to include the effect of reflections. This framework combines the formulations of an amplifier and an oscillator and permits us to quantify and illustrate the operation of a realistic device which is neither an ideal amplifier nor an ideal oscillator.

With the exception of the first section which, as indicated, is a review of Pierce's TWT theory, most of the material presented in this chapter has been developed in recent years as part of an effort to develop high power traveling-wave amplifiers. The first experiments on high power TWT performed at Cornell University [Shiffler (1989)] indicated that 100MW of power at 8.76GHz can be achieved before the system oscillates. Although no rf break-down was observed, the fact that the input is no longer isolated from the output, allows waves to be reflected backwards and this feedback could cause the system to oscillate. In order to isolate the input from the output the TWT was split in two sections separated by a sever (waveguide made of lossy material which operates below cut-off). The second set of experiments on a two stage high power TWT indicated that power levels in excess of 400 MW are achievable with no indication of rf break-down [Shiffler (1991)]. In this case however the spectrum of output frequencies was 300 MHz wide and a significant amount of power (up to 50 %) was measured in asymmetric sidebands. The latter observation was investigated theoretically [Schächter (1991)] and it was concluded that it is a result of amplified noise at frequencies selected by the interference of the two waves bouncing between the ends of the last stage. In fact we have shown [Schächter and Nation (1992)] that what we call amplifier and oscillator are the two extreme of possible operation regimes and any practical device operates somewhere in between, according to the degree of control we have on the reflection process. It was suggested to eliminate the problem using the transit-time isolation method. This method was successfully demonstrated [Kuang (1993)] experimentally and power levels of 200 MW were achieved at 9GHz. The spectrum of the output signal was less than 50 MHz wide and the pass-band of the periodic structure was less than 200 MHz.

4.1 Semi-Infinite Structure: Pierce-Like Theory

In the previous chapter it was justified to decouple the two groups of solutions described by the dispersion relation in (3.5.8) because the propagating electromagnetic modes have a phase velocity larger than c, whereas the space-charge waves have a phase velocity which is of the order of v_0. In principle, it is possible to slow down the phase velocity of the electromagnetic wave, in the absence of the beam, below c and then, the waves may become coupled – it is there where resonance occurs. One possibility to slow down the phase velocity, which will be considered throughout this chapter, is to load the waveguide with a dielectric material. In Fig. 4.1 we illustrate schematically the dispersion curve of an electromagnetic wave which propagates in a dielectric loaded waveguide. The space-charge wave intersects the former's curve at resonance.

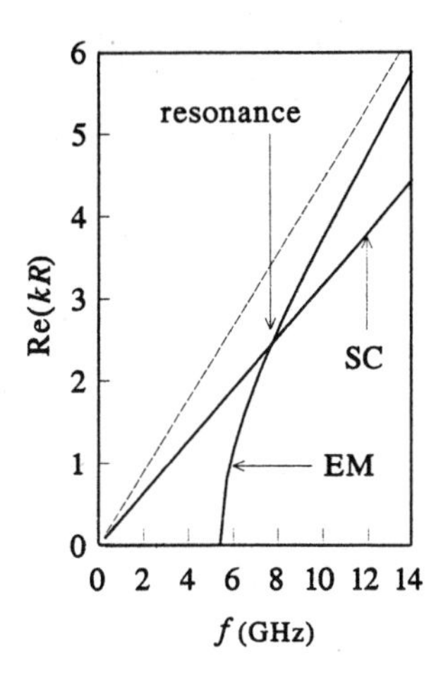

Fig. 4.1. Solution of the dispersion relation of the electromagnetic and space charge waves. The resonance is at the intersection of the two curves. The upper curve is the asymptote of the electromagnetic mode: $k = \omega\sqrt{\varepsilon_r}/c$

4.1.1 Dielectric Filled Waveguide

As a first step we shall assume full overlap between the beam and the dielectric. Although, in general no such overlap is permissible, this model will be used to explain in a simple way the quantities which describe the beam-wave interaction in a slow-wave structure. A realistic but somewhat more complex picture will be presented in the next sub-section. The dispersion relation of the TM_{0s} modes in the presence of a dielectric material (ε_r) is given by

$$\left(\varepsilon_r\frac{\omega^2}{c^2} - k_s^2\right)\left[1 - \frac{\omega_p^2}{\varepsilon_r\gamma^3(\omega - v_0k_s)^2}\right] = \frac{p_s^2}{R^2}, \qquad (4.1.1)$$

k_s represents the wavenumber of the sth mode and p_s is the zero of the zero order Bessel function of the first kind i.e., $\mathrm{J}_0(p_s) \equiv 0$. For simplicity it is assumed that only the first mode, $s = 1$, is excited and in the absence of the beam the solution of the electromagnetic wave is given by

$$k_1^{(0)} = \sqrt{\varepsilon_r\frac{\omega^2}{c^2} - \frac{p_1^2}{R^2}}, \qquad (4.1.2)$$

where R is the radius of the waveguide. The solution of the dispersion relation with the beam present is further assumed to have the form

$$k_1 = k_1^{(0)} + \delta k. \qquad (4.1.3)$$

Substituting this relation in (4.1.1), assuming that the beam effect on the distribution of the electromagnetic field is small on the scale of one wavelength i.e.,

$$|\delta k| \ll k_1^{(0)}, \tag{4.1.4}$$

neglecting the beam effect on the wave which propagates anti-parallel to the beam and finally assuming that

$$\frac{p_1^2}{R^2} \gg 2k_1^{(0)}|\delta k|, \tag{4.1.5}$$

we obtain the following simplified version of the dispersion relation

$$\delta k(\delta k - \Delta k)^2 = -K_0^3 \equiv -\frac{1}{2}\frac{p_1^2}{R^2}\frac{\omega_{\rm p}^2}{\varepsilon_{\rm r} k_1 v_0^2 \gamma^3}. \tag{4.1.6}$$

In this expression K_0 is the coupling wavenumber and $\Delta k \equiv \omega/v_0 - k_1^{(0)}$ represents the slip between the beam and the electromagnetic wave; this is more easily observed when dividing Δk by ω since

$$\frac{\Delta k}{\omega} = \frac{1}{v_0} - \frac{1}{v_{\rm ph}}. \tag{4.1.7}$$

The dispersion relation in (4.1.1) is a fourth order polynomial whereas (4.1.6) is a third order polynomial since the effect of the beam on the reflected wave was neglected. This dispersion relation was initially presented by Pierce (1947) in the context of beam-wave interaction in a helix – and it will be referred to hereafter as the Pierce's approach. Soon afterwards Chu and Jackson (1948) presented the formulation based on full electromagnetic field analysis. In both cases the dynamics of the beam was non-relativistic since the regime of operation at the time did not require relativistic analysis.

A third order polynomial has explicit analytic solution [see Abramowitz and Stegun (1968) p. 17]. Two of its roots are complex provided that

$$q \equiv \Delta k + \frac{3}{4^{1/3}} K_0 > 0, \tag{4.1.8}$$

and then the imaginary part of δk is equal to

$$\begin{aligned} \mathrm{Im}\,(\delta k) &= \frac{\sqrt{3}}{2}\left[-q^3 - \frac{1}{4}K_0^3 + (K_0 q)^{3/2}\right]^{1/3} \\ &- \frac{\sqrt{3}}{2}\left[-q^3 - \frac{1}{4}K_0^3 - (K_0 q)^{3/2}\right]^{1/3}, \end{aligned} \tag{4.1.9}$$

which can be readily shown, by assuming that Δk and K_0 are *independent*, to have its maximum at

$$\Delta k \equiv \omega\left(\frac{1}{v_0} - \frac{1}{v_{\rm ph}}\right) = 0. \tag{4.1.10}$$

This is also the resonance condition (see Fig. 4.1) and it can be formulated as

$$v_{\rm ph} = v_0 \,, \tag{4.1.11}$$

which indicates that maximum growth rate occurs when the electron beam is synchronous with the wave. At resonance (4.1.6) has three solutions

$$\delta k_1 = -K_0 \,, \quad \delta k_2 = K_0 \left(\frac{1}{2} - j\frac{\sqrt{3}}{2} \right) , \quad \delta k_3 = K_0 \left(\frac{1}{2} + j\frac{\sqrt{3}}{2} \right) , \tag{4.1.12}$$

corresponding to the three waves which propagate in the forward direction. The first wave has a constant amplitude and its phase velocity is larger than v_0; the other two waves have a slower phase velocity and their amplitude vary in space. The third solution corresponds to a wave whose amplitude grows exponentially in space. The maximum spatial growth rate is therefore

$$\mathrm{Im}\,(\delta k)_{\max} = \frac{\sqrt{3}}{2} K_0 \,. \tag{4.1.13}$$

Further discussion of the interaction between electromagnetic waves and space-charge waves in slow-wave structures can be found in the early literature e.g., Pierce (1950), Slater (1950), Hutter (1960), Chodorow and Susskind (1964), or more recently in Gilmour (1986). At this point we wish to emphasize the difference between the operation of a traveling-wave tube in the relativistic and non-relativistic regime [Naqvi (1996)]. We have emphasized, in the context of (4.1.9–10), that maximum gain occurs at resonance if we assume that K_0 and Δk are independent. But clearly this is not generally the case since both quantities are velocity dependent as revealed by their definition in (4.1.6–7). Furthermore, the validity of (4.1.8–9) is limited to the close vicinity of the expansion point, therefore we should solve (4.1.1) with no additional approximations. The result is illustrated in Fig. 4.2 for two cases. In one case the phase velocity of the wave is $\beta_{\rm ph} = 0.3$ corresponding to a non-relativistic regime and in the other, $\beta_{\rm ph} = 0.9$. In both cases at "resonance" ($\beta = \beta_{\rm ph}$) the spatial growth rate was designed to be the same. We observe that at low velocities the predictions of the Pierce approach behave as expected but at higher energies the peak gain occurs at much lower velocity than anticipated by the model. This can be attributed to the $\gamma\beta$ dependence of the coupling coefficient (K_0). At low energies the coupling occurs in a relatively narrow range of velocities and as a result, the change in K_0 is small and the peak gain occurs at resonance. At higher energies, although the gain might have decreased if K_0 were constant, it actually increases because of the increase in K_0 due to its $1/\gamma\beta$ dependence. For a large deviation from resonance the slip will ultimately take over and the gain drops. Now to some further comments regarding the *kinematics* of the interaction:

Remark 1. From the resonance condition (4.1.11) we conclude that maximum gain is achieved at a frequency which is related to the geometric and mechanical parameters by

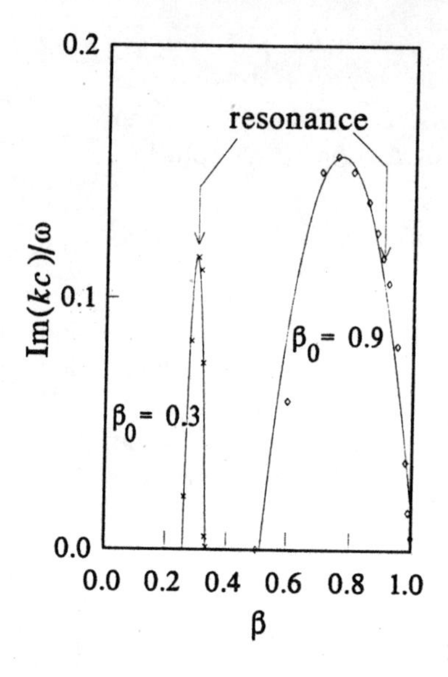

Fig. 4.2. Imaginary part of the wavenumber as a function of the beam velocity in two different cases. In one the wave is designed to propagate at low phase velocity ($\beta_{\rm ph} = 0.3$) and the second at high phase velocity ($\beta_{\rm ph} = 0.9$). In both cases the gain at resonance, $\beta = \beta_{\rm ph}$, was designed to be the same

$$\omega_{\rm r} = \frac{p_1 v_0}{R} \frac{1}{\sqrt{\varepsilon_{\rm r}(v_0/c)^2 - 1}} . \tag{4.1.14}$$

If we assume that at a frequency, ω_0, there is no longer growth, i.e., $q(\omega = \omega_0) = 0$, and ω_0 is only slightly apart from the resonance frequency $\omega_{\rm r}$ ($\omega_0 = \omega_{\rm r} + \delta\omega$ and $|\delta\omega| \ll \omega_{\rm r}$) then we can make a crude estimate the frequency range in which the ideal system under consideration will amplify. For this purpose we use the condition for an imaginary solution in (4.1.8) and the definition of Δk in (4.1.10): we substitute $\omega = \omega_{\rm r} + \delta\omega$ and $k_1 = k_1^{(0)} + v_{\rm gr}(\omega = \omega_{\rm r})\,\delta\omega$ thus

$$\frac{\omega_{\rm r} + \delta\omega}{v_0} - k_1^{(0)} - \delta\omega v_{\rm gr} + \frac{3}{4^{1/3}} K_0 = 0 . \tag{4.1.15}$$

We shall define the *interaction bandwidth*, $\delta\omega_i$, as equal $\delta\omega$:

$$\delta\omega_i \equiv \delta\omega = \frac{3}{4^{1/3}} K_0 v_{\rm gr} \frac{v_0}{v_0 - v_{\rm gr}} . \tag{4.1.16}$$

This result indicates that the interaction extends beyond the resonance frequency ($\omega_{\rm r}$) in an interval which is *linearly* proportional to the maximum spatial growth rate ($\mathrm{Im}\,(\delta k)_{\rm max}$):

$$\delta\omega_i = 3^{1/2} 2^{1/3} \frac{v_0}{v_0 - v_{\rm gr}} v_{\rm gr} \mathrm{Im}\,(\delta k)_{\rm max} . \tag{4.1.17}$$

Furthermore, the closer the group velocity is to the beam velocity, the broader the interaction and finally, note that this quantity is not dependent on the total interaction length.

Remark 2. The coupling coefficient K_0^3 can now be represented in terms of more familiar quantities. First we define the average current which flows along the waveguide as $I = en_0 v_0 \pi R^2$. Next we define the cross-section through which the wave propagates as $S_w = \pi R^2$. The interaction impedance defined in (2.3.29) was calculated for the present configuration in (2.3.30), reads $Z_{\rm int} = \eta_0 (p_1 c/\varepsilon_{\rm r} \omega R)^2/\beta_{\rm en}$. With these quantities we can express K_0^3 as

$$K_0^3 = \frac{1}{2}\frac{1}{S_w}\frac{\omega}{c}\frac{eIZ_{\rm int}}{mc^2}\frac{1}{(\gamma\beta)^3}. \quad (4.1.18)$$

In the linear regime, the linearity of K_0^3 in the interaction impedance is a general feature whenever the electrons interact with a TM mode and their oscillation is longitudinal. And so is its dependence on the normalized momentum of the particle, $\gamma\beta$. The last expression can also be formulated in terms of the energy velocity using (2.3.33) as

$$K_0^3 = \frac{1}{2}\frac{1}{S_w}\frac{\omega}{c}\frac{eI\eta_0}{mc^2}\frac{1}{(\gamma\beta)^3}\frac{1}{\varepsilon_{\rm int}\beta_{\rm en}}, \quad (4.1.19)$$

which indicates that the growth rate is inversely proportional to the energy velocity. Since in most cases of (our) interest this is equal to the goup velocity, we observe that if we substitute in the expression for the interaction bandwidth, the latter still decreases with the group velocity as $\delta\omega_i \propto {v_{\rm gr}}^{2/3}$.

Remark 3. The entire approach relies on a *linearized* hydrodynamic approximation. Which implies that the deviation from the initial *average* energy is small i.e.,

$$\gamma \gg |\delta\gamma|. \quad (4.1.20)$$

Later we shall adopt the single particle equation of motion for description of the electron dynamics and it will be shown that an *individual* particle can have energy which is more than twice the *average* initial energy. Nevertheless, the average energy modulation of all particles can still be relatively small – namely the relation in (4.1.20) still holds.

Now to the *dynamics* of the interaction. Let us consider a system of length d and assume that we know the value of the field at the input i.e., $E_z(r, z = 0) = E_0 {\rm J}_0(p_1 r/R)$. Furthermore, at this location the beam is not modulated yet thus, $\delta v(z = 0) = 0$ and $\delta n(z = 0) = 0$. According to the three modes we found previously [(4.1.12)] we can write the solution for E_z as:

$$E_z(r, z, \omega) = {\rm J}_0\left(p_1 \frac{r}{R}\right) {\rm e}^{-jk_1^{(0)} z}\left[E_1 {\rm e}^{-j\delta k_1 z} + E_2 {\rm e}^{-j\delta k_2 z} + E_3 {\rm e}^{-j\delta k_3 z}\right]. \quad (4.1.21)$$

The three conditions above determine three sets of algebraic equations:

$$E_1 + E_2 + E_3 = E_0\,,$$
$$\frac{E_1}{\omega/v_0 - k_1^{(0)} - \delta k_1} + \frac{E_2}{\omega/v_0 - k_1^{(0)} - \delta k_2} + \frac{E_3}{\omega/v_0 - k_1^{(0)} - \delta k_3} = 0\,,$$
$$\frac{E_1(k_1^{(0)} + \delta k_1)}{(\omega/v_0 - k_1^{(0)} - \delta k_1)^2} + \frac{E_2(k_1^{(0)} + \delta k_2)}{(\omega/v_0 - k_1^{(0)} - \delta k_2)^2} + \frac{E_3(k_1^{(0)} + \delta k_3)}{(\omega/v_0 - k_1^{(0)} - \delta k_3)^2} = 0\,, \quad (4.1.22)$$

and in principle, we can now solve for E_1, E_2 and E_3 such that we can determine the total electromagnetic field at the output ($z = d$). For the sake of simplicity we shall limit our discussion to the solution near resonance (where according to the Pierce approach the gain reaches its maximum). The three δk's are of the same order of magnitude so we can estimate that $|E_1| \simeq |E_2| \simeq |E_3| \simeq E_0/3$. Therefore, the z component of the electric field is

$$E_z(r,z,\omega) \simeq \frac{E_0}{3} \mathrm{J}_0\left(p_1 \frac{r}{R}\right) \mathrm{e}^{-jk_1^{(0)} z} \times \left[\mathrm{e}^{-j\delta k_1 z} + \mathrm{e}^{-j\delta k_2 z} + \mathrm{e}^{-j\delta k_3 z}\right]. \quad (4.1.23)$$

According to the three solutions in (4.1.12) the first δk is always real therefore its amplitude is constant, the amplitude of the second decays exponentially, and the third grows in space. The local gain is the ratio between the local amplitude and the amplitude at the input therefore is given by

$$\begin{aligned} G(z) &\equiv \frac{|E(z)|}{|E(0)|} \\ &= \frac{1}{3}\left|\mathrm{e}^{-j\delta k_1 z} + \mathrm{e}^{-j\delta k_2 z} + \mathrm{e}^{-j\delta k_3 z}\right| \\ &= \frac{1}{3}\left|\mathrm{e}^{3jK_0 z/2} + \mathrm{e}^{-\sqrt{3}K_0 z/2} + \mathrm{e}^{\sqrt{3}K_0 z/2}\right|; \end{aligned} \quad (4.1.24)$$

note that the whole expression was devided by $\mathrm{e}^{-jK_0 z/2}$ which corresponds to $\mathrm{Re}(\delta k_2)$ and $\mathrm{Re}(\delta k_3)$. The expression in (4.1.24) is illustrated in Fig. 4.3 where we present $G(z)$ in dB for several values of $K_0 d$. Although one of the solutions grows exponentially in the first part of the interaction region, the local gain is zero – this effect is referred to in literature as *spatial lethargy* since it takes some space for the exponentially growing wave to become dominant. We can estimate this lethargy length d_l by determining the location where the amplitude of the first and third waves combined, reach the value at the input i.e., $|\exp(3jK_0 d_\mathrm{l}/2) + \exp(\sqrt{3}K_0 d_\mathrm{l}/2)|/3 = 1$. This equation can be solved numerically and the result is

$$d_\mathrm{l} = \frac{1.412}{K_0}\,; \quad (4.1.25)$$

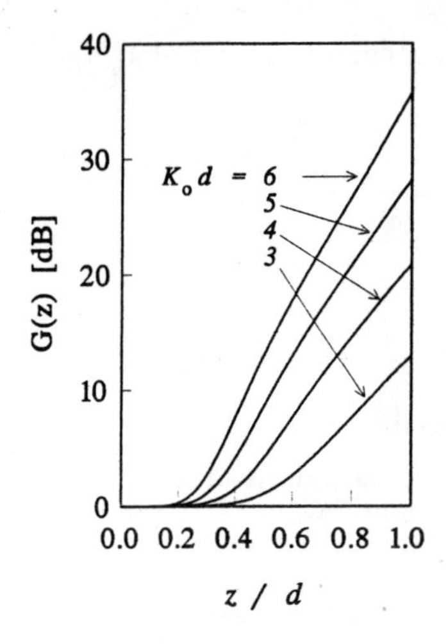

Fig. 4.3. The way the gain develops in space for different values of the coupling coefficient K_0

the second wave decays exponentially therefore it is neglected in this calculation. We shall further discuss this effect in the context of the macro-particle approach in Sect. 4.3. At the end of the interaction region we can also neglect the mode which has a constant amplitude (assuming that the gain is large enough) and the total gain (at $z = d$) is defined as

$$\mathrm{gain}_{\mathrm{dB}} = 20 \log_{10} \left[\frac{1}{3} \mathrm{e}^{\sqrt{3} K_0 d/2} \right] . \tag{4.1.26}$$

The total gain and the lethargy length are related by

$$\frac{d_{\mathrm{l}}}{d} = \frac{1.412\sqrt{3}}{2} \ln^{-1} \left[3 \times 10^{\mathrm{gain(dB)}/20} \right] . \tag{4.1.27}$$

The effect is evident in Fig. 4.3 and using the relations above we found for $K_0 d = 3$ the lethargy length is 0.47 d while the gain is 13 dB. For twice this growth ($K_0 d = 6$ which can be achieved by increasing the current) the lethargy length is 0.23 d and the gain is 35 dB.

4.1.2 Partially Filled Waveguide

Although in the previous model, the beam and the dielectric were occupying the entire space of the waveguide, has its tutorial merit, it is impractical with regards to the generation of radiation. In general no beam-dielectric overlap is permitted namely, there has to be a significant distance between the electron beam and the structure which slows down the wave. In this sub-section we shall consider the interaction between a beam of radius R_{b} and a wave which propagates in a waveguide partially filled with dielectric material (ε_{r}). The dielectric occupies the region between $R_{\mathrm{d}} < r < R$ where R is the radius of the waveguide and $R_{\rm d} > R_{\rm b}$ – see Fig. 4.4. The system is semi-infinitely long, thus no reflections occur.

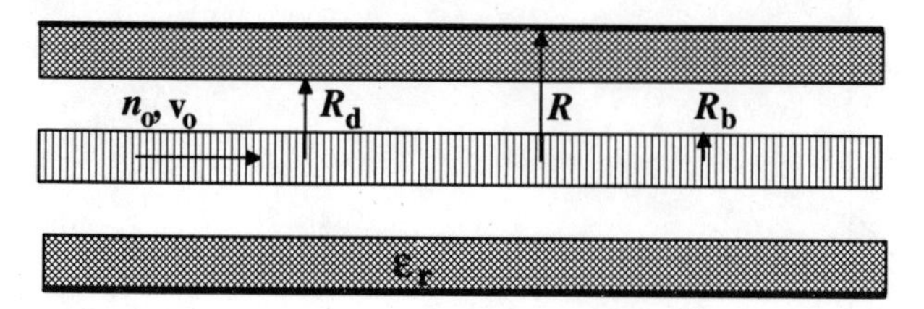

Fig. 4.4. A dielectric loaded waveguide. The radius of the guide is denoted by R. $R_{\rm d}$ and $R_{\rm b}$ stand for the the dielectric inner radius and beam's radius, respectively

As a first stage we shall consider the electromagnetic problem in the absence of the electrons. A $\rm TM_{01}$ mode is assumed to propagate along the waveguide and it is described by the z component of the magnetic vector potential, which in the vacuum gap $(0 < r < R_{\rm d})$ reads:

$$A_z(r,z,\omega) = A_0 {\rm I}_0(\Gamma r){\rm e}^{-jkz}\,, \tag{4.1.28}$$

where $\Gamma^2 = k^2 - (\omega/c)^2$. In the dielectric material $(R_{\rm d} < r < R)$ the magnetic vector potential is given by:

$$A_z(r,z,\omega) = B_0 T_0(\kappa r){\rm e}^{-jkz}\,, \tag{4.1.29}$$

where $\kappa^2 = \varepsilon_{\rm r}(\omega/c)^2 - k^2$, ${\rm I}_0(\xi)$ is the zero order modified Bessel function of the first kind and

$$T_0(\kappa r) \equiv {\rm J}_0(\kappa r)Y_0(\kappa R) - Y_0(\kappa r){\rm J}_0(\kappa R)\,. \tag{4.1.30}$$

The electromagnetic field in the vacuum gap is

$$\begin{aligned} H_\phi(r,z,\omega) &= -A_0\frac{1}{\mu_0}\Gamma {\rm I}_1(\Gamma r){\rm e}^{-jkz}\,, \\ E_r(r,z,\omega) &= -A_0\frac{c^2k}{\omega}\Gamma {\rm I}_1(\Gamma r){\rm e}^{-jkz}\,, \\ E_z(r,z,\omega) &= -A_0\frac{c^2}{j\omega}\Gamma^2 {\rm I}_0(\Gamma r){\rm e}^{-jkz}\,. \end{aligned} \tag{4.1.31}$$

In a similar way in the dielectric material,

$$\begin{aligned} H_\phi(r,z,\omega) &= B_0\frac{1}{\mu_0}\kappa T_1(\kappa r){\rm e}^{-jkz}\,, \\ E_r(r,z,\omega) &= B_0\frac{c^2k}{\omega\varepsilon_{\rm r}}\kappa T_1(\kappa r){\rm e}^{-jkz}\,, \\ E_z(r,z,\omega) &= B_0\frac{c^2}{j\omega\varepsilon_{\rm r}}\kappa^2 T_0(\kappa r){\rm e}^{-jkz}\,, \end{aligned} \tag{4.1.32}$$

where

$$T_1(\kappa r) \equiv \mathrm{J}_1(\kappa r)Y_0(\kappa R) - Y_1(\kappa r)\mathrm{J}_0(\kappa R)\,. \tag{4.1.33}$$

In order to determine the wavenumber k we now impose the boundary conditions at $r = R_\mathrm{d}$: the continuity of H_ϕ implies

$$-A_0\Gamma \mathrm{I}_1(\Gamma R_\mathrm{d}) = B_0\kappa T_1(\kappa R_\mathrm{d})\,, \tag{4.1.34}$$

whereas the continuity of E_z

$$-A_0\Gamma^2 \mathrm{I}_0(\Gamma R_\mathrm{d}) = B_0\frac{1}{\varepsilon_\mathrm{r}}\kappa^2 T_0(\kappa R_\mathrm{d})\,. \tag{4.1.35}$$

From these two equations the dispersion relation of the passive (subscript $_\mathrm{pa}$) device reads

$$D_\mathrm{pa}(\omega,k) \equiv \varepsilon_\mathrm{r}\mathrm{I}_0(\theta_d)T_1(\chi_d) - \frac{\chi_d}{\theta_d}\mathrm{I}_1(\theta_d)T_0(\chi_d) = 0\,, \tag{4.1.36}$$

where $\theta_d = \Gamma R_\mathrm{d}$ and $\chi_d = \kappa R_\mathrm{d}$. Figure 4.5 illustrates a solution of this dispersion relation [line *(a)*]. For comparison, two other dispersion relations are plotted: line *(b)* represents the empty waveguide, whereas line *(c)* corresponds to a waveguide filled with the same dielectric. From the dispersion relation we observe that at low frequencies (long wavelength) the mode behaves as if no dielectric exists. At high frequencies the mode is primarily confined by the dielectric slab.

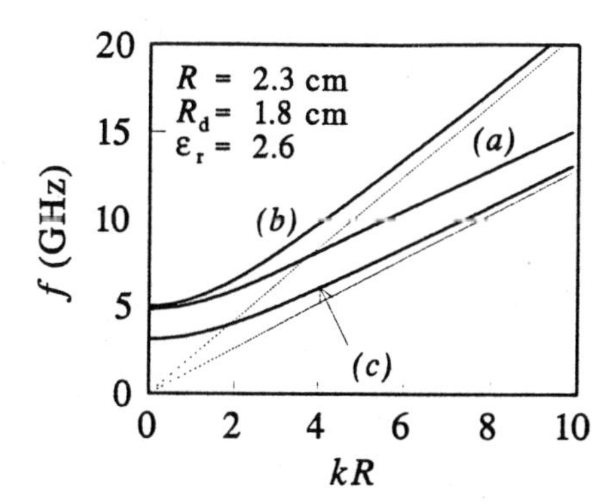

Fig. 4.5. (**a**) Solution of the dispersion relation in (4.1.36).(**b**) represents the empty waveguide, whereas (**c**) corresponds to a waveguide filled with dielectric

Next we shall consider the effect of the electron beam. In this case the magnetic vector potential in the beam region $0 < r < R_\mathrm{b}$ is given by

$$A_z(r,z,\omega) = A_0\mathrm{I}_0(\Lambda r)\mathrm{e}^{-jkz}\,,$$

$$\Lambda^2 = \Gamma^2\left[1 - \frac{\omega_\mathrm{p}^2}{\gamma^3(\omega - v_0 k)^2}\right]\,, \tag{4.1.37}$$

with the plasma frequency, $\omega_{\rm p}$, defined in (3.4.18). Accordingly, the electromagnetic field reads

$$H_\phi(r,z,\omega) = -A_0 \frac{1}{\mu_0} \Lambda \mathrm{I}_1(\Lambda r) \mathrm{e}^{-jkz} \,,$$

$$E_r(r,z,\omega) = -A_0 \frac{c^2 k}{\omega} \Lambda \mathrm{I}_1(\Lambda r) \mathrm{e}^{-jkz} \,,$$

$$E_z(r,z,\omega) = -A_0 \frac{c^2}{j\omega} \Gamma^2 \mathrm{I}_0(\Lambda r) \mathrm{e}^{-jkz} \,. \tag{4.1.38}$$

In the vacuum gap between the beam and the dielectric ($R_{\rm b} < r < R_{\rm d}$) the potential is

$$A_z(r,z,\omega) = \Big[B_0 \mathrm{I}_0(\Gamma r) + C_0 \mathrm{K}_0(\Gamma r)\Big] \mathrm{e}^{-jkz} \,, \tag{4.1.39}$$

while the electromagnetic field is

$$H_\phi(r,z,\omega) = -\frac{1}{\mu_0} \Gamma \Big[B_0 \mathrm{I}_1(\Gamma r) - C_0 \mathrm{K}_1(\Gamma r)\Big] \mathrm{e}^{-jkz} \,,$$

$$E_r(r,z,\omega) = -\frac{c^2 k}{\omega} \Gamma \Big[B_0 \mathrm{I}_1(\Gamma r) - C_0 \mathrm{K}_1(\Gamma r)\Big] \mathrm{e}^{-jkz} \,,$$

$$E_z(r,z,\omega) = -\frac{c^2}{j\omega} \Gamma^2 \Big[B_0 \mathrm{I}_0(\Gamma r) + C_0 \mathrm{K}_0(\Gamma r)\Big] \mathrm{e}^{-jkz} \,. \tag{4.1.40}$$

In the dielectric the expression is similar to (4.1.29):

$$A_z(r,z,\omega) = D_0 T_0(\kappa r) \mathrm{e}^{-jkz} \,, \tag{4.1.41}$$

and the electromagnetic field

$$H_\phi(r,z,\omega) = D_0 \frac{1}{\mu_0} \kappa T_1(\kappa r) \mathrm{e}^{-jkz} \,,$$

$$E_r(r,z,\omega) = D_0 \frac{c^2 k}{\omega \varepsilon_{\rm r}} \kappa T_1(\kappa r) \mathrm{e}^{-jkz} \,,$$

$$E_z(r,z,\omega) = D_0 \frac{c^2}{j\omega \varepsilon_{\rm r}} \kappa^2 T_0(\kappa r) \mathrm{e}^{-jkz} \,. \tag{4.1.42}$$

In order to determine the dispersion relation we now impose the boundary conditions at $r = R_{\rm b}$ and $r = R_{\rm d}$. Continuity of H_ϕ implies at $r = R_{\rm b}$:

$$A_0 \Lambda \mathrm{I}_1(\Lambda R_{\rm b}) = \Gamma \Big[B_0 \mathrm{I}_1(\Gamma R_{\rm b}) - C_0 \mathrm{K}_1(\Gamma R_{\rm b})\Big] \,, \tag{4.1.43}$$

while at $r = R_{\rm d}$:

$$-\Gamma \Big[B_0 \mathrm{I}_1(\Gamma R_{\rm d}) - C_0 \mathrm{K}_1(\Gamma R_{\rm d})\Big] = D_0 \kappa T_1(\kappa R_{\rm d}) \,. \tag{4.1.44}$$

Similarly the continuity of the longitudinal component of the electric field (E_z) implies at $r = R_b$:

$$A_0(-\Gamma^2)\mathrm{I}_0(\Lambda R_\mathrm{b}) = (-\Gamma^2)\left[B_0\mathrm{I}_0(\Gamma R_\mathrm{b}) + C_0\mathrm{K}_0(\Gamma R_\mathrm{b})\right], \tag{4.1.45}$$

and at $r = R_\mathrm{d}$:

$$(-\Gamma^2)\left[B_0\mathrm{I}_0(\Gamma R_\mathrm{d}) + C_0\mathrm{K}_0(\Gamma R_\mathrm{d})\right] = D_0\frac{1}{\varepsilon_\mathrm{r}}\kappa^2 T_0(\kappa R_\mathrm{d}). \tag{4.1.46}$$

These are four homogeneous equations and the non-trivial solution is determined from the condition that the determinant of the corresponding matrix is zero. Thus the dispersion equation of the active (subscript $_\mathrm{act}$) system is

$$D_\mathrm{act}(\omega, k) \equiv D_\mathrm{pa}(\omega, k) + D_\mathrm{beam}(\omega, k) = 0. \tag{4.1.47}$$

The first term is the dispersion relation of the passive system and the second (D_beam) represents the beam effect:

$$D_\mathrm{beam}(\omega, k) = \frac{\Gamma\mathrm{I}_1(\Gamma R_\mathrm{b})\mathrm{I}_0(\Lambda R_\mathrm{b}) - \Lambda\mathrm{I}_0(\Gamma R_\mathrm{b})\mathrm{I}_1(\Lambda R_\mathrm{b})}{\Gamma\mathrm{K}_1(\Gamma R_\mathrm{b})\mathrm{I}_0(\Lambda R_\mathrm{b}) + \Lambda\mathrm{K}_0(\Gamma R_\mathrm{b})\mathrm{I}_1(\Lambda R_\mathrm{b})} \times \left[\varepsilon_\mathrm{r}\mathrm{K}_0(\theta_d)T_1(\chi_d) + \frac{\chi_d}{\theta_d}\mathrm{K}_1(\theta_d)T_0(\chi_d)\right]. \tag{4.1.48}$$

The solution of the dispersion relation in the beam absence is denoted by $k^{(0)}(\omega)$ and the effect of the beam will be represented by a deviation, δk, from this value namely, $k = k^{(0)} + \delta k$. If we now expand the first term in (4.1.47) around $k^{(0)}$, we find

$$D_\mathrm{pa}(\omega, k^{(0)}) + \delta k\left[\frac{\partial D_\mathrm{pa}(\omega, k)}{\partial k}\right]_{k=k^{(0)}} = \delta k\left[\frac{\partial D_\mathrm{pa}(\omega, k)}{\partial k}\right]_{k=k^{(0)}}, \tag{4.1.49}$$

since by the definition of $k^{(0)}$, the dispersion function $D_\mathrm{pa}(\omega, k^{(0)})$ is identically zero. Regarding the second term in the dispersion relation of the active system (4.1.47), we can readily check that for $\omega_\mathrm{p} = 0$ the beam term D_beam is zero. We shall consider only the correction of the first order in the term $\omega_\mathrm{p}^2/(\omega - v_0 k)^2\gamma^3$, therefore we can neglect the beam effect in the denominator of D_beam which after a first order Taylor expansion implies

$$D_\mathrm{beam}(\omega, k^{(0)}) \simeq \frac{\omega_\mathrm{p}^2}{2\gamma^3(\omega - v_0 k)^2}\left[\theta_b^2\left(\mathrm{I}_0^2(\theta_b) - \mathrm{I}_1^2(\theta_b)\right)\frac{\chi_d}{\theta_d^2}\frac{T_0(\chi_d)}{\mathrm{I}_0(\theta_d)}\right]_{k=k^{(0)}}. \tag{4.1.50}$$

In the process of evaluating this expression the following relation was used $\mathrm{I}_0(x)\mathrm{K}_1(x) + \mathrm{I}_1(x)\mathrm{K}_0(x) = 1/x$ and $\theta_b = \Gamma R_\mathrm{b}$. Equations (4.1.49–50) can be substituted in the active dispersion relation (4.1.47) and the result written in an identical form as (4.1.6). The beam-wave coupling is determined by

$$K_0^3 = \frac{1}{2}\frac{eI\eta_0}{mc^2}\frac{1}{(\gamma\beta)^3}\frac{1}{\pi R^2}\left[\mathrm{I}_0^2(\theta_b) - \mathrm{I}_1^2(\theta_b)\right]_{k=k^{(0)}}$$
$$\times \left[\left(\frac{R}{R_\mathrm{d}}\right)^2 \frac{\chi_d T_0(\chi_d)}{\mathrm{I}_0(\theta_d)} \left(\frac{\partial D_\mathrm{pa}}{\partial k}\right)^{-1}\right]_{k=k^{(0)}} . \qquad (4.1.51)$$

The effect of the radius in the case of a pencil beam is now clearly revealed since the coupling coefficient K_0^3 is

$$K_0^3(R_\mathrm{b}) = K_0^3(0)\left[\mathrm{I}_0^2(\theta_b) - \mathrm{I}_1^2(\theta_b)\right]_{k=k^{(0)}} . \qquad (4.1.52)$$

The modified Bessel function of the first kind is a monotonic function thus for $R_\mathrm{b} \to 0$ the coupling coefficient K_0^3 has its minimum and grows with increasing beam radius. This fact can be readily understood bearing in mind that the slow wave which interacts with the electrons decays exponentially from the dielectric surface inward. The larger the radius of the beam, the stronger the electric field it encounters, therefore the coupling is stronger. Further discussion on the interaction in a dielectric loaded waveguide was presented by Walsh (1987).

4.2 Finite Length Effects

All the examples of beam-wave interaction presented so far, such as resistive wall instability – Sect. 3.5.3, two-beam instability – Sect. 3.5.4 and traveling-wave interaction in the last two sub-sections, disregard the possibility of reflections from the output end. In practice, there are several causes for reflections to occur: *(i)* the characteristic impedance of an electromagnetic wave in the interaction region differs from the impedance of the input and output waveguides. *(ii)* Even if at a given frequency this impedance mismatch can be tuned, at others, reflections may dominate and control the interaction process. *(iii)* In any interaction scheme the electromagnetic energy has to be decoupled from the beam. This decoupling process is always associated with some kind of discontinuity, therefore in this regard, reflections are an inherent part of the interaction.

In order to illustrate the effect of reflections, we first consider the extreme where the interaction is dominated by reflections - which is the case in an oscillator. It will be shown that in zero order the amplitude of the electromagnetic field in an oscillator is constant in space but varies in time whereas in an amplifier, it is constant in time but it varies in space. The oscillator analysis is followed by an investigation of the effect of reflections on an amplifier and the section concludes with some remarks on the interaction in an extended slow-wave cavity.

4.2.1 Oscillator

The simplest configuration we can conceive for an oscillator is a section of an amplifier with two reflecting mirrors at the two ends. There are two major differences between an amplifier and an oscillator: *(i)* in an amplifier the frequency is set externally whereas the wavenumber is determined by the waveguide and the interaction. Consequently, *(ii)* the amplitude of the wave varies in space but at a given location it is constant in time. The opposite holds for an oscillator; the wavenumber is set by the cavity (mirrors) and the frequency is determined internally by both cavity and interaction. The amplitude is constant in space (at a given moment) and it varies in time. Therefore, if we assume that the distance between the two mirrors is d, then from the condition that $E_r(r, z = 0, d) = 0$ we conclude that

$$k = \nu \frac{\pi}{d}, \tag{4.2.1}$$

where ν is an integer which labels each longitudinal mode. In the remainder we shall assume that the interaction is only with one of these modes ($\nu = 1$). In the beam absence the resonant frequency is

$$\omega_{\mathrm{r}} = \frac{c}{\sqrt{\varepsilon_{\mathrm{r}}}} \sqrt{\frac{p_1^2}{R^2} + \frac{\pi^2}{d^2}}, \tag{4.2.2}$$

and the beam introduces a small deviation $\delta\omega = \omega - \omega_{\mathrm{r}}$ which is a solution of the following dispersion relation

$$\delta\omega(\delta\omega - \Delta\omega)^2 = \Omega_0^3 \equiv \frac{1}{2}\left(\frac{p_1 c}{R}\right)^2 \frac{\omega_{\mathrm{p}}^2}{\omega_{\mathrm{r}} \varepsilon_{\mathrm{r}}^2 \gamma^3}, \tag{4.2.3}$$

where similarly to the amplifier case Ω_0 is the coupling frequency and $\Delta\omega = v_0\pi/d - \omega_{\mathrm{r}}$ is the slip. The main difference here is that in contrast to (4.1.6) the right-hand side is *positive* thus the same analysis of the third order polynomial can be applied for (4.2.3). We shall not repeat it but rather present the important results. The maximum temporal growth rate occurs at resonance and it is given by

$$\delta\omega_1 = \Omega_0, \quad \delta\omega_2 = -\frac{1}{2}\Omega_0(1 - j\sqrt{3}), \quad \delta\omega_3 = -\frac{1}{2}\Omega_0(1 + j\sqrt{3}). \tag{4.2.4}$$

The first solution corresponds to a wave whose phase velocity is larger than the average velocity of the electrons (v_0). The other two have phase velocities smaller than v_0. The second solution represents a wave which is decaying in time whereas the third grows.

Comparing the dispersion relation for the amplifier (4.1.6) with the dispersion relation for the oscillator (4.2.3) we find that the relation between Ω_0 and K_0 is given by $\Omega_0^3 = K_0^3 v_{\mathrm{gr}} v_0^2$ which implies that the connection between the maximum *temporal growth rate* $[\mathrm{Im}\,(\delta\omega)_{\max}]$ in an oscillator and the maximum *spatial growth rate* in an amplifier $[\mathrm{Im}\,(\delta k)_{\max}]$ is

$$\frac{\mathrm{Im}\,(\delta\omega)_{\max}}{\mathrm{Im}\,(\delta k)_{\max}}\frac{1}{c} = \left(\beta_{\mathrm{gr}}\beta^2\right)^{1/3} . \tag{4.2.5}$$

As in the amplifier case, we can define the *temporal lethargy* as the period of time, τ_l, during which the electromagnetic energy in the oscillator starts to grow exponentially relative to the situation before the beam was launched (here we have assumed that initially the cavity was filled with electromagnetic energy):

$$\tau_l = \frac{1.412}{\Omega_0} . \tag{4.2.6}$$

If we consider a system which is characterized by $K_0 d = 4$, $d = 20\,\mathrm{cm}$, $\beta = 0.9$ and $\beta_{\mathrm{gr}} = 0.2$ then the temporal lethargy is 0.4 nsec, therefore on the scale of a pulse of hundreds of nanoseconds this is negligible. However if $K_0 d = 0.04$ (keeping all other parameters the same) the temporal lethargy is 40 nsec which becomes significant.

In most cases of interest the cavity is not initially filled with electromagnetic energy and the mirrors/walls are not ideal reflectors. As a result, the electromagnetic signal has to build up from noise and at the same time to overcome ohmic loss (to the walls) or radiation loss (through extraction ports/mirrors). All the combined loss mechanisms cause a decay in the electromagnetic field which, as a zero order, can be described by the variation in time of the energy which in turn, can be assumed to be proportional to the total amount of energy stored at a given time i.e.,

$$\frac{\mathrm{d}}{\mathrm{d}t}W + \frac{2}{\tau_{\mathrm{loss}}}W = 0 . \tag{4.2.7}$$

On the other hand, in Sect. 2.5.3 it was shown that a single particle excites a variety of electromagnetic waves in a cavity. When a uniform distribution of particles is injected into a cavity, waves of different frequencies and phases are generated and absorbed in the same time. This is *noise* induced by the beam in the cavity. For a *coherent* signal to develop from noise it is necessary that the beam-wave interaction will exceed some threshold which can be expressed in terms of the injected current. This amount is determined from the condition that the growth due to the interaction at least cancels the decay due to loss mechanisms i.e.,

$$\mathrm{e}^{\sqrt{3}\Omega_0 t/2}\mathrm{e}^{-t/\tau_{\mathrm{loss}}} \geq 1 . \tag{4.2.8}$$

Using this expression we can explicitly write for the *threshold current*

$$\begin{aligned} I_{\mathrm{th}} &= \frac{16}{3\sqrt{3}}\frac{mc^2}{e\eta_0}\frac{S_w}{(c\tau_{\mathrm{loss}})^2}\frac{\gamma^3\beta}{\omega_{\mathrm{r}}\tau_{\mathrm{loss}}}\varepsilon_{\mathrm{int}} \\ &= 3.08 I_e \gamma^3 \beta \frac{S_w}{(c\tau_{\mathrm{loss}})^2(\omega_{\mathrm{r}}\tau_{\mathrm{loss}})}\varepsilon_{\mathrm{int}} . \end{aligned} \tag{4.2.9}$$

Note that this current is proportional to the dielectric coefficient of the interaction and it scales as γ^3.

4.2.2 Gain and Bandwidth Considerations

One of the main assumptions in the analysis of the beam-wave interaction in an amplifier so far was that there are no reflections from the output end of the structure. In the previous sub-section we examined the dramatic change in the characteristics of the beam-wave interaction as reflections are deliberately introduced causing temporal rather than spatial growth. Based on the (pure) electromagnetic analysis presented in Chap. 2 the assumption of zero reflections in an amplifier is not justified in general since the wavelengths of the electromagnetic wave (and thus the characteristic impedance) in the interaction region and the extraction region are different. If discontinuities in the characteristic impedance occur they generate reflected waves. The reflected waves interfere (constructively or destructively) with the incoming waves to generate transmission patterns which were discussed in Sect. 2.5.1. According to this picture there are frequency ranges for which the transmission coefficient has a maximum or minimum. When the beam is present the situation is somewhat more complex since, in addition to the regular electromagnetic mode which can propagate, there are also space-charge waves which carry energy. However, in the case of a sufficiently long system such that at the output end the exponentially growing mode is dominant we can still assume only two waves bouncing between the input and output ends.

The starting point is similar to what was presented in Sect. 2.5.1: consider a waveguide of radius R is filled with a dielectric material according to

$$\varepsilon_{\rm r}(z) = \begin{cases} 1 & -\infty < z < 0\,, \\ \varepsilon_{\rm r} & 0 < z < d\,, \\ 1 & d < z < \infty\,. \end{cases} \tag{4.2.10}$$

A wave is launched from $z \to -\infty$ toward the discontinuity at $z = 0$ and for sake of simplicity we shall assume that this wave is composed of a single mode (TM_{01} i.e., $s = 1$). The z component of the magnetic vector potential in the first region ($-\infty < z < 0$) is given by

$$A_z(r, -\infty < z < 0, \omega) = \mathrm{J}_0\left(p_1 \frac{r}{R}\right)\left[A_{\rm in}\mathrm{e}^{-jk_1 z} + A_{\rm ref}\mathrm{e}^{jk_1 z}\right], \tag{4.2.11}$$

where $A_{\rm in}$ is the amplitude of the incoming wave and $A_{\rm ref}$ represents the amplitude of the reflected wave because of the discontinuity; $k_1 = \sqrt{(\omega/c)^2 - (p_1/R)^2}$. Between the two discontinuities at $z = 0$ and $z = d$ the solution has a similar form

$$A_z(r, 0 < z < d, \omega) = \mathrm{J}_0\left(p_1 \frac{r}{R}\right)\left[A\mathrm{e}^{-jKz} + B\mathrm{e}^{jk_2 z}\right], \tag{4.2.12}$$

where $k_2 = \sqrt{\varepsilon_r(\omega/c)^2 - (p_1/R)^2}$. The term $A e^{-jKz}$ is an "effective" wave which represents all three modes we discussed in Sect. 4.1. Finally in the third region there is no reflected wave thus

$$A_z(r, d < z < \infty, \omega) = \mathrm{J}_0\left(p_1 \frac{r}{R}\right) A_{\mathrm{tr}} \mathrm{e}^{-jk_1(z-d)} . \tag{4.2.13}$$

The four, as yet unknown, amplitudes A_{ref}, A_{tr}, A and B are determined by imposing the boundary conditions at $z = 0, d$:

$$Z_1 \left[A_{\mathrm{in}} - A_{\mathrm{ref}}\right] = Z_2 [A - B] , \tag{4.2.14}$$

$$A_{\mathrm{in}} + A_{\mathrm{ref}} = A + B , \tag{4.2.15}$$

$$Z_2 [A \,\mathrm{e}^{-j\bar{\psi}_+} - B \mathrm{e}^{j\bar{\psi}_-}] = Z_1 A_{\mathrm{tr}} \tag{4.2.16}$$

and

$$A \,\mathrm{e}^{-j\bar{\psi}_+} + B \mathrm{e}^{j\bar{\psi}_-} = A_{\mathrm{tr}} . \tag{4.2.17}$$

In these expressions the following definitions were used:

$$Z_1 \equiv \eta_0 \sqrt{1 - \left(\frac{p_1 c}{\omega R}\right)^2} , \quad Z_2 \equiv \eta_0 \frac{1}{\varepsilon_r} \sqrt{\varepsilon_r - \left(\frac{p_1 c}{\omega R}\right)^2} , \tag{4.2.18}$$

$\bar{\psi}_+ \equiv Kd$ represents the phase and amplitude variation of the effective wave as it propagates from $z = 0$ to d and $\bar{\psi}_- = k_2 d$ represents the phase shift of the backward wave. Note that the effect of the beam was neglected in the impedance terms.

The transmission coefficient is defined as $\tau = A_{\mathrm{tr}}/A_{\mathrm{in}}$ and is given by

$$\tau = \frac{4 Z_1 Z_2 \mathrm{e}^{-j(\bar{\psi}_+ - \bar{\psi}_-)}}{\mathrm{e}^{j\bar{\psi}_-}(Z_1 + Z_2)^2 - \mathrm{e}^{-j\bar{\psi}_+}(Z_1 - Z_2)^2} . \tag{4.2.19}$$

Before we consider the beam effect on the transmission coefficient let us examine the passive device namely when no beam is present. The wavenumbers in this case are the same thus $\bar{\psi}_+ = \bar{\psi}_- = \bar{\psi} = k_2 d$ and the transmission coefficient reads

$$\tau = \frac{4 Z_1 Z_2}{\mathrm{e}^{j\bar{\psi}}(Z_1 + Z_2)^2 - \mathrm{e}^{-j\bar{\psi}}(Z_1 - Z_2)^2} . \tag{4.2.20}$$

The peaks of the transmission coefficient occur when $2k_2^{(0)} d = 2n\pi$ and the valleys at $2k_2 d = n\pi$. According to the first relation, the "distance" (Δk_2) between two peaks is $(\Delta k_2) d = \pi$. Using the definition of the group velocity, v_{gr} in (2.3.24), the "distance" Δf, between two peaks is approximately given by

$$(\Delta f)_{\mathrm{pa}} \simeq \frac{v_{\mathrm{gr}}}{2d} . \tag{4.2.21}$$

The subscript $_{\rm pa}$ indicates that this is the bandwidth of a *passive* device. In general, the bandwidth of a given peak is the difference between the two frequencies for which the transmission coefficient is half (-3 dB) of its peak value. Since there are cases where the total height of the peak is less than 3 dB, we define in this case the bandwidth as the distance between two peaks (or two bottom points). If the impedance mismatch is much larger and the difference between the peak and bottom values is much larger then the bandwidth is explicitly dependent on both impedances. At least in what concerns the operation of a traveling-wave tube the trend is to work with minimum reflections, therefore the definition in (4.2.21) is sufficient for our purpose.

Our next step is to calculate the bandwidth of an *active* device. For this purpose we direct our attention back to the expression for the transmission coefficient in (4.2.19). It was mentioned above that the effect of the beam is effectively represented by $e^{-j\bar{\psi}_+}$. This is the ratio between the amplitude of the wave at the output and input. According to the simplified interaction model we developed in Sect. 4.1 we can represent this ratio as

$$e^{-j\bar{\psi}_+} \simeq \frac{1}{3} e^{-j(k_2+\delta k_3)d} , \tag{4.2.22}$$

where δk_3 is that solution of (4.1.6) which has a positive imaginary part as presented in (4.1.12) for the resonance case. In this expression it has been tacitly assumed that this growing wave is dominant.

In this effective representation we have two contributions: one is the real part of the wavenumber and the other is its imaginary part. We shall consider first the effect of the *real part*. We already indicated that, without electrons, the peaks in the transmission coefficient are separated by $2k_2 d = 2\pi$ ($n = 0, \pm 1, ..$). In a similar way when the beam is injected they occur at $[2k_2 + \mathrm{Re}(\delta k_3)]d = 2\pi n$. Consequently, the frequency shift, δf, in the location of the peak due to the interaction is

$$\delta f \simeq -\frac{1}{4\pi} v_{\rm gr} \mathrm{Re}(\delta k_3) , \tag{4.2.23}$$

The maximum frequency shift is expected to occur at maximum gain, namely at resonance hence:

$$\delta f_{\rm max} \simeq -\frac{1}{4\pi\sqrt{3}} v_{\rm gr} \mathrm{Im}\,(\delta k_3) . \tag{4.2.24}$$

As an example, consider coupling coefficient $K_0 = 30\,\mathrm{m}^{-1}$ and group velocity of 0.5 c; the anticipated frequency shift is 180 MHz. For a group velocity 10 times smaller the frequency shift drops to 18 MHz. Here it is important to emphasize two aspects: *(i)* The effect of the beam on the reflection process at the end of the extended cavity was ignored. This is not always justified since the capacitive effects at the ends may become significant and consequently the impedances, which we assumed to be virtually frequency independent, may vary significantly causing an additional frequency shift. *(ii)* When we mention here "frequency shift" what is meant is that the frequency where maximum gain occurs, shifts from one frequency to another but the system operates all along in a *linear* regime, namely the frequency at the output is identical with that at the input and only the frequency where maximum transmission occurs, varies because of the interaction. Next we shall examine the effect of the *imaginary part* of the wavenumber on the transmission coefficient.

The transmission coefficient near the peak and close to resonance is given by

$$|\tau| \simeq \frac{4Z_1Z_2 e^{\sqrt{3}K_0 d/2}}{(Z_1+Z_2)^2 - (Z_2-Z_1)^2 e^{\sqrt{3}K_0 d/2}}, \tag{4.2.25}$$

where K_0 has been defined in (4.1.6). In the relatively close vicinity of a peak we may approximate the transmission coefficient for the case when no beam is present by:

$$|\tau_{\mathrm{pa}}| \simeq \frac{f_1}{\sqrt{(f-f_0)^2+f_1^2}}, \tag{4.2.26}$$

where f_0 is the frequency where the peak is located and, $2\sqrt{3}f_1$ is the bandwidth of the peak which according to the definition in (4.2.21) equals $(2\sqrt{3}f_1 =)v_{\mathrm{gr}}/2d$. When the beam is present, the wave is amplified by a gain factor, $g(f)$ which at resonance reads $g(f_0) \simeq \frac{1}{3}e^{\sqrt{3}K_0 d/2}$. By analogy with the expression in (4.2.25) we can write the transmission coefficient for an active system as

$$|\tau_{\mathrm{act}}| \simeq \frac{f_1 g(f)}{\sqrt{(f-f_0-\delta f_{\mathrm{max}})^2\, g^2(f)+f_1^2}}. \tag{4.2.27}$$

The bandwidth is the difference between the two frequencies at which $|\tau_{\mathrm{act}}|$ reaches 1/2 of its peak value. Ignoring the frequency shift, these two frequencies are a solution of

$$(f-f_0)^2 = 4f_1^2 g^{-2}(f_0) - f_1^2 g^{-2}(f). \tag{4.2.28}$$

Next we assume that the interaction bandwidth [see (4.1.17)] is much broader than f_1, therefore the right-hand side in the last expression can be approximated with $3f_1^2 g^{-2}(f_0)$. This result indicates that the bandwidth of an active (and high gain) system $(\Delta f)_{\mathrm{act}}$ is related to the gain and passive device bandwidth, $(\Delta f)_{\mathrm{pa}}$, by

$$(\Delta f)_{\mathrm{act}} = \frac{(\Delta f)_{\mathrm{pa}}}{g(f_0)}, \tag{4.2.29}$$

or

$$(\Delta f)_{\mathrm{act}} = (\Delta f)_{\mathrm{pa}} 10^{-\frac{\mathrm{gain[dB]}}{20}}. \tag{4.2.30}$$

This result indicates that the product *bandwidth* $\times$ *gain* is constant. For example, a gain of 25 dB in a system whose passive bandwidth is 200 MHz causes the bandwidth of the active device to be 11 MHz.

4.2.3 Interaction in an Extended Cavity

If we examine the condition for the occurrence of the peaks in the last section and the resonance condition for cavity creation (see Sect. 4.2.1) – we find that the two are identical. In fact we have indicated in Sect. 2.5.1 that the denominator of the transmission coefficient determines the resonance frequencies of the system. Whether these frequencies are real or imaginary depends basically on whether electromagnetic energy can leave the system either as a propagating wave [Davis (1994)] or via a dissipative (ohmic loss). An additional insight on the nature of the process can be achieved if the transmission coefficient of a system with two discontinuities (three characteristic impedances – see Fig. 2.3 in Sect. 2.5.1) is represented in terms of the local reflection and transmission coefficients. For this purpose let us define the transmission coefficient from the first region ($-\infty < z < 0$) to the second ($0 < z < d$) by τ_{12} as

$$\tau_{12} = \frac{2Z_1}{Z_1 + Z_2}, \tag{4.2.31}$$

where Z_1 and Z_2 are the characteristic impedances in each one of the regions. The reflection from the first region when a wave impinges from the second region is denoted by ϱ_{21} and is given by

$$\varrho_{21} = \frac{Z_2 - Z_1}{Z_1 + Z_2}, \tag{4.2.32}$$

and correspondingly the wave reflected from the third section when the wave impinges from the second is

$$\varrho_{23} = \frac{Z_2 - Z_3}{Z_3 + Z_2}; \tag{4.2.33}$$

in a similar way, the wave which is transmitted into the third section in this case is

$$\tau_{23} = \frac{2Z_2}{Z_3 + Z_2}. \tag{4.2.34}$$

Using this notation in addition to the phase and amplitude advance as described in the previous sub-section ($\mathrm{e}^{-j\bar{\psi}_+}$ and $\mathrm{e}^{j\bar{\psi}_-}$), we find for the transmission coefficient of the active system [(4.2.19)] the following expression

$$\tau_{\mathrm{act}} = \tau_{12} \frac{\mathrm{e}^{-j\bar{\psi}_+}}{1 - \varrho_{21}\varrho_{23}\mathrm{e}^{-j\bar{\psi}_+}\mathrm{e}^{-j\bar{\psi}_-}} \tau_{23}. \tag{4.2.35}$$

Using this notation we can now emphasize several aspects of the finite length effect:

Remark 1. The transmission coefficient of the active system depends on the ability to couple the power into the system (τ_{12}), the gain and the reflection process in the interaction region (the middle term) and on the ability to extract the power out of the system (τ_{23})

Remark 2. The middle term denominator includes all the information about the effect of reflections on the interaction process and in addition, it provides us with a criterion regarding transition to oscillation. As in the case of the empty cavity the eigen-frequencies of the system are determined by the zeros of the denominator i.e.,

$$1 - \varrho_{21}\varrho_{23}\mathrm{e}^{-j\bar{\psi}_+}\mathrm{e}^{-j\bar{\psi}_-} = 0\,. \tag{4.2.36}$$

From this expression we conclude that the necessary condition for oscillation is

$$g(f)|\varrho_{23}||\varrho_{21}| \geq 1\,. \tag{4.2.37}$$

The physical interpretation of this expression is the following: consider a wave of an amplitude 1 at the input end of the interaction region. As it traverses the system the wave is amplified according to the gain in the system, $g(f)$. At the output end it is partially reflected ($|\varrho_{23}|$) and it undergoes an additional reflection ($|\varrho_{21}|$) at the input. If the amplitude after this last reflection is larger than unity the amplitude will continue to grow in time after each round trip thus the system will oscillate. In order to illustrate the effect let us consider two systems: (*i*) $|\varrho_{23}| = |\varrho_{21}| = 0.1$, for which case the maximum gain before oscillation occurs is $20|\log_{10}(0.1 \times 0.1)| = 40\,\mathrm{dB}$. *(ii)* The other case of interest represents a situation in which one end (typically the input) is effectively short circuited thus $|\varrho_{21}| = 1.0$ and the second has a reasonably good transition such that the effective reflection coefficient is $|\varrho_{23}| = 0.05$; the maximum gain before oscillation in this case is $20|\log_{10}(0.05 \times 1.0)| = 26\,\mathrm{dB}$.

Remark 3. If the system does not operate in a regime which is close to oscillation, it is possible to write the transmission coefficient of the active system, in the following form:

$$\begin{aligned}\tau_{\mathrm{act}} &= \tau_{12}\mathrm{e}^{-j\bar{\psi}_+}\left[\sum_{n=0}^{\infty}\left(\varrho_{21}\varrho_{23}\mathrm{e}^{-j\bar{\psi}_+}\mathrm{e}^{-j\bar{\psi}_-}\right)^n\right]\tau_{23} \\ &= \tau_{12}\mathrm{e}^{-j\bar{\psi}_+}\left[1 + \varrho_{21}\varrho_{23}\mathrm{e}^{-j\bar{\psi}_+}\mathrm{e}^{-j\bar{\psi}_-} + \left(\varrho_{21}\varrho_{23}\mathrm{e}^{-j\bar{\psi}_+}\mathrm{e}^{-j\bar{\psi}_-}\right)^2 \cdots\right]\tau_{23}\,.\end{aligned} \tag{4.2.38}$$

In the framework of this notation it is tacitly assumed that the electron pulse is infinitely long and there are an infinite number of reflections (as the number of terms in the sum). This is obviously not the case in practice and only a limited number of terms has to be considered according to the pulse length and the time it takes the signal to complete one round trip.

Remark 4. If there are fluctuations in the current or voltage the expression in (4.2.38) is more adequate for generalization purposes than (4.2.25). Let us denote the total round trip amplitude and phase shift by $\bar{R} = \varrho_{21}\varrho_{23}e^{-j\bar{\psi}_+}e^{-j\bar{\psi}_-}$; using this notation, the transmission coefficient of the active system reads

$$\tau_{\rm act} = \tau_{12}e^{-j\bar{\psi}_+}\left[1 + \bar{R} + \bar{R}^2 + \bar{R}^3\cdots\right]\tau_{23}\,. \tag{4.2.39}$$

If the current varies along the pulse, then $\bar{R}(I)$ is a function of the current and the natural generalization will be

$$\tau_{\rm act} = \tau_{12}e^{-j\bar{\psi}_+}\left[1 + \bar{R}(I_1) + \bar{R}(I_1)\bar{R}(I_2) + \bar{R}(I_1)\bar{R}(I_2)\bar{R}(I_3)\cdots\right]\tau_{23}\,, \tag{4.2.40}$$

where I_ν indicates the average current in the course of the νth reflection. We shall return to this subject when we discuss the generalized formulation of an amplifier and an oscillator.

4.2.4 Backward-Wave Oscillator

In the type of structures on which we have based our model so far the wave and the energy it carries flow both in the same direction. Therefore, if the input of an amplifier is at $z = 0$ and the beam flows in the positive direction, then power is converted from the beam to the wave. Consequently, comparing the power at a location d far enough from the input we will find that $P(z = d) > P(z = 0)$. Furthermore, the wave at $z = 0$ does not "know" that it is going to be amplified since there is no reflected wave or in other words there is no feedback to provide this information. On the other hand, in an oscillator, the mirrors at both ends together with the structure itself provide a feedback which cause the amplitude at the input to follow the amplitude at the output such that approximately the amplitude of the wave is constant in space.

Imagine now a situation in which the wave propagates in the positive direction, but the energy flows in the opposite direction – this is exactly the case in periodic structures which will be discussed in the next chapter. The information regarding the interaction is carried by the wave opposite to the beam and in fact the input and the output trade places: the input in such a case is at $z = d$ and the output is at $z = 0$. In order to quantify our statements we shall start from the expression for the interaction wavenumber as presented in (4.1.19) and since it was assumed that the energy velocity and the phase velocity are parallel we can now consider a situation in which $\beta_{\rm en}$ is negative and so is K_0^3. As a result we get, instead of (4.1.6),

$$\delta k(\delta k - \Delta k)^2 = K_0^3 , \tag{4.2.41}$$

where the only difference is that the right-hand side is positive (as in the oscillator). The solution at resonance is different and it reads

$$\begin{aligned}
\delta k_1 &= K_0 , \\
\delta k_2 &= -K_0 \left(\frac{1}{2} - j\frac{\sqrt{3}}{2} \right) , \\
\delta k_3 &= -K_0 \left(\frac{1}{2} + j\frac{\sqrt{3}}{2} \right) .
\end{aligned} \tag{4.2.42}$$

As in the traveling-wave amplifier the wave propagates in the forward direction, therefore similar to (4.1.21), we can write

$$\begin{aligned}
E_z(r,z,\omega) =& \mathrm{J}_0 \left(p_1 \frac{r}{R} \right) \mathrm{e}^{-jk_1^{(0)}(z-d)} \\
& \times \left[E_1 \mathrm{e}^{-j\delta k_1 (z-d)} + E_2 \mathrm{e}^{-j\delta k_2 (z-d)} + E_3 \mathrm{e}^{-j\delta k_3 (z-d)} \right] .
\end{aligned} \tag{4.2.43}$$

Since at the input ($z = d$) the beam is assumed to be uniform (not bunched) and the initial amplitude is E_0, the boundary conditions imply

$$\begin{aligned}
& E_1 + E_2 + E_3 = E_0 , \\
& \frac{E_1}{\omega/v_0 - k_1^{(0)} - \delta k_1} + \frac{E_2}{\omega/v_0 - k_1^{(0)} - \delta k_2} + \frac{E_3}{\omega/v_0 - k_1^{(0)} - \delta k_3} = 0 , \\
& \frac{E_1 (k_1^{(0)} + \delta k_1)}{(\omega/v_0 - k_1^{(0)} - \delta k_1)^2} + \frac{E_2 (k_1^{(0)} + \delta k_2)}{(\omega/v_0 - k_1^{(0)} - \delta k_2)^2} \\
& \qquad + \frac{E_3 (k_1^{(0)} + \delta k_3)}{(\omega/v_0 - k_1^{(0)} - \delta k_3)^2} = 0 ;
\end{aligned} \tag{4.2.44}$$

as in the traveling-wave tube case these three equations determine the amplitudes E_1, E_2 and E_3. At the output, the third solution is dominant; thus the gain is given by

$$G \equiv \frac{|E(0)|}{|E(d)|} \simeq \frac{1}{3} \mathrm{e}^{\sqrt{3} K_0 d/2} . \tag{4.2.45}$$

Although the right-hand side is identical to the traveling-wave amplifier result, the fact that in this case the feedback is inherent in the interaction process and is not dependent on load impedance, makes the backward device substantially less sensitive to the load. Furthermore, Carmel (1989) has shown experimentally that the presence of a stationary background plasma (gas) can improve substantially the efficiency of the system. In fact, several years before that Carmel (1973) had shown that high power microwave radiation can be generated by a backward-wave oscillator driven by a relativistic, high current, electron beam.

4.3 Macro-Particle Approach

The hydrodynamic approximation is adequate for the description of the interaction in the linear regime when we wish to consider the variation in the *average* dynamic variables – density and velocity fields. As we approach saturation, the spread in the velocity and density field becomes significant and the validity of the hydrodynamic approximation becomes questionable. In order to solve the problem we have to adopt a more fundamental approach – which is based on the solution of the single-particle equation of motion.

There are at least three ways to develop the simplified set of equations which describe the interaction between electrons and a wave. All three have the one particle equation of motion in common and assume that the basic form of the solution of the electromagnetic field is preserved. The three methods differ in the way the equation which describes the amplitude and the phase of the electromagnetic field is developed. One possibility is to start from the non-homogeneous wave equation for the magnetic vector potential, the second method is to start from the wave equation for E_z and in the third method the starting point is Poynting's theorem. Throughout this text we shall use either the first (Chap. 6) or the third. It is the latter which will be used in this section.

4.3.1 Simplified Set of Equations

The starting point is Poynting's theorem:

$$\nabla \cdot \mathbf{S} + \frac{\partial}{\partial t} W = -\mathbf{J} \cdot \mathbf{E} . \tag{4.3.1}$$

Assuming that the walls of the system are made of an ideal metal, then all the power flux flows in the z direction thus we can integrate over the cross-section (πR^2) of the system:

$$\begin{aligned} &\frac{\partial}{\partial z} 2\pi \int_0^R \mathrm{d}r\, r\, S_z(r,z,t) + \frac{\partial}{\partial t} 2\pi \int_0^R \mathrm{d}r\, r\, W(r,z,t) \\ &= -2\pi \int_0^R \mathrm{d}r\, r\, J_z(r,z,t)\, E_z(r,z,t) . \end{aligned} \tag{4.3.2}$$

We assume that a very strong magnetic field confines the electron motion to the z direction; therefore the only non-zero component of the current density is longitudinal. Furthermore, the system is assumed to operate in the linear regime and it oscillates at a single frequency ω. For the present purposes we average out over one period of the wave $T = 2\pi/\omega$ and if we assume that there is no reflected wave and consequently, there is no change in the amount of electromagnetic energy stored in the system, then (4.3.2) reads

$$\frac{\mathrm{d}}{\mathrm{d}z}\left[2\pi \int_0^R \mathrm{d}r\, r \frac{1}{T}\int_0^T \mathrm{d}t S_z(r,z,t)\right]$$
$$= -2\pi \int_0^R \mathrm{d}r\, r \frac{1}{T}\int_0^T \mathrm{d}t\, J_z(r,z,t)\, E_z(r,z,t)\,. \tag{4.3.3}$$

The first term is the total average power which propagates along the system:

$$P(z) = 2\pi \int_0^R \mathrm{d}r\, r \frac{1}{T}\int_0^T \mathrm{d}t S_z(r,z,t)\,, \tag{4.3.4}$$

and according to the definition of the interaction impedance for a very thin pencil beam it is given by

$$P(z) = \frac{\pi R^2}{Z_{\mathrm{int}}}\frac{1}{T}\int_0^T \mathrm{d}t E_z^2(R_{\mathrm{b}},z,t)\,. \tag{4.3.5}$$

The factor two difference between this equation and (2.3.29) is due to the fact that in the latter, the field has already been averaged on time. The principal assumption of the current approach is that at a given frequency and at a given location, the same interaction impedance which relates the total average power to the longitudinal electric field in vacuum, relates the same quantities when the beam is also present. Consequently, Poynting's theorem now reads

$$\pi R^2 \frac{\mathrm{d}}{\mathrm{d}z}\left[\frac{1}{Z_{\mathrm{int}}}\frac{1}{T}\int_0^T \mathrm{d}t E_z^2(R_{\mathrm{b}},z,t)\right]$$
$$= -2\pi \frac{1}{T}\int_0^T \mathrm{d}t\, E_z(R_{\mathrm{b}},z,t)\int_0^{R_{\mathrm{b}}} \mathrm{d}r\, r\, J_z(r,z,t)\,, \tag{4.3.6}$$

where again we used the thin beam approximation namely, the transverse variations of the electric field are negligible across the beam thickness, therefore the electric field was extracted from the integral in the right-hand side of the equation.

Within the framework of the single particle description, the current density of an azimuthally symmetric flow of electrons is given by

$$J_z(r,z,t) = -e\sum_i v_i(t)\delta\Big[z - z_i(t)\Big]\frac{1}{2\pi r}\delta\Big[r - r_i(t)\Big]\,; \tag{4.3.7}$$

$z_i(t)$ and $r_i(t)$ are the longitudinal and the radial location of the ith electron at a time t. With this definition of the current density the radial integration is straightforward and it reads

$$\frac{\mathrm{d}}{\mathrm{d}z}\left[\frac{1}{T}\int_0^T \mathrm{d}t E_z^2(R_{\mathrm{b}},z,t)\right]$$
$$= \frac{eZ_{\mathrm{int}}}{\pi R^2}\frac{1}{T}\int_0^T \mathrm{d}t\, E_z(R_{\mathrm{b}},z,t)\sum_{i=1}^N v_i(t)\delta\Big[z - z_i(t)\Big]\,. \tag{4.3.8}$$

The second main assumption in this approach is that the effect of the beam on the (single mode) distribution of the electric field is only longitudinal. In other words, if in the beam absence, the longitudinal electric field in the beam region was given by $E_z(R_b, z, t) = E_0 \cos(\omega t - kz - \theta_0)$ where the amplitude (E_0) and the phase (θ_0) are constant, in the presence of the beam the same component reads

$$E_z(R_b, z, t) = E(z) \cos\Big[\omega t - kz - \theta(z)\Big] \tag{4.3.9}$$

and both the amplitude and the phase are allowed to vary in the longitudinal direction. We proceed now by performing the time integration on both sides of (4.3.8). In the left-hand side, the integration over the trigonometric functions is straightforward and in the right-hand side, we take advantage of the Dirac delta function, thus

$$\frac{d}{dz}\left[\frac{1}{2}E^2(z)\right] = \frac{eZ_{int}N}{\pi R^2 T} E(z)\Big\langle \cos\Big[\omega\tau_i(z) - kz - \theta(z)\Big]\Big\rangle , \tag{4.3.10}$$

where $\tau_i(z)$ is defined as

$$\tau_i(z) = \tau_i(0) + \int_0^z d\zeta \frac{1}{v_i(\zeta)} , \tag{4.3.11}$$

and it represents the time it takes the ith electron to reach the point z. $v_i(z)$ is the velocity of the ith electron at z, N is the total number of electrons in one period (T) of the wave and $\langle\cdots\rangle \equiv N^{-1}\sum_{i=1}^{N}\cdots$. We can now identify eN/T as the instantaneous current i.e., $I = eN/T$.

It is convenient at this point to adopt a complex notation namely, $\bar{E}(z) \equiv E(z)\,e^{-j\theta(z)}$, which permits us to write (4.3.10) as

$$\begin{aligned}\frac{1}{2}\frac{d}{dz}E^2(z) &= \frac{1}{2}\frac{d}{dz}\Big[\bar{E}(z)\bar{E}^*(z)\Big] = \frac{1}{2}\Big[\bar{E}(z)\frac{d}{dz}\bar{E}^*(z) + \bar{E}^*(z)\frac{d}{dz}\bar{E}(z)\Big] \\ &= \frac{IZ_{int}}{\pi R^2}\frac{1}{2}\Big[\bar{E}(z)\Big\langle e^{j\chi_i(z)}\Big\rangle + \bar{E}^*(z)\Big\langle e^{-j\chi_i(z)}\Big\rangle + \Big] , \end{aligned} \tag{4.3.12}$$

where $\chi_i(z) = \omega\tau_i(z) - kz$. The last expression can also be written as

$$\bar{E}(z)\Big[\frac{d}{dz}\bar{E}^*(z) - \frac{IZ_{int}}{\pi R^2}\langle e^{j\chi_i(z)}\rangle\Big] + \bar{E}^*(z)\Big[\frac{d}{dz}\bar{E}(z) - \frac{IZ_{int}}{\pi R^2}\langle e^{-j\chi_i(z)}\rangle\Big] = 0 \tag{4.3.13}$$

and since it has to be satisfied for any $\bar{E}(z)$, we conclude that

$$\frac{d}{dz}\bar{E}(z) - \frac{IZ_{int}}{\pi R^2}\Big\langle e^{-j\chi_i(z)}\Big\rangle = 0 , \tag{4.3.14}$$

which describes the dynamics of the amplitude and phase of the electromagnetic field and its dependence on the distribution of particles.

The next step is to simplify the equation of motion of the electrons. Since the motion is in one dimension it is more convenient to use the single particle energy conservation as introduced in (3.1.5). Using the explicit expression for the electric field in (4.3.9) and following the motion of the electron in space we have

$$\frac{\mathrm{d}}{\mathrm{d}z}\gamma_i(z) = -\frac{e}{mc^2}\frac{1}{2}\left[\bar{E}(z)\mathrm{e}^{j\chi(z)} + \mathrm{c.c.}\right]. \tag{4.3.15}$$

It is more convenient to present these two equations [(4.3.14) and (4.3.15)] using a normalized notation. For this purpose we normalize z to the length of the interaction region and define $\zeta = z/d$ as a normalized coordinate. The normalized (complex) amplitude of the longitudinal component of the electric field in the region of the beam is

$$a(\zeta) = \frac{e\bar{E}(z)d}{mc^2}, \tag{4.3.16}$$

and the coupling coefficient α is

$$\alpha = \frac{eIZ_{\mathrm{int}}}{mc^2}\frac{d^2}{\pi R^2}. \tag{4.3.17}$$

Using this notation the variation in space of the normalized amplitude is given by

$$\frac{\mathrm{d}}{\mathrm{d}\zeta}a(\zeta) = \alpha\left\langle \mathrm{e}^{-j\chi_i(\zeta)}\right\rangle, \tag{4.3.18}$$

and the single particle energy conservation reads

$$\frac{\mathrm{d}}{\mathrm{d}\zeta}\gamma_i(\zeta) = -\frac{1}{2}\left[a(\zeta)\mathrm{e}^{j\chi_i(\zeta)} + \mathrm{c.\,c.}\right]. \tag{4.3.19}$$

To complete the description of the particles' dynamics we have to determine the dynamics of the phase term χ_i. According to its definition and (4.3.11) we find

$$\frac{\mathrm{d}}{\mathrm{d}\zeta}\chi_i(\zeta) = \Omega\frac{1}{\beta_i} - K, \tag{4.3.20}$$

where $K = kd$ and $\Omega = \omega d/c$. The last three equations form a closed set of equations which describe the interaction.

Before we proceed to solutions of this set of equations for a practical system, it will be shown that the approximations involved do not affect the global energy conservation. This is readily obtained by averaging the single particle energy conservation (4.3.19) and substituting the equation for the complex normalized amplitude (4.3.18):

$$\frac{\mathrm{d}}{\mathrm{d}\zeta}\left[\langle\gamma_i\rangle + \frac{1}{2\alpha}|a(\zeta)|^2\right] = 0. \tag{4.3.21}$$

In addition, we will show how this set of first order differential equations leads to the same solution we found using the hydrodynamic approximation. To retrieve this limit we take twice the derivative of the amplitude equation in (4.3.18). After the first derivative we obtain

$$\frac{\mathrm{d}^2}{\mathrm{d}\zeta^2}a(\zeta) = -j\alpha\left\langle\left(\frac{\Omega}{\beta_i} - K\right)\mathrm{e}^{-j\chi_i(\zeta)}\right\rangle , \tag{4.3.22}$$

and after the second

$$\frac{\mathrm{d}^3}{\mathrm{d}\zeta^3}a(\zeta) = -j\alpha\Omega\left\langle\frac{1}{(\beta_i\gamma_i)^3}\mathrm{e}^{-j\chi_i(\zeta)}\frac{\mathrm{d}}{\mathrm{d}\zeta}\gamma_i\right\rangle - \alpha\left\langle\left(\frac{\Omega}{\beta_i} - K\right)^2\mathrm{e}^{-j\chi_i(\zeta)}\right\rangle . \tag{4.3.23}$$

Next we substitute the explicit expression for the single-particle energy conservation from (4.3.19) and consider only the slow varying term. The result is:

$$\left[\frac{\mathrm{d}^3}{\mathrm{d}\zeta^3} - j\frac{1}{2}\alpha\Omega\left\langle\frac{1}{(\beta_i\gamma_i)^3}\right\rangle\right]a(\zeta) = -\alpha\left\langle\left(\frac{\Omega}{\beta_i} - K\right)^2\mathrm{e}^{-j\chi_i(\zeta)}\right\rangle . \tag{4.3.24}$$

The differential equation on the left-hand side is equivalent to the third order polynomial obtained using the hydrodynamic approximation. According to this expression, if the variation in the momentum is small, the spatial growth rate is given by the imaginary part of the root of the characteristic polynomial:

$$\mathrm{Im}\,(k) = \frac{\sqrt{3}}{2}\left[\frac{1}{2}\frac{eIZ_{\mathrm{int}}}{mc^2}\frac{1}{\pi R^2}\frac{\omega}{c}\left\langle\frac{1}{(\beta_i\gamma_i)^3}\right\rangle\right]^{1/3} , \tag{4.3.25}$$

and this is identical with the result in (4.1.13). The right-hand side term in (4.3.24) represents the driving term in the system. If at the input the phase and the velocity of the particles are completely uncorrelated then its contribution is zero. As the interaction progresses in space, the phase and the velocity of the particles become correlated and its contribution increases.

4.3.2 Phase-Space Distribution: Linear Regime

We shall now investigate the beam-wave interaction using this set of simplified equations (4.3.18–20). The slow-wave structure consists of a dielectric loaded waveguide which is 20 cm long. The system is driven by a 850 kV, 450 A electron pencil beam. In addition, a wave is launched at the input. The longitudinal component of the electric field at the beam location is assumed to be 1 MV/m. For a practical solution of the equations of motion we divide the entire ensemble of electrons into 64 clusters equally populated with electrons. The internal distribution in each one of these clusters is assumed to remain unchanged along the interaction process. Figure 4.6 illustrates the way the gain and the efficiency

$$\eta(\%) \equiv \frac{\left\langle \gamma(z=0) \right\rangle - \left\langle \gamma(z) \right\rangle}{\left\langle \gamma(z=0) \right\rangle - 1} \times 100\,, \tag{4.3.26}$$

vary along the system. As in the hydrodynamic model we observe first the "buid-up" region where the gain is effectively zero, followed by a region where the gain (in dB) increases linearly. The efficiency in this case is less than 10% which means that the *average energy* has dropped by less than 10%. This is, on average, what one can expect from a single-stage traveling-wave tube (TWT) without special intervention. Both the lethargy and the linear gain section are in reasonable agreement with the regular Pierce approach.

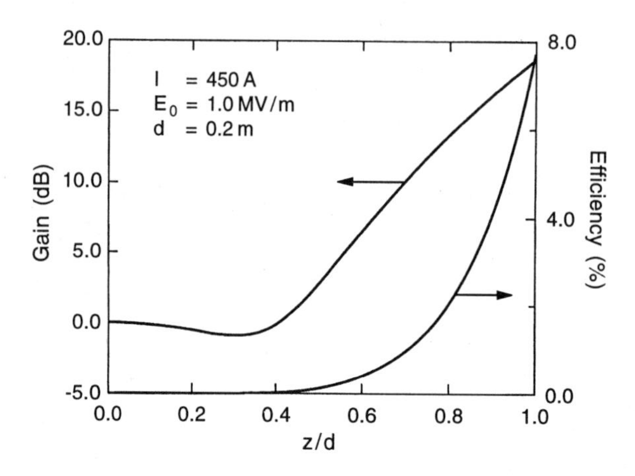

Fig. 4.6. Gain and efficiency along the interaction region. The upper curve represents the gain and the lower one shows the efficiency

We shall now exploit the present formalism to investigate more systematically the interaction process. Figure 4.7 illustrates the way the phase-space distribution evolves along the interaction region. At the entrance $z = 0.0\,\mathrm{d}$, the clusters are uniformly distributed in the domain $-\pi < \chi < \pi$ and $|\gamma - 2.665| < 0.005$. After crossing 20% of the interaction region, the electrons in-phase with the wave were decelerated while those in anti-phase ($\chi = \pm\pi$) are accelerated. As the electrons advance to $z = 0.4\,\mathrm{d}$, the bunching process continues and the electrons' energy spread is now $\pm 6\%$ around the initial average value. At this stage the bottom point of the distribution starts to be shifted towards $\chi = \pi$. This is also the point where the collective effect becomes dominant and the gain starts to grow exponentially. The two processes mentioned above (increase of the energy spread and distribution shift) continue as electrons advance towards $z = d$. At $z = 0.6\,d$ the energy spread is already $\pm 10\%$ and the bottom point of the distribution has slipped 0.8 radians from $\chi = 0$. In the last 20% of the interaction region

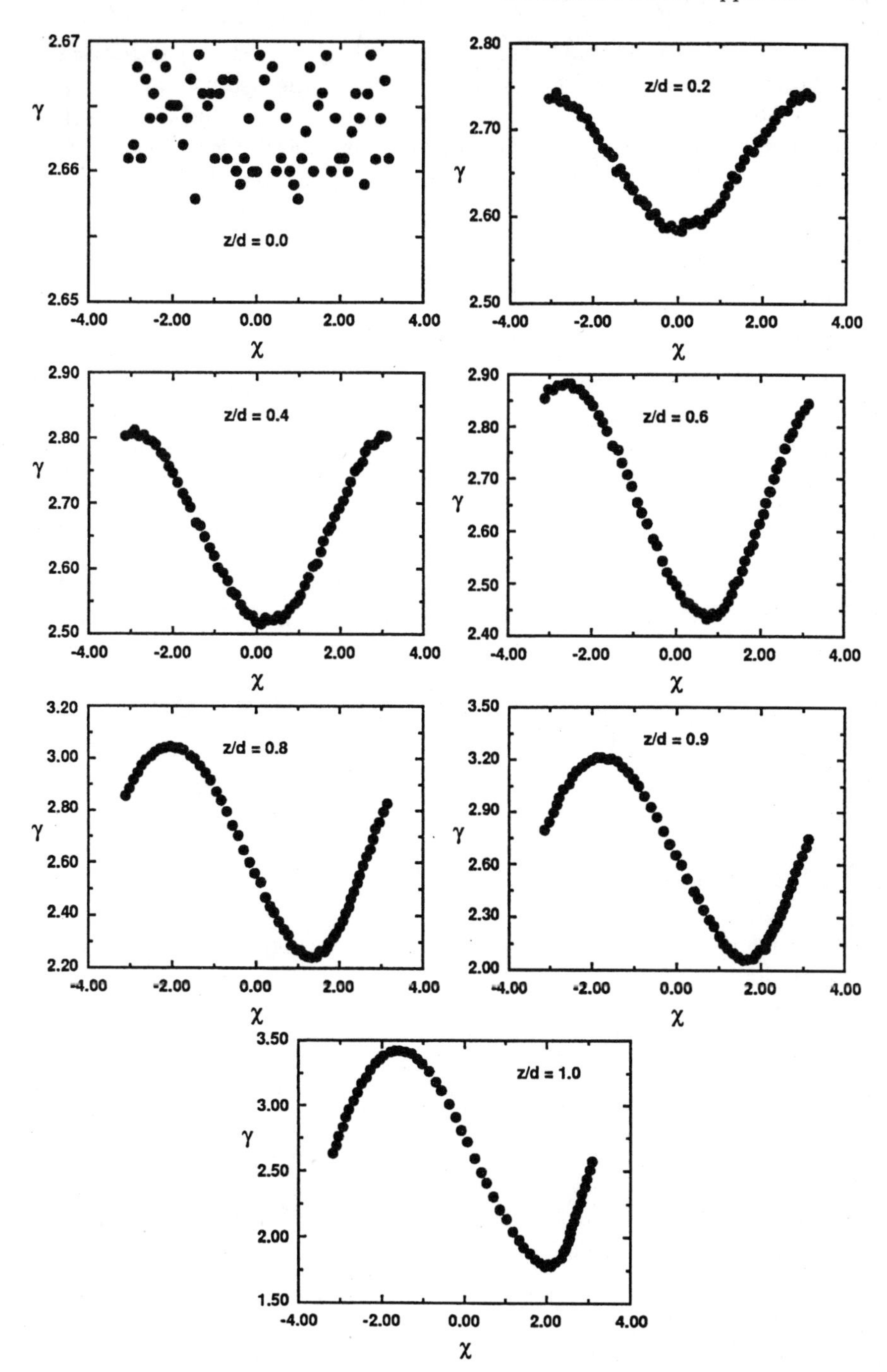

Fig. 4.7. Phase-space distribution at various locations along the interaction region

the electrons are strongly bunched and the energy spread ("peak-to-peak") is actually larger than the average kinetic energy of the electrons at the input ($\gamma_{\max} = 3.5$, $\gamma_{\min} = 1.75$). This is a remarkable result bearing in mind that the efficiency (and thus the change in the average energy) is less than 10%.

The slow and fast electrons have a completely different (relative) weight on the interaction process. According to (4.3.25) the spatial growth rate is proportional to

$$\left[\left\langle \frac{1}{(\beta_i \gamma_i)^3} \right\rangle\right]^{1/3}, \tag{4.3.27}$$

therefore the low momentum electrons have a much larger effect on the interaction process than the fast ones.

In order to have a more quantitative measure of the energy spread we can determine its variation as a function of the other parameters. The first step is to define the energy spread as

$$\Delta\gamma = \sqrt{\langle\gamma^2\rangle - \langle\gamma\rangle^2}\,. \tag{4.3.28}$$

Next if we multiply the single particle energy conservation equation (4.3.21) by γ_i and average over the entire ensemble we obtain

$$\frac{\mathrm{d}}{\mathrm{d}\zeta}\langle\gamma_i^2\rangle = -\left[a(\zeta)\left\langle \gamma_i \mathrm{e}^{j\chi_i(\zeta)}\right\rangle + \mathrm{c.\,c.}\right]. \tag{4.3.29}$$

In a similar way, the variation in space of the square of the average energy is given by

$$\frac{\mathrm{d}}{\mathrm{d}\zeta}\langle\gamma_i\rangle^2 = -\left[a(\zeta)\langle\gamma_i\rangle\left\langle \mathrm{e}^{j\chi_i(\zeta)}\right\rangle + \mathrm{c.\,c.}\right]. \tag{4.3.30}$$

If we now subtract (4.3.30) from (4.3.29) and use the definition of the energy spread we find

$$\begin{aligned}\frac{\mathrm{d}}{\mathrm{d}\zeta}\Delta\gamma^2 = &-a(\zeta)\left[\left\langle \gamma_i \mathrm{e}^{j\chi_i(\zeta)}\right\rangle - \langle\gamma_i\rangle\left\langle \mathrm{e}^{j\chi_i(\zeta)}\right\rangle\right] \\ &- a^*(\zeta)\left[\left\langle \gamma_i \mathrm{e}^{-j\chi_i(\zeta)}\right\rangle - \langle\gamma_i\rangle\left\langle \mathrm{e}^{-j\chi_i(\zeta)}\right\rangle\right].\end{aligned} \tag{4.3.31}$$

This expression indicates that the energy spread is controlled by two principal quantities: (*i*) the amplitude of the radiation field which is obvious since the latter determines the modulation. But this is not sufficient since (*ii*) the phase and energy of all particles have to be correlated i.e.,

$$\left|\langle\gamma_i \mathrm{e}^{j\chi_i(\zeta)}\rangle - \langle\gamma_i\rangle\langle \mathrm{e}^{j\chi_i(\zeta)}\rangle\right| > 0\,, \tag{4.3.32}$$

in order to cause any variation of the energy spread. Otherwise even for a large amplitude of the radiation field, the change in the energy spread is zero.

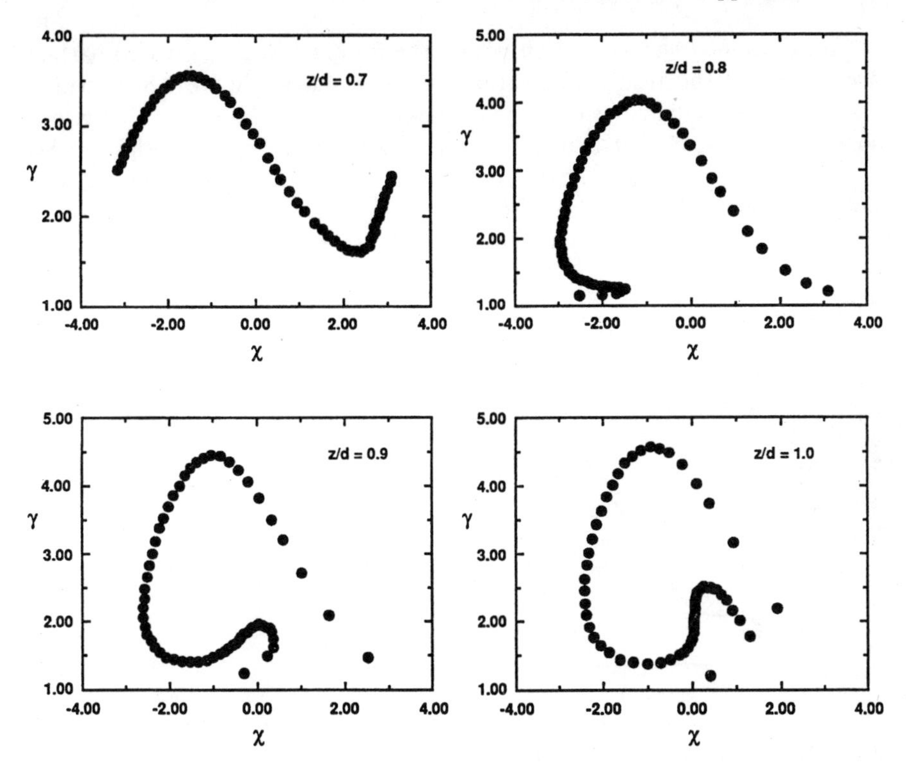

Fig. 4.8. Phase-space distribution at the end of the interaction region

4.3.3 Phase-Space Distribution: Saturation

The next step is to investigate the saturation process. For this purpose we increase the total length by 50% such that now $d = 30$ cm. Saturation occurs when the electrons start to "absorb" energy from the wave. This will happen when the electrons which were initially decelerated reach the point of π phase with the wave and they start to be accelerated. To examine the conditions for saturation we shall first resort to the hydrodynamic model. Consider the longitudinal component of the electric field which is a self consistent solution of the interaction process near resonance: $E(z) \simeq E_0 \cos[\omega t - z(\omega/v_0 - K_0/2)]\mathrm{e}^{\sqrt{3}K_0 z/2}$. The amplitude of the oscillation can be estimated by substituting in the equation of motion: $mv_0(v_0K_0)^2\gamma^3\delta z = eE_0\mathrm{e}^{\sqrt{3}z/2}$. Accordingly the saturation length is defined by $\frac{1}{2}K_0\delta z(z = d_{\mathrm{sat}}) = \pi$ hence

$$\frac{d_{\mathrm{sat}}}{d} = \frac{1}{\sqrt{3}K_0 d/2} \ln\left[2\pi \frac{K_0 d(\gamma\beta)^3}{a_0}\right] . \tag{4.3.33}$$

For the present parameters $d_{\mathrm{sat}}/d = 0.7$.

We examined the saturation within the framework of the macro-particle approach and the result is illustrated in Figs. 4.8–9. Figure 4.8 illustrates the phase-space distribution only in the last 33% of the interaction region since in the first 67% it is identical with what we presented in Fig. 4.9. We observe that the bottom of the distribution ($z = 0.7\,\mathrm{d}$) is almost at the π point. Beyond this point slow electrons are accelerated. This is accompanied by a decrease in the gain and efficiency as illustrated in Fig. 4.9.

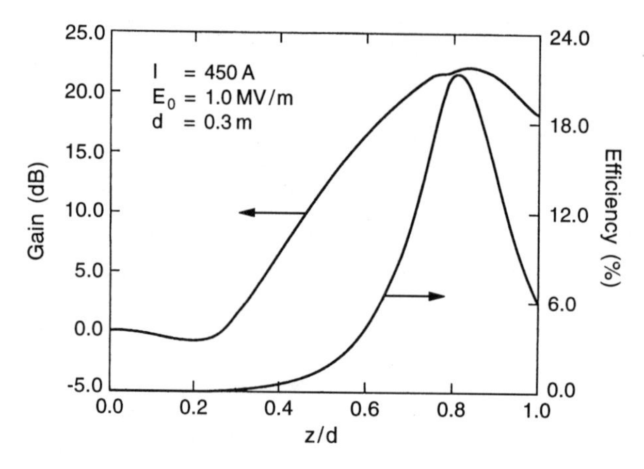

Fig. 4.9. Gain and efficiency in a 50% longer device

4.3.4 Interaction in a Slowly Tapered Structure

In order to increase the efficiency (and consequently the gain) we have to compensate for the energy lost by the electrons. The velocity drop associated with this process and the collective effect itself, cause the phase shift we presented above. In order to adjust the relative phase between the wave and the slow electrons we can taper the slow-wave structure. In terms of the dielectric loaded waveguide this can be done by changing the dielectric coefficient along the z axis or changing the radius of the waveguide and/or dielectric slab. Let us now examine the case when $\varepsilon_{\mathrm{r}}(z)$ varies in space. We shall assume that this variation is small such that

$$\frac{\omega}{c} \gg \left|\frac{d\varepsilon_{\mathrm{r}}(z)}{dz}\right|\frac{1}{\varepsilon_{\mathrm{r}}(z)}\,. \tag{4.3.34}$$

Subject to this condition the equations which describe the dynamics of the amplitude and the particles' dynamics remain unchanged and only the phase equation becomes

$$\frac{\mathrm{d}}{\mathrm{d}\zeta}\chi_i(\zeta) = \frac{\Omega}{\beta_i(\zeta)} - K(\zeta)\,, \tag{4.3.35}$$

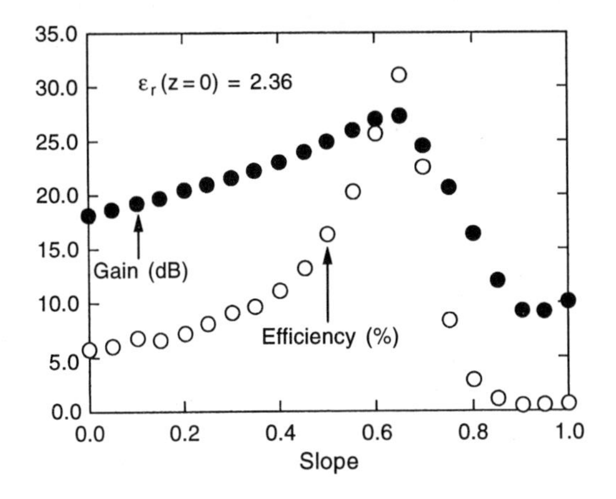

Fig. 4.10. Gain and efficiency as function of the slope of the dielectric coefficient as defined in (4.3.36)

since the normalized wavenumber is $K^2(\zeta) = \varepsilon_r(\zeta)\Omega^2 - p_1^2(d/R)^2$. To simplify the analysis even further we assume a linear variation in space of the dielectric coefficient ε_r namely

$$\varepsilon_r(\zeta) = \varepsilon_r(0) + C_1\zeta\,. \tag{4.3.36}$$

We may now ask what is the optimal value of the slope C_1, given the initial electromagnetic field and the beam characteristics, to obtain maximum efficiency and gain. We expect such an optimal value to occur from inspection of Fig. 4.9: When the phase velocity is constant, at $z = 0.8d$ the phase shift is such that they are in anti-phase. Gradually slowing down the phase velocity, by increasing ε_r, we may push the saturation beyond $z = d$. Increasing ε_r too much could cause the wave to be too slow and the beam-wave coupling is weak and consequently, the system does not reach saturation, thus remaining in the linear regime without extracting maximum energy from the beam.

For the parameters mentioned above we found that the peak occurs at $C_1 = 0.65$ and the efficiency was increased from the 6% in the uniform case to 31% as illustrated in Fig. 4.10. This increase in efficiency is accompanied by 10 dB increase in gain. Figure 4.11 illustrates the gain and the efficiency for a slope which is somewhat below the optimal value $C_1 = 0.6$. Comparing to the uniform case (Fig. 4.10) the saturation point was shifted from $z = 0.8d$ to $z = 0.9d$.

4.3.5 Noise

One of the disadvantages of the hydrodynamic approximation is that the beam is conceived as a fluid and as such, the particle character of the electron is lost. As a result, for evaluation of noise effects one has to postulate velocity or density fluctuations – see discussion by Haus (1959). In the present approach the individual character of the particles is preserved. If a single electron is launched into a slow-wave structure which at a given frequency has

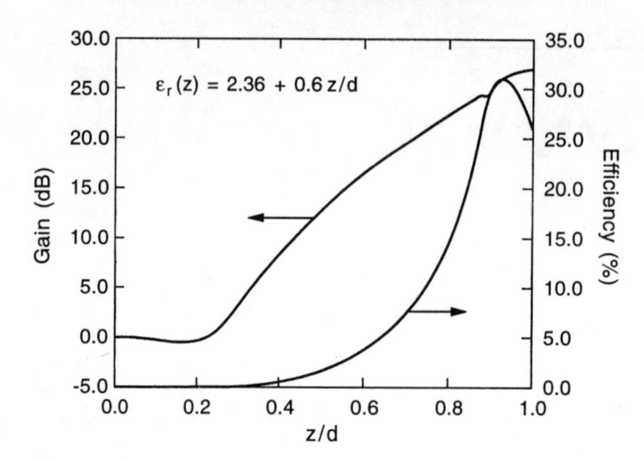

Fig. 4.11. Variation of gain and efficiency in space for linear tapering somewhat below the optimal value

an interaction impedance $Z_{\rm int}$, then the variation in the amplitude $E(z)$ is given by

$$\frac{\rm d}{{\rm d}z} E_i(z) = \frac{eZ_{\rm int}}{\pi R^2 T} {\rm e}^{-j[\chi_i(0) + \int_0^z {\rm d}z' (\omega/v_i(z') - k)]} . \tag{4.3.37}$$

We ignore now the effect of the radiation field on the particles such that in the phase term we can take v_i as constant. At the output of the structure the electric field is

$$E_i(d) = \frac{eZ_{\rm int}}{\pi R^2 T} d{\rm e}^{-j\chi_i(0)} {\rm e}^{-j[\omega/v_i(z') - k]d/2} \,{\rm sinc}[(\omega/v_i - k)d/2] ; \tag{4.3.38}$$

here ${\rm sinc}(x) = \sin(x)/x$. For a uniform distribution of electrons the average electric field at the output is zero due to the random phase of the particles relative to the wave: $\langle {\rm e}^{-j\chi_i(0)} \rangle = 0$. Nevertheless, each such electron emits spontaneous radiation whose power level at the output is given by

$$P_i = \frac{\pi R^2}{2Z_{\rm int}} \left| E_i(d) \right|^2 . \tag{4.3.39}$$

The *total average power* of spontaneous radiation emitted by a uniform beam is

$$P_{\rm sp} = \left[\frac{1}{2} Z_{\rm int} I \frac{e}{T} \right] \frac{d^2}{\pi R^2} \,{\rm sinc}^2 \left[\left(\frac{\omega}{v_0} - k \right) \frac{d}{2} \right] . \tag{4.3.40}$$

This power is linearly proportional to the number of electrons (since $I = eN/T$).

4.3.6 Super-Radiant Emission

If for a *uniform* beam the power emitted was proportional to the number of electrons N, in the case of a *pre-bunched* beam, the emitted power is proportional to N^2. In order to show that we shall again ignore the effect of the radiation field on the electrons. Contrary to the previous case where the low level of emitted power justifies completely this assumption, in this case it is no longer justified and for an adequate solution one has to take into account the variation in the electrons' phase-space distribution. Nevertheless, in order to have a zero order estimate we shall ignore the effect of the radiation on the electrons. In the framework of this approximation the amplitude equation reads

$$\frac{\mathrm{d}}{\mathrm{d}z}E(z) = \frac{IZ_{\mathrm{int}}}{\pi R^2}\left\langle \mathrm{e}^{-j[\chi_i(0)+z(\omega/v_i-k)]}\right\rangle ; \tag{4.3.41}$$

the phase $(\chi_i(0))$ and the energy γ_i are correlated and for simplicity we shall consider a cold bunch (very small energy spread) which has a phase distribution $-\pi < -\chi_0 < \chi_i(0) < \chi_0 < \pi$, hence

$$E(d) = \frac{IZ_{\mathrm{int}}}{\pi R^2}\,\mathrm{sinc}(\chi_0)\mathrm{e}^{-j(\omega/v_0-k)d/2}\,\mathrm{sinc}\left[\left(\frac{\omega}{v_0}-k\right)\frac{d}{2}\right] . \tag{4.3.42}$$

Note that at the limit $\chi_0 = \pi$ we obtain the result of a uniformly distributed beam namely, $E(d) = 0$. As we shall see a significant amount of the kinetic energy of the electrons is transferred to the radiation field, therefore we introduce now a correction to the assumption that v remains unchanged. If the efficiency is 100%, the velocity at the output is zero so we shall consider the average value of $\mathrm{e}^{-j(\omega/v_0-k)d/2}\,\mathrm{sinc}[(\omega/v_0-k)d/2]$ at the input $(v = v_0)$ and output $(v = 0)$; the result is $\frac{1}{2}\mathrm{e}^{-j(\omega/v_0-k)d/2}\,\mathrm{sinc}[(\omega/v_0-k)d/2]$. The power emitted in this case is

$$P_{\mathrm{sr}} \simeq \left[\frac{1}{2}Z_{\mathrm{int}}I^2\right]\frac{d^2}{\pi R^2}\,\mathrm{sinc}^2\left[\left(\frac{\omega}{v_0}-k\right)\frac{d}{2}\right]\,\mathrm{sinc}^2\chi_0\frac{1}{4} . \tag{4.3.43}$$

For $\chi_0 = \pi/2$, $\beta = 0.92$, $I = 400\,\mathrm{A}, d = 20\,\mathrm{cm}$, $R = 1.0\,\mathrm{cm}$, $Z_{\mathrm{int}} = 250\,\Omega$ the maximum power radiated is of the order of 250 MW which is more than 70% of the instantaneous power carried by the beam (0.85 MV $\times$ 400 A $\simeq$ 340 MW). This crude estimate of the super-radiant emission is related to spontaneous radiation

$$P_{\mathrm{sr}} \propto N\,\mathrm{sinc}^2\chi_0 P_{\mathrm{sp}} , \tag{4.3.44}$$

which emphasizes our statement at the beginning of this sub-section regarding the factor N difference between the power emitted in the two processes.

4.3.7 Resonant Particle Model

In Sect. 4.3.2 we found that as electrons lose energy to the wave, their velocity is decreased and therefore they slip from the resonance condition. We have also shown, using a very simple model, that this effect can be corrected if the slow-wave structure is tapered. Let us now examine this process in a more systematic way. Our goal is to determine how the structure should vary in space in order to extract maximum energy from a given distribution of electrons and a given input field. As stated, this requirement (in general) will be very difficult to meet; however we can solve the problem for a limited set of distributions. In particular, we can solve for a very narrow phase-space distribution which can be approximated by a single macro-particle and this solution gives us a crude design as for how the structure should vary in space. With such a design we can release somewhat the constraint on the initial particle distribution and address more practical problems.

The equations which describe the dynamics of a system which consists of a single macro particle and an electromagnetic wave are given by

$$\begin{aligned} \frac{\mathrm{d}}{\mathrm{d}\zeta} a_{\mathrm{r}}(\zeta) &= \alpha\, \mathrm{e}^{-j\chi_{\mathrm{r}}(\zeta)}, \\ \frac{\mathrm{d}}{\mathrm{d}\zeta} \gamma_{\mathrm{r}}(\zeta) &= -\frac{1}{2}\left[a_{\mathrm{r}}(\zeta)\, \mathrm{e}^{j\chi_{\mathrm{r}}(\zeta)} + \mathrm{c.c.}\right], \\ \frac{\mathrm{d}}{\mathrm{d}\zeta} \chi_{\mathrm{r}}(\zeta) &= \frac{\Omega}{\beta_{\mathrm{r}}(\zeta)} - K(\zeta). \end{aligned} \tag{4.3.45}$$

The coupling coefficient α, is considered to be constant but the wave number in the phase term is allowed to vary (the important variations are assumed to be controlled by the phase term). We further assume that (*i*) the electrons are ideally bunched such that they form a single macro-particle which (*ii*) remain "glued" together along the entire interaction region. (*iii*) The initial velocity of the macro-particle is equal to the phase velocity of the wave. The problem is to determine the necessary variation in the wavenumber of the slow-wave structure in order to keep the macro-particle in resonance along the entire interaction region. This last condition can be mathematically formulated as

$$\frac{\mathrm{d}}{\mathrm{d}\zeta} \chi_{\mathrm{r}}(\zeta) = \frac{\Omega}{\beta_{\mathrm{r}}(\zeta)} - K(\zeta) = 0; \tag{4.3.46}$$

subscript r indicates the resonant particle. Because of this condition this is called the *resonant particle model.* From the first term in this equation we conclude that the phase is constant (χ_{r}) along the interaction region, therefore the integration of the amplitude equation is straightforward

$$a_{\mathrm{r}}(\zeta) = a(0) + \alpha\zeta \mathrm{e}^{-j\chi_{\mathrm{r}}}. \tag{4.3.47}$$

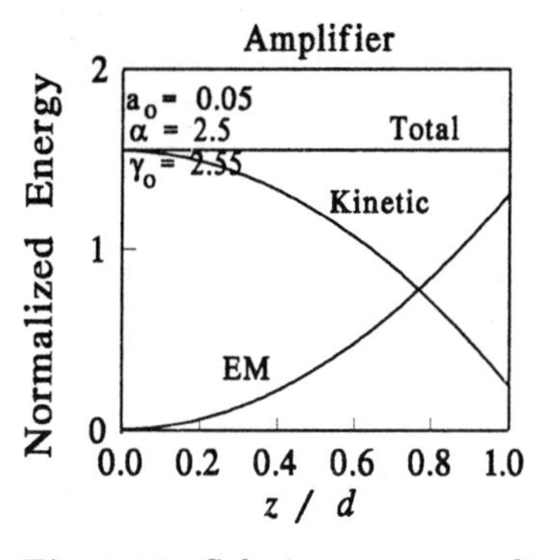

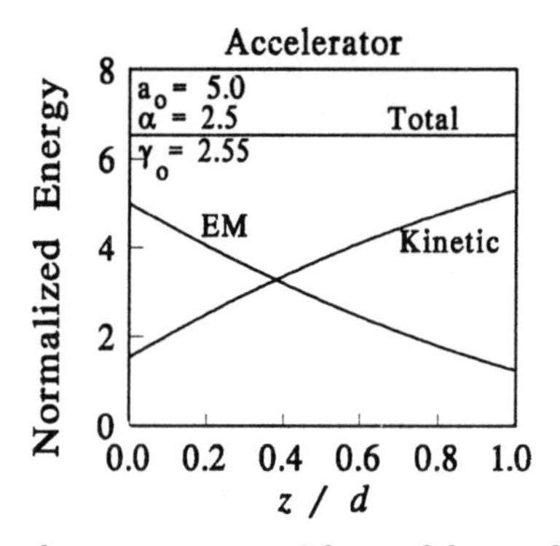

Fig. 4.12. Solution corresponding to the resonant particle model: amplifier left frame and accelerator right frame. In practice the loading (change in the EM energy) is small in a linear accelerator as further discussed in Chap. 8

Our natural next step is to substitute this expression into the single particle equation of motion; the result is

$$\gamma_{\mathrm{r}}(\zeta) = \gamma_{\mathrm{r}}(0) - \zeta a(0) \cos \chi_{\mathrm{r}} - \alpha \frac{\zeta^2}{2} . \tag{4.3.48}$$

In order to determine the wavenumber of the structure, the last result is substituted in (4.3.46) and obtain

$$K(\zeta) = \Omega \frac{\gamma_{\mathrm{r}}(\zeta)}{\sqrt{\gamma_{\mathrm{r}}^2(\zeta) - 1}} . \tag{4.3.49}$$

According to the initial phase of the resonant particle with the wave the latter will gain ($\chi_{\mathrm{r}} = \pi$) or lose ($\chi_{\mathrm{r}} = 0$) energy and the system operates as an accelerator or amplifier respectively. In either one of the two cases, the solution above satisfies the global energy conservation i.e.,

$$\gamma_{\mathrm{r}}(\zeta) - 1 + \frac{1}{2\alpha} \left| a_{\mathrm{r}}(\zeta) \right|^2 = \gamma_{\mathrm{r}}(0) - 1 + \frac{1}{2\alpha} \left| a_{\mathrm{r}}(0) \right|^2 . \tag{4.3.50}$$

In an amplifier the normalized kinetic energy ($\gamma_{\mathrm{r}} - 1$) at the input is much larger than the electromagnetic energy per particle ($|a_{\mathrm{r}}(0)|^2/2\alpha$). The situation is reversed in an accelerator – see Fig. 4.12. However, in practice the loading effect (change in the electromagnetic energy) in the case of an accelerator is usually small. As an example consider a system which operates as an amplifier at 10 GHz, its length is 2 m and the normalized coupling coefficient is $\alpha = 2.5$. The initial energy of the electrons corresponds to $\gamma_{\mathrm{r}} = 2.55$ and the input radiation power corresponds to an initial amplitude of $a_{\mathrm{r}}(0) = 0.05$. According to (4.3.47–48) the energy at the output (for $\chi_{\mathrm{r}} = 0$) will correspond to $\gamma_{\mathrm{r}} = 1.25$ thus the efficiency is more than 80%.

The resonant particle model, as presented above, is obviously an idealization of a realistic system which has a finite spread on the initial phase distribution. It is used for the design stage when it is required to calculate the parameters of the structure i.e., the variation in space of the wave number in (4.3.49). In a realistic system, the electrons are not "glued" together therefore they spread. We formulate next the equations for the deviations from the ideal model: $\delta a(\zeta) \equiv a(\zeta) - a_{\rm r}(\zeta)$ represents the change in the amplitude of the radiation field. In a similar way $\delta\gamma_i(\zeta) \equiv \gamma_i(\zeta) - \gamma_{\rm r}(\zeta)$ and $\delta\chi_i(\zeta) \equiv \chi_i(\zeta) - \chi_{\rm r}$. These deviations satisfy the following set of equations:

$$\begin{aligned}
\frac{\mathrm{d}}{\mathrm{d}\zeta}\delta a(\zeta) &= \alpha \mathrm{e}^{-j\chi_{\rm r}} \left[\langle \mathrm{e}^{-j\delta\chi_i(\zeta)} \rangle - 1 \right] , \\
\frac{\mathrm{d}}{\mathrm{d}\zeta}\delta\gamma_i(\zeta) &= -\frac{1}{2} \left\{ \mathrm{e}^{j\chi_{\rm r}} \left[[a_{\rm r}(\zeta) + \delta a(\zeta)] \mathrm{e}^{j\delta\chi_i(\zeta)} - a_{\rm r}(\zeta) \right] + \mathrm{c.c.} \right\} , \\
\frac{\mathrm{d}}{\mathrm{d}\zeta}\delta\chi_i(\zeta) &= \Omega \left[\frac{1}{\beta_i(\zeta)} - \frac{1}{\beta_{\rm r}(\zeta)} \right] .
\end{aligned} \tag{4.3.51}$$

It is instructive to examine this set of equations in a regime where these deviations are small

$$\begin{aligned}
\frac{\mathrm{d}}{\mathrm{d}\zeta}\delta a(\zeta) &= -j\alpha \mathrm{e}^{-j\chi_{\rm r}} \langle \delta\chi_i(\zeta) \rangle , \\
\frac{\mathrm{d}}{\mathrm{d}\zeta}\delta\gamma_i(\zeta) &= -\frac{1}{2} \left[a_{\rm r}(\zeta) j\delta\chi_i(\zeta) + \delta a(\zeta) \mathrm{e}^{j\chi_{\rm r}} + \mathrm{c.c.} \right] , \\
\frac{\mathrm{d}}{\mathrm{d}\zeta}\delta\chi_i(\zeta) &= \frac{\Omega}{[\gamma_{\rm r}(\zeta)\beta_{\rm r}(\zeta)]^3} \delta\gamma_i(\zeta) .
\end{aligned} \tag{4.3.52}$$

It is evident from the first two equations that if the average phase distribution $\langle \delta\chi_i(\zeta) \rangle$ vanishes or is very small, the system behaves as if driven by a single macro particle. The third equation indicates that the phase distribution tends to spread as the momentum of the electrons decreases, diminishing in the process the energy conversion. On the other hand, if the bunch is being accelerated, the phase deviations are much smaller and, as will be shown in Chap. 8, the bunch is actually compressed.

The next step is to examine the operation of a realistic system which has the same parameters as in the example at the beginning ot the previous paragraph. The simulation is based on the solution of (4.3.51) and is performed as follows: we take 10240 macro-particles uniformly distributed in the range $|\delta\gamma_i(0) = \gamma_i(0) - \gamma_{\rm r}(0)| < \gamma_{\rm r}(0)/80$ and $|\delta\chi_i(0)| < \pi/36$. Figure 4.13 illustrates the phase-space plot at four different locations along the interaction region. First frame illustrates the bunch at the input. After 70 cm the bunch lost about 10% of its momentum and it still maintains its shape. After another 70 cm the bunch has lost a total of about 30% of its initial momentum. However at this point the distribution spread in phase space becomes fairly large and towards the output the spread is large. We observe that this is basically a long line which is cut by the way we chose to present our data (Mod(π)). A more convenient way to examine this process is to overlay the first three

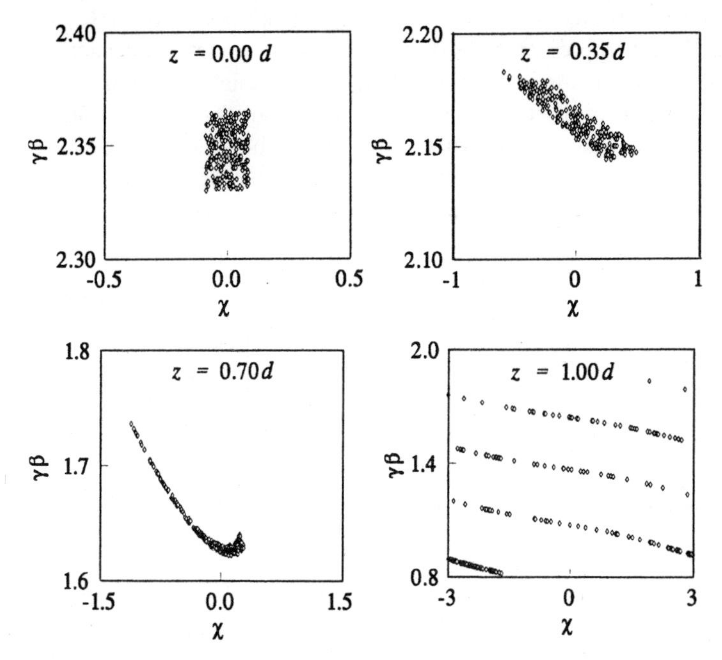

Fig. 4.13. Phase-space distribution in a tapered structure designed based on the resonant particle model

frames on one plot. The result is presented in Fig. 4.14. Here we clearly observe the significant increase in the phase distribution (in accordance with our previous discussion).

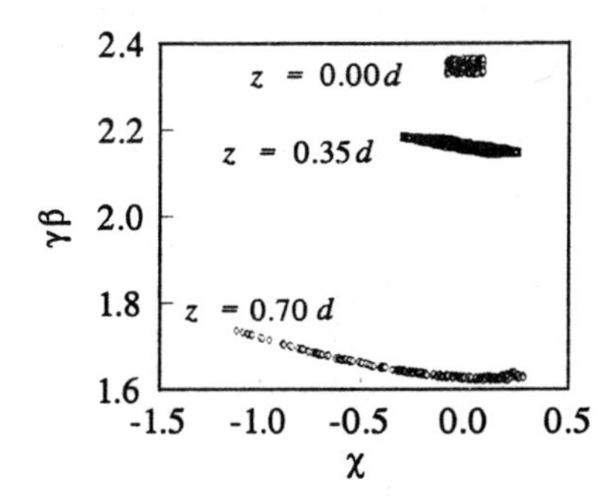

Fig. 4.14. Evolution of phase-space distribution

Finally, we present in Fig. 4.15 the gain and the efficiency (left frame) and in the right the average energy of the electrons and their energy spread. Several characteristics are evident: *(i)* The gain starts to grow in space immediately without the *spatial lethargy* required in an amplifier driven by a uniform beam. This result is obvious bearing in mind that the wave at the input does not have to bunch the beam – the latter is already bunched. The gain is about 33 dB in comparison to 34 dB predicted by the resonant particle model and 16 dB using a uniform structure. *(ii)* The efficiency is more than 65% comparing to more than 80% predicted by the resonant particle model or a typical 10% in a uniform structure. *(iii)* From the energy spread we conclude that the bunch maintains its shape for about 70% of the interaction region. Beyond this point there is a significant increase in the energy spread.

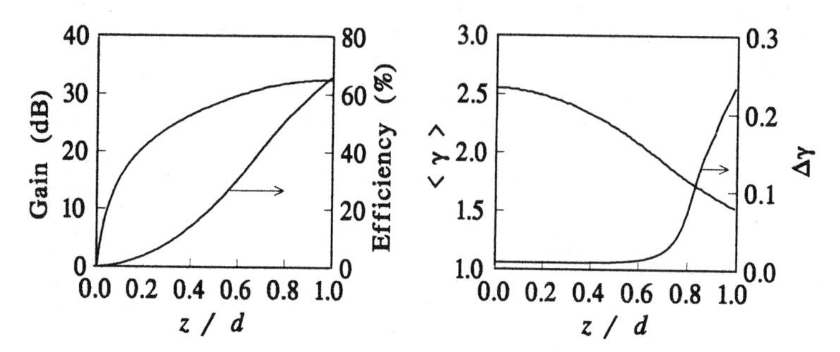

Fig. 4.15. Gain and efficiency variation along the interaction region (left frame). Average energy and energy spread in the right frame

4.4 Complex Interaction Impedance

The formulation so far assumed that the interaction impedance is real and is uniform in the interaction region, however the power is in general a complex quantity and the structure may vary in space. Consequently, we shall now formulate the beam-wave interaction without these two constraints. The discussion is limited to the linear regime, therefore a steady state regime is considered.

Our starting point is the complex Poynting's theorem (2.1.32) which in its integral form reads

$$\frac{\mathrm{d}}{\mathrm{d}z}P(z) + 2j\omega\left[W_{\mathrm{m}}(z) - W_{\mathrm{e}}(z)\right] = \int \mathrm{d}a\left[\frac{1}{2}\mathbf{J}^{*}(\mathbf{r})\cdot\mathbf{E}(\mathbf{r})\right], \tag{4.4.1}$$

where

$$\begin{aligned} W_{\mathrm{m}}(z) &= \frac{\mu_0}{4}\int \mathrm{d}a|\mathbf{H}(\mathbf{r})|^2, \\ W_{\mathrm{e}}(z) &= \frac{\varepsilon_0}{4}\int \mathrm{d}a|\mathbf{E}(\mathbf{r})|^2, \\ P(z) &= \int \mathbf{da}\cdot\left[\frac{1}{2}\mathbf{E}(\mathbf{r})\times\mathbf{H}^{*}(\mathbf{r})\right], \end{aligned} \tag{4.4.2}$$

represent the average energy per unit length stored in the magnetic and electrical field correspondingly. $P(z)$ is the complex power in the system and da is an area element perpendicular to the z direction. The real part of $P(z)$ is indicative of the average power flow in the system whereas its imaginary counterpart represents the energy stored in the system. As in the previous section, the motion is assumed to be limited to the longitudinal direction, therefore

$$\frac{\mathrm{d}}{\mathrm{d}z}P(z) + 2j\omega\Big[W_{\mathrm{m}}(z) - W_{\mathrm{e}}(z)\Big] = \frac{1}{2}IE(z)\langle \mathrm{e}^{j\omega\tau_i(z)}\rangle\,, \tag{4.4.3}$$

where $\tau_i(z)$ was defined in (4.3.11),

$$E(z) \equiv \frac{2}{R_{\mathrm{b}}^2}\int_0^{R_{\mathrm{b}}} \mathrm{d}r\, r\, E_z(r,z) \simeq E_z(R_{\mathrm{b}}, z)\,, \tag{4.4.4}$$

is the effective electric field which acts on the electrons and I is the instantaneous current at the input in one period of the wave.

We limit our discussion to the linear regime; thus the power must be quadratic in the components of the electromagnetic field. In particular, assuming local interaction, the local power has to be quadratic in the effective field present at the location where electrons will be injected i.e., $P(z) \propto |E(z)|^2$. Therefore, it is now convenient to introduce the quantity $Y(z)$ which is the interaction admittance

$$P(z) = \frac{1}{2}Y^*(z)|E(z)|^2\pi R^2\,, \tag{4.4.5}$$

and since in general $P(z)$ is a complex function, so is $Y(z)$ i.e.,

$$Y(z) \equiv Y_r(z) + jY_i(z)\,. \tag{4.4.6}$$

Following the same logic we can express the total magnetic energy per unit length at a given location as a quadratic function on the effective electric field $[E(z)]$ thus

$$W_{\mathrm{m}}(z) = \frac{\varepsilon_0}{4}\varepsilon_{\mathrm{m}}(z)|E(z)|^2\pi R^2\,, \tag{4.4.7}$$

and in a similar way

$$W_{\mathrm{e}}(z) = \frac{\varepsilon_0}{4}\varepsilon_{\mathrm{e}}(z)|E(z)|^2\pi R^2\,. \tag{4.4.8}$$

In contrast to $Y(z)$ which is a complex quantity, $\varepsilon_{\mathrm{m}}(z)$ and $\varepsilon_{\mathrm{e}}(z)$, are real defined functions. Using this notation we can write

$$\frac{\mathrm{d}}{\mathrm{d}z}\left[Y^*|E|^2\right] + j\omega\varepsilon_0\left[\varepsilon_{\mathrm{m}} - \varepsilon_{\mathrm{e}}\right]|E|^2\pi R^2 = \frac{I}{\pi R^2}E\langle \mathrm{e}^{j\omega\tau_i}\rangle\,. \tag{4.4.9}$$

This expression can be split into its real and imaginary components as follows:

$$\frac{\mathrm{d}}{\mathrm{d}z}\left[Y_r|E|^2\right] = \frac{I}{\pi R^2}|E|\Big\langle \cos[\psi + \omega\tau_i]\Big\rangle ,$$

$$-\frac{\mathrm{d}}{\mathrm{d}z}\left[Y_i|E|^2\right] + \omega\varepsilon_0\left[\varepsilon_{\mathrm{m}} - \varepsilon_{\mathrm{e}}\right]|E|^2 = \frac{I}{\pi R^2}|E|\Big\langle \sin[\psi + \omega\tau_i]\Big\rangle , \qquad (4.4.10)$$

where we used the fact that $E = |E|\mathrm{e}^{j\psi}$ and it was tacitly assumed that the effective area where the wave propagates πR^2 does not vary in space. If this is not the case, the definitions of Y, ε_{m} and ε_{e} can be changed to include this effect.

The relations in (4.4.10) can be further simplified if we realize that Y and $\Delta\varepsilon \equiv \varepsilon_{\mathrm{m}} - \varepsilon_{\mathrm{e}}$ are related. In a linear regime Y is independent of $|E|$ and, as indicated in the past, the interaction parameters are functions only of the frequency and geometry. In particular, it will be assumed that Y and $\Delta\varepsilon$ are current independent. In order to determine the relation mentioned above we consider the homogeneous case ($I = 0$), for which (4.4.10) read

$$2Y_r\frac{\mathrm{d}}{\mathrm{d}z}E_0 + E_0\frac{\mathrm{d}}{\mathrm{d}z}Y_r = 0 ,$$

$$-2Y_i\frac{\mathrm{d}}{\mathrm{d}z}E_0 - E_0\frac{\mathrm{d}}{\mathrm{d}z}Y_i + \omega\varepsilon_0\Delta\varepsilon E_0 = 0 , \qquad (4.4.11)$$

E_0 denoting the homogeneous field. Next we substitute $\mathrm{d}E_0/\mathrm{d}z$ from the first equation into the second and assuming that E_0 is not zero we find

$$\omega\varepsilon_0\Delta\varepsilon = Y_r\frac{\mathrm{d}}{\mathrm{d}z}\left[\frac{Y_i}{Y_r}\right] . \qquad (4.4.12)$$

We next substitute this relation in (4.4.10) which when written again in a complex form reads

$$\frac{\mathrm{d}}{\mathrm{d}z}\left[Y_r|E|^2\right] = \frac{I}{\pi R^2}|E|\frac{Y_r}{Y}\mathrm{e}^{-j\psi}\langle \mathrm{e}^{-j\omega\tau_i}\rangle . \qquad (4.4.13)$$

Since in our notation the left-hand side of this equation is real, the right-hand side of (4.4.13) contains all the information about the phase, ψ which has to satisfy $\psi + \arctan[\langle\sin(\omega\tau_i)\rangle \,/\, \langle\cos(\omega\tau_i)\rangle] + \arctan[Y_i/Y_r] = \pi n$, and consequently

$$\frac{\mathrm{d}}{\mathrm{d}z}\psi = -\frac{\omega}{c}\frac{1}{\tilde{\beta}} - \kappa_y , \qquad (4.4.14)$$

where

$$\tilde{\beta}^{-1} \equiv \frac{\langle\beta_i^{-1}\cos\omega\tau_i\rangle\langle\cos\omega\tau_i\rangle + \langle\beta_i^{-1}\sin\omega\tau_i\rangle\langle\sin\omega\tau_i\rangle}{\langle\cos^2\omega\tau_i\rangle + \langle\sin^2\omega\tau_i\rangle} ,$$

$$\kappa_y \equiv |Y|^{-2}\left[Y_r\frac{\mathrm{d}}{\mathrm{d}z}Y_i - Y_i\frac{\mathrm{d}}{\mathrm{d}z}Y_r\right] . \qquad (4.4.15)$$

This finally allows us to determine the equation for the complex amplitude of the electromagnetic field which is

$$\frac{\mathrm{d}}{\mathrm{d}z}\left[\sqrt{Y_r}E\right] + j\sqrt{Y_r}E\left[\frac{\omega}{c}\frac{1}{\tilde{\beta}} + \kappa_y\right] = \frac{1}{2}\frac{I}{\pi R^2}\frac{\sqrt{Y_r}}{Y}\langle \mathrm{e}^{-j\omega\tau_i}\rangle . \qquad (4.4.16)$$

In order to complete the description of the systems dynamics we recall that

$$\frac{\mathrm{d}}{\mathrm{d}z}\gamma_i = -\frac{e}{mc^2}\frac{1}{2}\left[E\mathrm{e}^{j\omega\tau_i} + \mathrm{c.c.}\right] , \qquad (4.4.17)$$

and according to its definition, the phase term satisfies

$$\frac{\mathrm{d}}{\mathrm{d}z}\tau_i = \frac{1}{v_i} . \qquad (4.4.18)$$

These three equations determine the dynamics of electrons and electromagnetic field for the case when the interaction impedance (admittance) is complex and it varies in space. Note that no a-priori assumption has been made regarding the form of the electromagnetic wave.

Before we conclude this section there are three aspects which should be emphasized. Firstly, the energy conservation associated with this set of equations is obtained by averaging (4.4.17) and, substituting (4.4.16), the result is

$$\frac{\mathrm{d}}{\mathrm{d}z}\left[\langle\gamma_i\rangle + \frac{e\pi R^2}{mc^2 I}Y_r|E|^2\right] = 0 . \qquad (4.4.19)$$

This clearly indicates that energy exchange is possible only when there is a non-zero real component to the interaction admittance (impedance). Secondly, when comparing this result (4.4.19) with the similar result in (4.3.21) we observe that there is a factor 1/2 difference. It occurs since in the present case, fast oscillation of the electromagnetic field is allowed whereas in Sect. 4.3.1 this oscillation was averaged out. Thirdly, contrary to the equations in the previous section, this set can also describe the propagation of a space-charge wave in a uniform waveguide below cut-off. In this case $Y_r = \mathrm{const} = 0$ and $Y_i = \mathrm{const} \neq 0$ thus

$$\left[\frac{\mathrm{d}}{\mathrm{d}z} + j\frac{\omega}{c}\frac{1}{\tilde{\beta}}\right]E = -j\frac{1}{2}\frac{I}{\pi R^2}\frac{1}{Y_i}\langle \mathrm{e}^{-j\omega\tau_i}\rangle , \qquad (4.4.20)$$

which represents the space-charge waves and according to (4.4.19) there is no energy exchange.

4.5 Amplifier and Oscillator: A Unified Approach

In Sect. 4.3 we formulated the interaction in an amplifier based on the single-particle equation of motion and ignoring variation in time i.e., reflections. The next step is to include these effects in the analysis. The motivation for this generalization was introduced already in Sect. 4.2.1 where we discussed the effect of reflections within the framework of the hydrodynamic model and it was shown that one manifestation of their effect is the product *gain* $\times$ *bandwidth* which was proved to be constant. Another consequence of a wave being reflected is amplitude variations which occur at the input, according to the time it takes the reflected wave to traverse the distance between the output and input end. Thus, when reflections are not negligible, the assumption of no time variations in a realistic amplifier is not justified.

The opposite situation occurs in oscillators. "Mirrors" at the two ends impose the variation in space of the electromagnetic field. Thus the field amplitude is considered constant in space and the beam-wave interaction determines the temporal growth rate. But the beam which enters the system is presumably unbunched therefore it will take some space for this beam to become bunched. If so, the modulation amplitude is expected to vary in space and consequently, the amplitude of the radiation field will vary in space. As before, this is in contradiction to the initial assumption.

In order to emphasize even further the difference between an amplifier and an oscillator we recall that within the framework of the hydrodynamic model, the beam-wave interaction was formulated in terms of a dispersion relation

$$D_{\rm act}(k,\omega) = 0\,. \tag{4.5.1}$$

In an ideal amplifier we assumed that there are *no variations in time of the amplitude* thus the frequency is set for us by an external generator ($\omega = \omega_0$) and we have to determine the variation in space represented by a set of k's which can be complex and they are a solution of:

$$D_{\rm act}(k,\omega = \omega_0) = 0\,. \tag{4.5.2}$$

This was one "extreme" among the regimes of beam-wave interaction. The opposite extreme happens in an ideal oscillator where we assume that there are *no variations in space of the amplitude* since the wavenumber k is determined by the separation of the mirrors (d) i.e., $k = \pi n/d$ where n is an integer, and we have to determine the variation in time represented by a set of frequencies which can be complex and they are a solution of:

$$D_{\rm act}(k = \pi n/d,\omega) = 0\,. \tag{4.5.3}$$

We shall now include the role of reflections on the beam-wave interaction and in this process we generalize the formulation which will allow us to derive the operation of an amplifier or an oscillator from one set of equations.

4.5.1 Simplified Set of Equations

As in the previous case the starting point is Poynting's theorem

$$\nabla \cdot \mathbf{S} + \frac{\partial}{\partial t} W = -\mathbf{J} \cdot \mathbf{E} \,. \tag{4.5.4}$$

Assuming that the walls of the system are made of an ideal metal, all the power flux flows in the z direction therefore we can integrate over the cross-section of the system:

$$\frac{\partial}{\partial z}\left[2\pi \int_0^R \mathrm{d}r\, r\, S_z(r,z,t)\right] + \frac{\partial}{\partial t}\left[2\pi \int_0^R \mathrm{d}r\, r\, W(r,z,t)\right]$$
$$= -2\pi \int_0^R \mathrm{d}r\, r\, J_z(r,z,t)\, E_z(r,z,t) \,. \tag{4.5.5}$$

The first term,

$$P(z,t) = 2\pi \int_0^R \mathrm{d}r\, r\, S_z(r,z,t) \,, \tag{4.5.6}$$

is the total instantaneous power which flows in the system and

$$W(z,t) = 2\pi \int_0^R \mathrm{d}r\, r\, W(r,z,t) \,, \tag{4.5.7}$$

represents the total instantaneous energy per unit length stored in the system. As before, it is assumed that the oscillation is longitudinal and the transverse variations in the electric field are negligible on the scale of the beam thickness. Thus, for a thin pencil beam of radius R_b, the right-hand term of (4.5.5) reads

$$-2\pi\, E_z(R_\mathrm{b},z,t) \int_0^R \mathrm{d}r\, r\, J_z(r,z,t) \,. \tag{4.5.8}$$

Now, since the current density is given by

$$J_z(r,z,t) = -e \sum_i v_i(t)\delta\Big[z - z_i(t)\Big]\frac{1}{2\pi r}\delta\Big[r - r_i(t)\Big] \,, \tag{4.5.9}$$

the integration over the transverse coordinate becomes trivial by virtue of the Dirac delta function. As a result, (4.5.8) reads

$$eE_z(R_\mathrm{b},z,t)\sum_{i=1}^{N} v_i(t)\delta\Big[z - z_i(t)\Big] \,, \tag{4.5.10}$$

where it was tacitly assumed that there is no transverse motion. According to our assumptions and definitions so far, Poynting's theorem is given by

$$\frac{\partial}{\partial z}P(z,t) + \frac{\partial}{\partial t}W(z,t) = eE_z(R_\mathrm{b},z,t)\sum_{i=1}^{N} v_i(t)\delta\left[z - z_i(t)\right] . \tag{4.5.11}$$

When reflections are non-negligible, the longitudinal electric field which acts on the electrons has two components: one which is propagating parallel to the electrons and another which is propagating anti-parallel:

$$\begin{aligned} E_z(R_\mathrm{b},z,t) =& E_+(z,t)\cos[\omega t - kz - \psi_+(z,t)] \\ &+ E_-(z,t)\cos[\omega t + kz - \psi_-(z,t)] . \end{aligned} \tag{4.5.12}$$

It will be further assumed that the amplitudes and the phases (E_+, E_-, ψ_+ and ψ_-) vary slowly comparing to the trigonometric function i.e., :

$$\begin{aligned} &\left|\frac{\partial}{\partial t}E_\pm\right| \ll \omega\left|E_\pm\right| , \quad \left|\frac{\partial}{\partial z}E_\pm\right| \ll k\left|E_\pm\right| , \\ &\left|\frac{\partial}{\partial z}\psi_\pm\right| \ll k\left|\psi_\pm\right| , \quad \left|\frac{\partial}{\partial t}\psi_\pm\right| \ll \omega\left|\psi_\pm\right| . \end{aligned} \tag{4.5.13}$$

Among the two waves, only the one propagating parallel has an average net effect, therefore the right-hand side of (4.5.11) simplifies to

$$eE_+(z,t)\cos\Big[\omega t - kz - \psi_+(z,t)\Big]\sum_{i=1}^{N} v_i(t)\delta\Big[z - z_i(t)\Big] . \tag{4.5.14}$$

Without loss of generality we can use the trigonometric properties of the cos function to write

$$\begin{aligned} =& eNE_+(z,t)\cos\psi_+(z,t)\Big\langle \cos\chi_i(t)v_i(t)\delta\Big[z - z_i(t)\Big]\Big\rangle \\ &+ eNE_+(z,t)\sin\psi_+(z,t)\Big\langle \sin\chi_i(t)v_i(t)\delta\Big[z - z_i(t)\Big]\Big\rangle , \end{aligned} \tag{4.5.15}$$

where

$$\chi_i(t) = \omega t - kz_i(t) , \tag{4.5.16}$$

is the phase of the ith particle relative to the wave at the time t and $\langle \cdots \rangle = \frac{1}{N}\sum_{i=1}^{N}\cdots$. The notation in (4.5.15) indicates that if the particles move with a velocity which is close to the phase velocity of the wave, all quantities are slow varying functions of z and t. This is in contrast to the left-hand side of the expression for Poynting theorem which consists of both slow and fast terms. Next we shall eliminate the contribution of the fast oscillation.

The fast variations of the electromagnetic field are determined by the angular frequency $\omega = 2\pi/T$. We can use the definition of the interaction impedance to average out these fast variations in the total power i.e.,

$$\bar{P}(z,t) = \frac{1}{T}\int_{t-T/2}^{t+T/2} \mathrm{d}t' P(z,t') = \frac{1}{T}\int_{t-T/2}^{t+T/2} \mathrm{d}t' \frac{\pi R^2}{Z_\mathrm{int}} E_z^2(R_\mathrm{b},z,t') ; \tag{4.5.17}$$

the result is

$$\bar{P}(z,t) = \frac{\pi R^2}{2Z_{\rm int}} \left[E_+^2(z,t) - E_-^2(z,t) \right] . \tag{4.5.18}$$

The cross term proportional to $(E_+ E_-)$ was neglected since it varies rapidly in space. In addition we took into account the fact that the power carried by the backward wave is in the opposite direction to that of the forward propagating wave. In a similar way the average energy stored per unit length is given by

$$\bar{W}(z,t) = \frac{1}{2}\pi R^2 \varepsilon_0 \varepsilon_{\rm int} \left[E_+^2(z,t) + E_-^2(z,t) \right] . \tag{4.5.19}$$

Thus we can write for the slow varying components of Poynting's theorem

$$\begin{aligned} &\frac{\partial}{\partial z}\left[E_+^2(z,t) - E_-^2(z,t)\right] + \frac{1}{v_{\rm en}}\frac{\partial}{\partial t}\left[E_+^2(z,t) + E_-^2(z,t)\right] \\ &= \frac{2eNZ_{\rm int}}{\pi R^2}\left[E_+(z,t)\cos\psi_+(z,t)\left\langle \cos\chi_i(t) v_i(t)\delta[z - z_i(t)]\right\rangle\right] \\ &+ \frac{2eNZ_{\rm int}}{\pi R^2}\left[E_+(z,t)\sin\psi_+(z,t)\left\langle \sin\chi_i(t) v_i(t)\delta[z - z_i(t)]\right\rangle\right] . \end{aligned} \tag{4.5.20}$$

Here we used (2.3.33) which relates the interaction impedance with the interaction dielectric coefficient and the energy velocity: $Z_{\rm int}\varepsilon_{\rm int} = \eta_0/\beta_{\rm en}$.

Before we proceed and simplify the amplitude equation, it will be more convenient to use a normalized notation. For this purpose, we examine the single-particle energy conservation ignoring the effect of the backward wave on the motion of the electrons,

$$\frac{\rm d}{{\rm d}t}\gamma_i = -\frac{e}{mc^2} v_i(t) E_+[z = z_i(t), t] \cos\left[\omega t - k z_i(t) - \psi_+(z = z_i(t), t)\right] ; \tag{4.5.21}$$

this justifies the normalization of the electric field according to

$$a_\pm(\zeta,\tau) = \frac{eE_\pm(z,t)d}{mc^2} {\rm e}^{-j\psi_\pm(z,t)} , \tag{4.5.22}$$

of a new spatial variable $\zeta = z/d$ where d is the total length of the interaction region, of a new time variable $\tau = tc/d$ and of the coupling coefficient α as

$$\alpha = \frac{eIZ_{\rm int}}{mc^2}\frac{d^2}{\pi R^2} \tag{4.5.23}$$

where I is the average current in one period of the wave (N is the number of particles in one period of the wave). With this notation (4.5.20–21) read

$$\begin{aligned} &\frac{\partial}{\partial \zeta}\left[|a_+(\zeta,\tau)|^2 - |a_-(\zeta,\tau)|^2\right] + \frac{1}{\beta_{\rm en}}\frac{\partial}{\partial \tau}\left[|a_+(\zeta,\tau)|^2 + |a_-(\zeta,\tau)|^2\right] \\ &= \alpha\left[a_+(\zeta,\tau)\left\langle {\rm e}^{j\chi_i(\tau)}\beta_i(\tau)\delta[\zeta - \bar{z}_i(t)]\right\rangle + {\rm c.c.}\right] , \end{aligned} \tag{4.5.24}$$

$$\frac{\mathrm{d}}{\mathrm{d}\tau}\gamma_i = -\beta_i(\tau)\frac{1}{2}\left[a_+[\zeta = \bar{z}_i(\tau), \tau]\mathrm{e}^{j\chi_i(\tau)} + \mathrm{c.c.}\right], \tag{4.5.25}$$

and

$$\frac{\mathrm{d}}{\mathrm{d}\tau}\chi_i(\tau) = \Omega - K\beta_i(\tau), \tag{4.5.26}$$

is the phase equation where $\Omega = \omega d/c$ and $K = kd$.

The amplitudes of the forward wave $[a_+(\zeta, \tau)]$ and backward wave $[a_-(\zeta, \tau)]$ are correlated at both ends of the structure by the reflection process. In the interaction region itself the two amplitudes are not coupled since we indicated that the electrons are interacting only with the forward wave. Consequently, the energy conservation associated with this wave reads

$$\left[\frac{\partial}{\partial\zeta} - \frac{1}{\beta_{\mathrm{en}}}\frac{\partial}{\partial\tau}\right]|a_-(\zeta, \tau)|^2 = 0, \tag{4.5.27}$$

which in turn implies that the equation for the amplitude of the forward wave is given by

$$\frac{\partial}{\partial\zeta}a_+(\zeta, \tau) + \frac{1}{\beta_{\mathrm{en}}}\frac{\partial}{\partial\tau}a_+(\zeta, \tau) = \alpha\left\langle \mathrm{e}^{-j\chi_i(\tau)}\beta_i(\tau)\delta\left[\zeta - \bar{z}_i(t)\right]\right\rangle. \tag{4.5.28}$$

In order to determine the effect of reflections we denote by a_0 the amplitude of the forward wave present in the system in the absence of the beam. The reflection process is represented by a scalar reflection coefficient at the input (ϱ_{in}) and output (ϱ_{out}) end. At any instant τ, the *change* in the forward wave amplitude is reflected from the output end towards the input according to

$$\left[a_+(\zeta = 1, \tau) - a_0\right]\varrho_{\mathrm{out}}\mathrm{e}^{-jK}. \tag{4.5.29}$$

As we indicated, the backward wave is not directly affected by the beam and therefore it propagates towards the input as if no beam were present. The time it takes this change in the energy (variation in the amplitude) to reach the input end is determined by the energy velocity of the cold structure. In our normalized notation this delay is $1/\beta_{\mathrm{en}}$. Thus after this delay, the change mentioned above undergoes an additional reflection - this time from the input end. The contribution of the reflection to the amplitude of the forward wave at the input end ($\zeta = 0$) is given by

$$a_+(\zeta = 0, \tau) - a_0 = \bar{\varrho}\left[a_+(\zeta = 1, \tau - 1/\beta_{\mathrm{en}}) - a_0\right], \tag{4.5.30}$$

where $\bar{\varrho} \equiv \varrho_{\mathrm{in}}\varrho_{\mathrm{out}}\mathrm{e}^{-2jK}$ is the feedback term of the passive (no beam) system. Let us now summarize the generalized set of equations which describe the dynamics of the field and the electrons when reflections are included:

$$
\begin{aligned}
&\left[\frac{\partial}{\partial \zeta} + \frac{1}{\beta_{\text{en}}}\frac{\partial}{\partial \tau}\right] a_+(\zeta, \tau) = \alpha\left\langle \mathrm{e}^{-j\chi_i(\tau)}\beta_i(\tau)\delta[\zeta - \bar{z}_i(t)]\right\rangle, \\
&\frac{\mathrm{d}}{\mathrm{d}\tau}\gamma_i(\tau) = -\beta_i(\tau)\frac{1}{2}\left[a_+[\zeta = \bar{z}_i(\tau), \tau]\mathrm{e}^{j\chi_i(\tau)} + \text{c.c.}\right], \\
&\frac{\mathrm{d}}{\mathrm{d}\tau}\chi_i(\tau) = \Omega - K\beta_i(\tau), \\
&a_+(\zeta = 0, \tau) - a_0 = \bar{\varrho}\left[a_+(\zeta = 1, \tau - 1/\beta_{\text{en}}) - a_0\right].
\end{aligned}
\tag{4.5.31}
$$

From this generalized formulation we can obtain the equations which were developed in the previous sections for an ideal amplifier.

4.5.2 Ideal Amplifier

In an ideal amplifier we expect no reflections and thus no time variations, of the amplitude ($\partial/\partial\tau \sim 0$). The amplitude equation is averaged on time and since the integration over the Dirac delta function is straightforward we have

$$
\frac{\mathrm{d}}{\mathrm{d}\zeta}a_+(\zeta) = \alpha\langle \mathrm{e}^{-j\chi_i(\zeta)}\rangle . \tag{4.5.32}
$$

Since no time variations are assumed, then according to the definition of $\mathrm{d}/\mathrm{d}\tau(\equiv \partial/\partial\tau + \beta_i\partial/\partial\zeta \sim \beta_i\partial/\partial\zeta = \beta_i\mathrm{d}/\mathrm{d}\zeta)$ it is sufficient to describe only the space variation and in this framework $\beta_i(\zeta)$ represents the velocity of the ith electron at ζ. Consequently, the equation of motion reads

$$
\frac{\mathrm{d}}{\mathrm{d}\zeta}\gamma_i = -\frac{1}{2}\left[a_+(\zeta)\mathrm{e}^{j\chi_i(\zeta)} + \text{c.c.}\right] . \tag{4.5.33}
$$

In a similar way

$$
\frac{\mathrm{d}}{\mathrm{d}\zeta}\chi_i(\zeta) = \frac{\Omega}{\beta_i} - K , \tag{4.5.34}
$$

and finally the reflections equation is identically satisfied since there are no reflections thus $\bar{\varrho} = 0$ and the amplitude at the input is always the same and it equals a_0.

4.5.3 Ideal Oscillator

For an ideal oscillator it is assumed that no variations in space occur and therefore the amplitude of the electric field experienced by the electrons does not depend on the location of any individual electron. As a result, we replace $a_+(\bar{z}_i(\tau), \tau)$, in the single particle energy conservation (4.5.31), with its value at the input - $a_+(0, \tau)$ - hence

$$
\frac{\mathrm{d}}{\mathrm{d}\tau}\gamma_i = -\beta_i(\tau)\frac{1}{2}\left[a_+(0, \tau)\mathrm{e}^{j\chi_i(\tau)} + \text{c.c.}\right] . \tag{4.5.35}
$$

The reflection coefficients from both ends are *unity* and the boundary conditions imply $K = \pi n$. As a result, $\bar{\varrho} = 1$ and the reflections equation (4.5.31) reads

$$a_+(0,\tau) = a_+(1, \tau - 1/\beta_{\rm en}) . \tag{4.5.36}$$

In order to determine the dynamics of the amplitude in an oscillator we average the amplitude equation [in (4.5.31)] over the interaction region:

$$\frac{1}{\beta_{\rm en}}\frac{\rm d}{{\rm d}\tau}a_+(0,\tau) = \alpha\Big\langle \beta_i(\tau){\rm e}^{-j\chi_i(\tau)}\Big\rangle , \tag{4.5.37}$$

and as in the case of an ideal amplifier, we calculate the normalized growth rate. For this purpose we take twice the derivative of (4.5.37), neglect terms which oscillate rapidly and obtain the following expression for $a_+(0,\tau)$

$$\begin{aligned} &\frac{{\rm d}^3 a_+}{{\rm d}\tau^3} + \frac{1}{2}\alpha'\left\langle \frac{1}{\gamma_i^3}\right\rangle \frac{{\rm d}a_+}{{\rm d}\tau} - \frac{1}{2}\alpha'\left\langle \frac{1}{\gamma_i^3}(\Omega - K\beta_i)\right\rangle a_+ \\ &\quad + \frac{1}{2}j\alpha' K\left\langle \frac{\beta_i}{\gamma_i^3}\right\rangle a_+ - \frac{3}{4}\alpha' a_+\left[\left\langle \frac{\beta_i}{\gamma_i^3}\,{\rm e}^{j\chi_i}\right\rangle a_+ + \left\langle \frac{\beta_i}{\gamma_i^3}\,{\rm e}^{-j\chi_i}\right\rangle a_+^*\right] \\ &= -\,\alpha'\langle \beta_i(\Omega - K\beta_i)^2\,{\rm e}^{-j\chi_i}\rangle , \end{aligned} \tag{4.5.38}$$

where $\alpha' = \alpha\beta_{\rm en}$. A simple evaluation of the growth rate in the linear regime of operation is possible by ignoring the fifth term (since it is non-linear); near resonance, where we expect the growth rate to be maximum, the third term is much smaller than the fourth. Finally, for relativistic electrons and long interaction length (such that $K \gg 1$) the second term is negligible relative to the fourth, therefore we have

$$\left[\frac{{\rm d}^3}{{\rm d}\tau^3} + \frac{1}{2}j\alpha' K\langle\frac{\beta_i}{\gamma_i^3}\rangle\right] a_+ = -\alpha'\left\langle \beta_i(\Omega - K\beta_i)^2\,{\rm e}^{-j\chi_i}\right\rangle . \tag{4.5.39}$$

Assuming that $\langle\frac{\beta_i}{\gamma_i^3}\rangle$ does not vary significantly in time, the normalized growth rate is

$$\bar{\omega} = \frac{\sqrt{3}}{2}\left[\frac{1}{2}\alpha\beta_{\rm en}K\left\langle\frac{\beta_i}{\gamma_i^3}\right\rangle\right]^{1/3} . \tag{4.5.40}$$

This expression is identical to that calculated in Sect. 4.2.1 developed using the hydrodynamic approximation.

4.5.4 Global Energy Conservation

Global energy conservation is obtained by multiplying the equation of motion (in (4.5.31)) by $\delta[\zeta-\bar{z}_i(\tau)]$ and averaging over the entire ensemble of particles. In the resulting expression,

$$\begin{aligned}&\left\langle \delta\Big[\zeta-\bar{z}_i(\tau)\Big]\frac{\mathrm{d}}{\mathrm{d}\tau}\gamma_i\right\rangle\\ &\quad= -\frac{1}{2}\Big[a_+[\zeta=\bar{z}_i(\tau),\tau]\Big\langle \beta_i(\tau)\mathrm{e}^{j\chi_i(\tau)}\delta[\zeta-\bar{z}_i(\tau)]\Big\rangle + \mathrm{c.c.}\Big]\,, \qquad (4.5.41)\end{aligned}$$

we substitute the equation for the amplitude (4.5.31) and get

$$\begin{aligned}&\left\langle \delta\Big[\zeta-\bar{z}_i(\tau)\Big]\frac{\mathrm{d}}{\mathrm{d}\tau}\gamma_i(\tau)\right\rangle\\ &\quad= -\frac{1}{2\alpha}\left[\frac{\partial}{\partial\zeta}|a_+(\zeta,\tau)|^2+\frac{1}{\beta_{\mathrm{en}}}\frac{\partial}{\partial\tau}|a_+(\zeta,\tau)|^2\right]. \qquad (4.5.42)\end{aligned}$$

In order to bring this last equation to a more familiar form we note that the left-hand side term, after differentiation by parts, reads

$$\frac{\mathrm{d}}{\mathrm{d}\tau}\Big\langle \delta[\zeta-\bar{z}_i(\tau)]\gamma_i\Big\rangle - \left\langle \gamma_i\frac{\mathrm{d}}{\mathrm{d}\tau}\delta[\zeta-\bar{z}_i(\tau)]\right\rangle, \qquad (4.5.43)$$

and the last term is zero by virtue of the definition of $\mathrm{d}/\mathrm{d}\tau$ namely

$$\begin{aligned}\frac{\mathrm{d}}{\mathrm{d}\tau}\delta[\zeta-\bar{z}_i(\tau)] &= \frac{\partial}{\partial\tau}\delta[\zeta-\bar{z}_i(\tau)]+\beta_i\frac{\partial}{\partial\zeta}\delta[\zeta-\bar{z}_i(\tau)]\\ &= \frac{\partial}{\partial\tau}\delta[\zeta-\bar{z}_i(\tau)]-\frac{\partial}{\partial\tau}\delta[\zeta-\bar{z}_i(\tau)]\\ &= 0\,. \qquad (4.5.44)\end{aligned}$$

Using the same definition we can write for (4.5.43)

$$\frac{\partial}{\partial\tau}\Big\langle \gamma_i\delta[\zeta-\bar{z}_i(\tau)]\Big\rangle + \frac{\partial}{\partial\zeta}\Big\langle \beta_i\gamma_i\delta[\zeta-\bar{z}_i(\tau)]\Big\rangle, \qquad (4.5.45)$$

which finally allows us to present (4.5.42) in the familiar form of a conservation law i.e.,

$$\begin{aligned}&\frac{\partial}{\partial\tau}\left[\Big\langle \gamma_i\delta[\zeta-\bar{z}_i(\tau)]\Big\rangle + \frac{1}{2\alpha\beta_{\mathrm{en}}}|a_+(\zeta,\tau)|^2\right]\\ &\quad+\frac{\partial}{\partial\zeta}\left[\Big\langle \beta_i\gamma_i\delta[\zeta-\bar{z}_i(\tau)]\Big\rangle + \frac{1}{2\alpha}|a_+(\zeta,\tau)|^2\right] = 0\,. \qquad (4.5.46)\end{aligned}$$

As in the hydrodynamic approximation, we can identify the average energy of an electron and its energy flux ($\langle\gamma_i\delta[\zeta-\bar{z}_i]\rangle$, $\langle\beta_i\gamma_i\delta[\zeta-\bar{z}_i]\rangle$ correspondingly) as well as the normalized electromagnetic power per particle $|a_+|^2/2\alpha$ and the normalized electromagnetic energy per particle $|a_+|^2/2\alpha\beta_{\mathrm{en}}$.

4.5.5 Reflections in an Amplifier

There are two processes which may cause significant time variations in an amplifier: saturation and reflections. Saturation occurs when the initial amplitude of the radiation field is large; we shall not consider here variation in time caused by saturation without reflections involved. An amplifier is generally designed to operate below the saturation level. However, if in the design process reflections are disregarded, then until the first reflection reaches the input the system will probably operate as designed. But as the first reflection adds to the initial amplitude it may bring the system to saturation.

In this sub-section we shall investigate the variation in time caused by reflections in an amplifier. The process is as follows: before the electron beam is injected in the structure the amplitude of the forward propagating wave is uniform in space and constant in time. Let us denote it by a_0. Ignoring effects associated with the pulse front, we may expect this amplitude to be amplified according to the equations determined previously for an ideal amplifier. The change in this amplitude is propagating with the energy velocity $v_{\rm en} \equiv c\beta_{\rm en}$ so it will take $t_{\rm fb} = d/v_{\rm en}$ to the amplified field to approach the input end. We denote by g_1 the first one-pass gain which is the ratio of the amplitude at the output of the interaction region to the input point. The contribution of reflections at the input will be denoted by b_ν – where ν is the index that numerates the bouncing process therefore it can be considered as a discrete (normalized) time variable. During the first period, the reflections have no contribution therefore $b_0 = 0$. The amplitude at $\zeta = 1$ is $g_1(a_0 + b_0)\mathrm{e}^{-jkd}$. Without the beam present the amplitude at this point is $a_0\mathrm{e}^{-jkd}$. Therefore only the difference is reflected. After an additional reflection from the input end we may write the contribution of the first reflection to the amplitude at the input as $b_1 = \bar{\varrho}\,[g_1(a_0 + b_0) - a_0]$. Before this reflection arrives ($t < 2t_{\rm fb}$) the amplitude at the input is constant and its value is a_0. Until the next reflection arrives the amplitude at the input has two contributions which are constant in time, one from the generator and the other from the reflection, i.e., $a_0 + b_1$.

After the ν'th reflection the amplitude at the input is $a_0 + b_\nu$ and at the output end $g_{\nu+1}(a_0+b_\nu)\mathrm{e}^{-jkd}$. As a result, the contribution of the reflections to the input amplitude after $\nu + 1$ steps is

$$b_{\nu+1} = \bar{\varrho}\,[g_{\nu+1}(a_0 + b_\nu) - a_0]\;. \tag{4.5.47}$$

This expression is a discrete formulation of the reflections equation introduced in (4.5.31); the process is summarized in Fig. 4.16. Note that at the limit of a very long pulse and a linear gain such that $g_\nu = g$ for any ν, we have for $b_\nu = a_0\bar{\varrho}(g - 1)/(1 - \bar{\varrho}g)$. It implies that the amplitude at the input reads

$$a_0 + b_\nu = a_0\frac{1 - \bar{\varrho}}{1 - \bar{\varrho}g}\,, \tag{4.5.48}$$

exactly as predicted by a linear (steady state) theory.

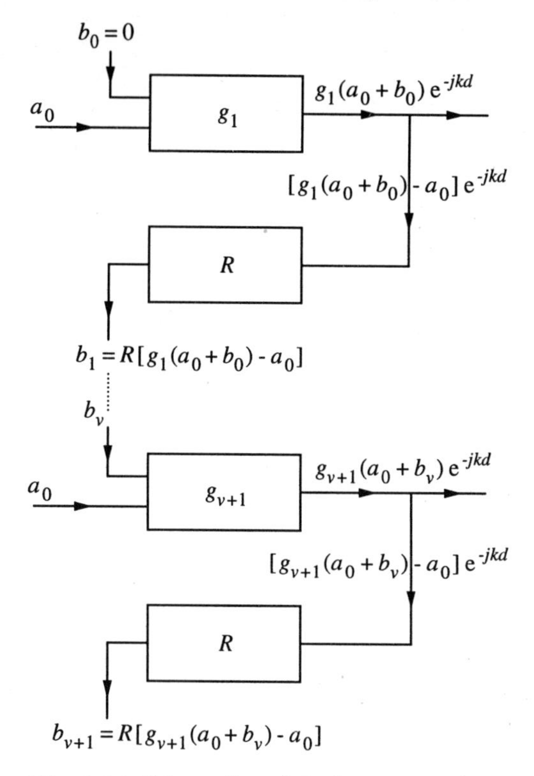

Fig. 4.16. Schematics of the beam-wave interaction in the presence of reflections

We shall consider now a simple system within the framework of the present analysis. For this purpose we examine an amplifier which is 20 cm long, operates at 8.8 GHz, driven by a 800 kV, 1 kA beam. In the absence of reflections the gain of the system is 32 dB for 80 kW at the input. Without loss of generality we chose both the reflection coefficients to be equal $\varrho_{\rm in} = \varrho_{\rm out} = \varrho$. The electrons' pulse is 100 ns long. In Fig. 4.17 we can see how the one-pass gain (squares) and the total gain (circles) are varying in time (ν indicating the index of the reflection i.e., $\nu = 1$ is reflection number one etc.). The total gain is the ratio between the accumulated amplitude, of the forward wave, at the output and the initial amplitude (before the beam was injected) at the input of the interaction region. For a small reflection coefficient, $\varrho = 0.1$, we observe that both gains are relatively stable. The fact that the total gain is smaller than the one-pass gain is not of particular significance at this point since this depends on the phase accumulated by the wave in its round trip. However, as the reflections are increased, the total amplitude at the input increases, saturation is reached and therefore the one-pass gain is systematically smaller than the total gain. There exists an intermediary point, $\varrho = 0.5$, where the system acts very unstable whereas at another, $\varrho = 0.7$, the system appears to be very stable in spite of the fact that the reflection is higher. This is a

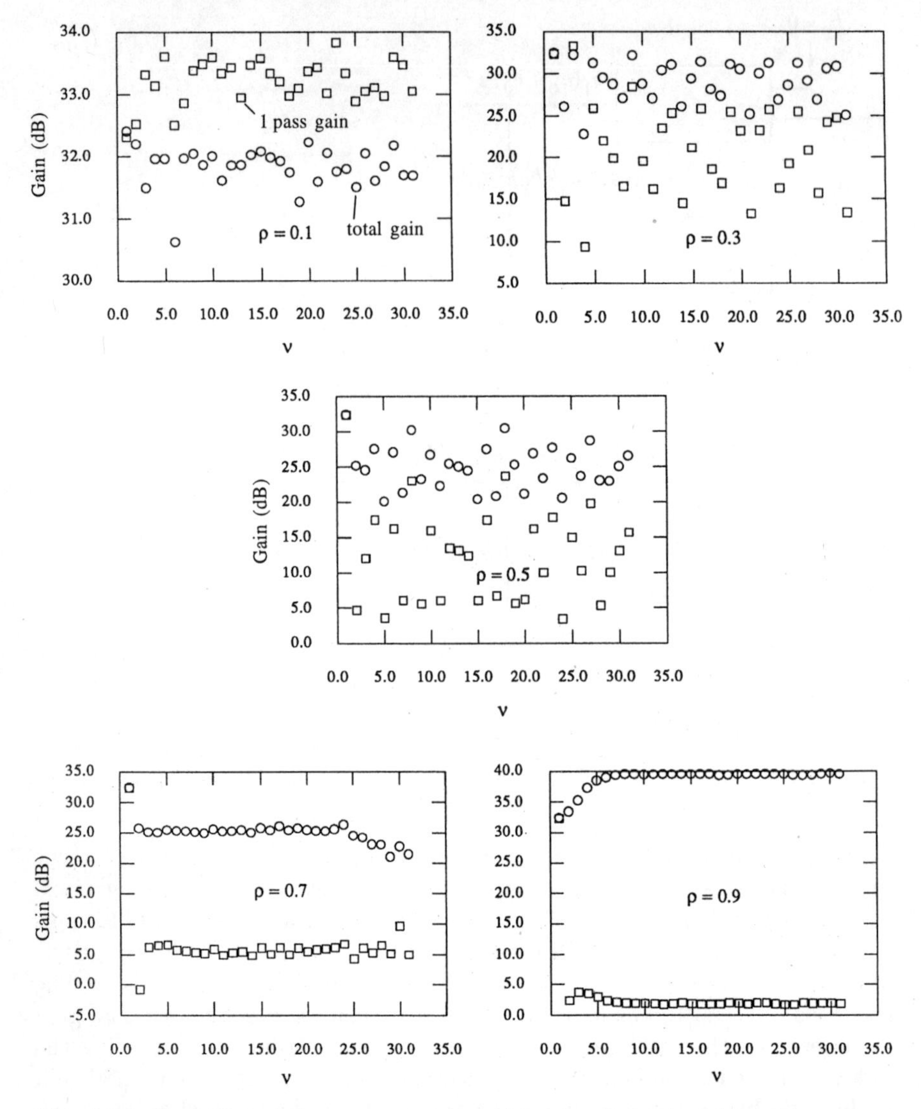

Fig. 4.17. Evolution of the one-pass gain and total gain in the interaction region

direct result of the phase dependence of the reflected amplitude. Ultimately at high reflection ($\varrho = 0.9$) the system reveals an immediate increase of the amplitude in time associated with practically zero one-pass gain, indicating that the system is operating as an oscillator. Note that whatever the reflection coefficient was, before the first reflection arrives, the one-pass gain and the total gain are equal.

In order to show the general influence of the reflection coefficient on the total gain and the one-pass gain we have averaged out these two quantities over the entire number of reflections for different values of the reflection coefficient. Figure 4.18 illustrates this result. We observe here that the average one-pass gain is monotonically decreasing when increasing the reflection coefficient. The average total gain is stable for small ϱ corresponding to a linear regime of operation; it slightly decreases for intermediary reflections – corresponding to saturation and it increases again when the reflection is so high that the system practically operates as an oscillator. Note that in this case the one-pass gain is practically zero.

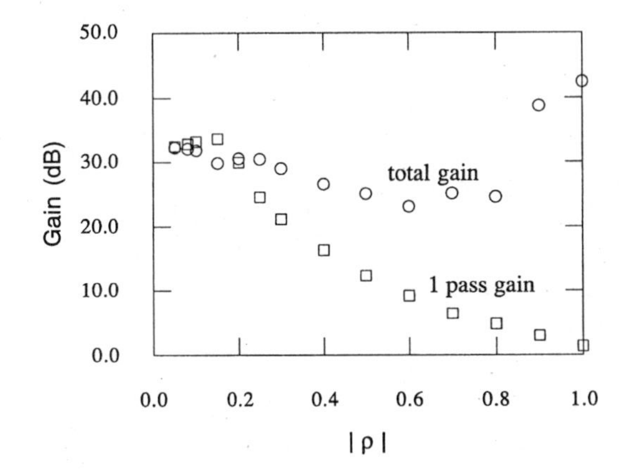

Fig. 4.18. Average one-pass gain and total gain as a function of the reflection coefficient

An additional insight of the physical process can be achieved by examining the spectrum of the signal as illustrated in Figs. 4.19–20. The power in each frequency component of the signal is normalized to the power in the central frequency (8.8 GHz). When the reflection is low ($\varrho < 0.15$) the power in all the other frequencies is 30 dB below the level of the main signal. For $\varrho = 0.2$ the eigen-frequencies of the "oscillator" are less than 15 dB below the central frequency. The power in the sidebands is increasing monotonically with the reflection coefficient ϱ, and at $\varrho = 0.4$ they dominate.

4.5.6 Spatial Variations in an Oscillator

Part of the energy in an oscillator is extracted by making the mirror(s) of a reflection coefficient smaller than unity. As a result, the amount of electromagnetic energy available for interaction with the electrons decreases. Since this power is extracted at the ends, it is revealed as an effective variation of the field amplitude. In order to illustrate the effect of the spatial variation on the operation of an oscillator we start by integrating the equation which describes the dynamics of the amplitude (4.5.24) over the entire length of the oscillator:

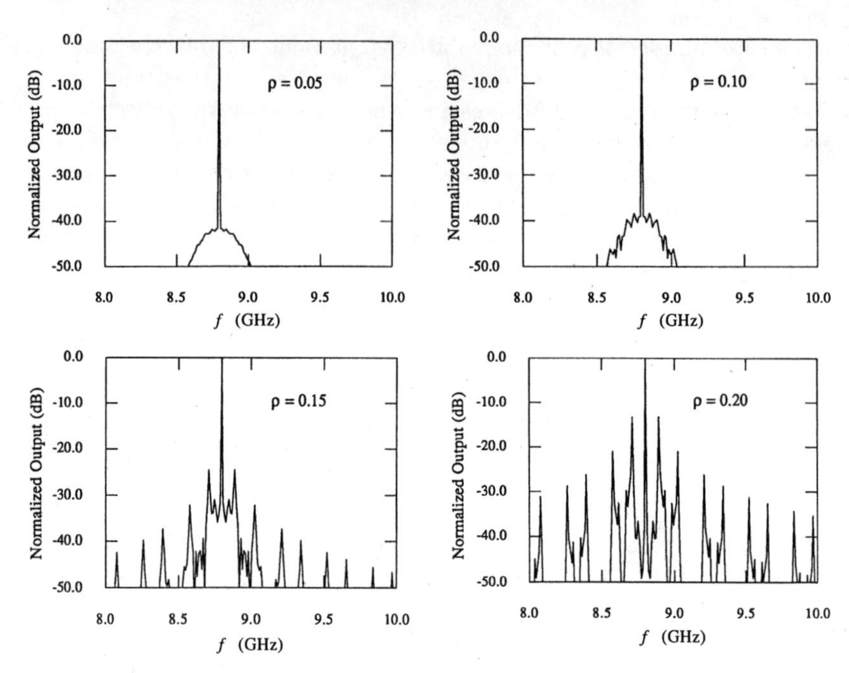

Fig. 4.19. Normalized output power spectrum for various reflection coefficients

$$\frac{\mathrm{d}}{\mathrm{d}\tau}|a_+(1,\tau)|^2 + \beta_{\mathrm{en}}\left[|a_+(1,\tau)|^2 - |a_+(0,\tau)|^2\right] = \alpha\beta_{\mathrm{en}}\langle\beta_i(\tau)\mathrm{e}^{-j\chi_i(\tau)}\rangle\,. \tag{4.5.49}$$

Next we substitute the reflections equation from (4.5.31). The result is

$$\frac{\mathrm{d}}{\mathrm{d}\tau}|a_+(1,\tau)|^2 + \beta_{\mathrm{en}}\left[|a_+(1,\tau)|^2 - |a_0(1-\bar{\varrho}) + \bar{\varrho}a_+(1,\tau-1/\beta_{\mathrm{en}})|^2\right]$$
$$= \alpha\tau_o\beta_{\mathrm{en}}\left\langle\beta_i(\tau)\mathrm{e}^{-j\chi_i(\tau)}\right\rangle\,. \tag{4.5.50}$$

Expanding in Taylor series with respect to $1/\beta_{\mathrm{en}}$ (this normalized characteristic time is assumed to be much shorter than the pulse duration) and assuming that $a_0 = 0$ we finally get

$$\left[\frac{\mathrm{d}}{\mathrm{d}\tau} + \beta_{\mathrm{en}}\frac{1-|\bar{\varrho}|^2}{1+|\bar{\varrho}|^2}\right]a_+(1,\tau) = \frac{\alpha\beta_{\mathrm{en}}}{1+|\bar{\varrho}|^2}\left\langle\beta_i(\tau)\mathrm{e}^{-j\chi_i(\tau)}\right\rangle\,. \tag{4.5.51}$$

This expression replaces (4.5.37) in the description of a non-ideal oscillator. Note that the second term on the left-hand side of (4.5.51) represents the "radiation" loss due to the finite transmission from both ends of the oscillator. This becomes even more evident from the expression for general energy conservation

$$\frac{\mathrm{d}}{\mathrm{d}\tau}\left[\langle\gamma_i\rangle + \frac{1}{2\alpha\beta_{\mathrm{en}}}(1+|\bar{\varrho}|^2)|a_+(1,\tau)|^2\right] = -\frac{1}{\alpha}(1-|\bar{\varrho}|^2)|a_+(1,\tau)|^2 \tag{4.5.52}$$

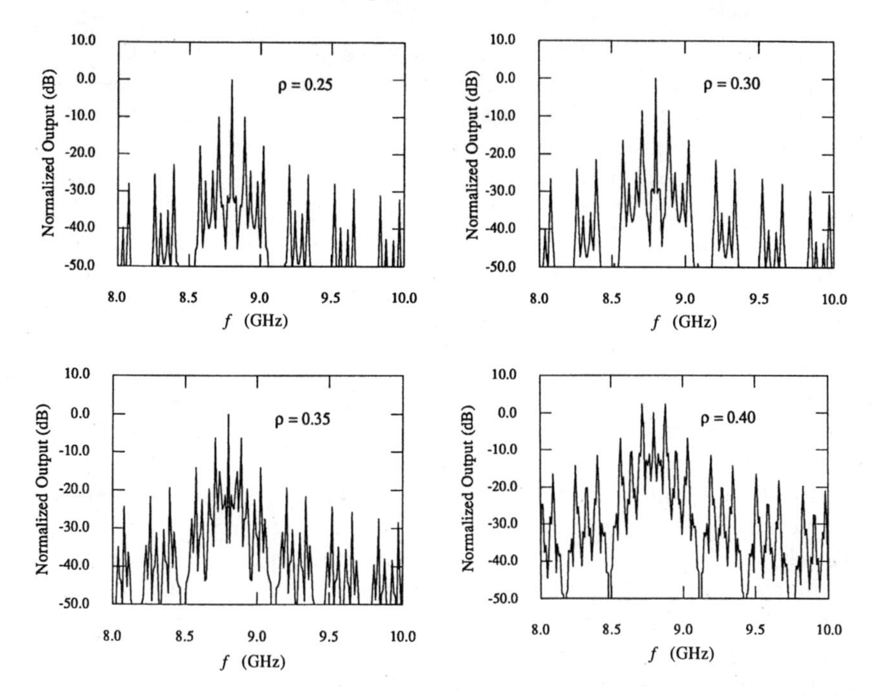

Fig. 4.20. Normalized output power spectrum for various reflection coefficients

as revealed by the right-hand side term.

The only source of energy in the oscillator is the beam and when the mirrors are ideal, all the kinetic energy converted in radiation power is confined to the volume of the oscillator. If part of this energy is allowed to flow out, then self-sustained oscillation is possible only if the current injected is above a threshold value which depends on the reflection coefficients. In order to determine the threshold current we first have to realize that the radiation loss is associated with an exponential decay with a coefficient (see (4.5.51)) $\beta_{\rm en}(1-|\bar{\varrho}|^2)/(1+|\bar{\varrho}|^2)$. For self-sustained oscillation this decay has to be compensated by the exponential increase due to the interaction – as determined in (4.5.40), i.e.,

$$\beta_{\rm en}\frac{1-|\bar{\varrho}|^2}{1+|\bar{\varrho}|^2} < \bar{\omega} = \frac{\sqrt{3}}{2}\left[\frac{1}{2}\alpha\beta_{\rm en}K\langle\frac{\beta_i}{\gamma_i^3}\rangle\right]^{1/3} . \tag{4.5.53}$$

Therefore, the condition for self-sustained oscillation can be formulated as

$$I > I_{\rm th} \equiv \frac{16}{3^{3/2}}\frac{mc^2}{eZ_{\rm int}\Omega}{\beta_{\rm en}}^2\beta^2\gamma^3\frac{\pi R^2}{d^2}\left[\frac{1-|\bar{\varrho}|^2}{1+|\bar{\varrho}|^2}\right]^3 . \tag{4.5.54}$$

Note that in case of "radiation losses" the threshold current is quadratic in the energy velocity, therefore the lower $\beta_{\rm en}$, the lower the current required for the system to oscillate.

Energy extracted from both ends is one mechanism responsible for spatial variation but it is not the only one. Another mechanism is associated with the fact that electrons entering the oscillator are unbunched and their build-up into bunches is not "immediate" in space but it takes some portion of the interaction length. After this transient region there will be no variations in space, provided that the system does not reach saturation - which will be not considered here. In order to illustrate this effect, we examine the same system as in the case of the amplifier; in this case however the input power $P_{\rm in}$ is zero, the pulse length is 50 nsec instead of 100 nsec in the amplifier and the "mirrors" at both ends have a reflection coefficient $\varrho = 0.9$. The entire pulse was assumed to consist of 35,000 macro-particles, 512 of those being at any time in the oscillator. In Fig. 4.21 we illustrate the phase space of these electrons which are in the interaction region. In the first 20% of the pulse duration there is not sufficient electromagnetic field built in the oscillator in order to affect significantly the electron's distribution (although there is a small increase in the momentum spread). After 40% of the pulse has passed, we clearly see the spatial transient in the interaction region. At this point in time the constant amplitude regime is achieved after about 20% of the total interaction length. The normalized momentum spread which at the beginning is less than 0.06 is now larger than 0.35. Later the bunches continue to grow – the momentum spread is further increased approaching 3 at the end of the electrons' pulse.

Before we conclude we wish to emphasize the difference between the two transients which occur in an oscillator. One is the temporal lethargy which we have discussed already and it is indicative of the time it takes for the exponential growth of the electromagnetic energy to become dominant. However the transient presented in Fig. 4.21 is a *spatial transient in an oscillator*. It is not a result of the three eigen-modes mentioned above since in an (ideal) oscillator these modes have a constant amplitude in space. As we mentioned above, this is a result of the finite length it takes the radiation field to bunch the "fresh" electrons.

The last two sub-sections indicate that the convenient picture of a traveling-wave tube operating either as an amplifier or as an oscillator is too simplistic. In fact we have shown that these two regimes are the extreme cases and any system operates somewhere in between corresponding to the reflection coefficients at both ends, the phase accumulated in one round trip and the gain. Furthermore, in the absence of reflections and saturation in an amplifier it is justified to assume that the amplitude of the electromagnetic wave remains constant in time. However, even a low reflection coefficient may affect the performance of an amplifier if the gain is high enough. When reflections were included in the analysis, the amplitude was shown to vary in time. The resulting spectrum revealed peaks at other frequencies. These peaks are symmetric to the initial frequency and their separation is determined by the feedback time, namely the energy velocity.

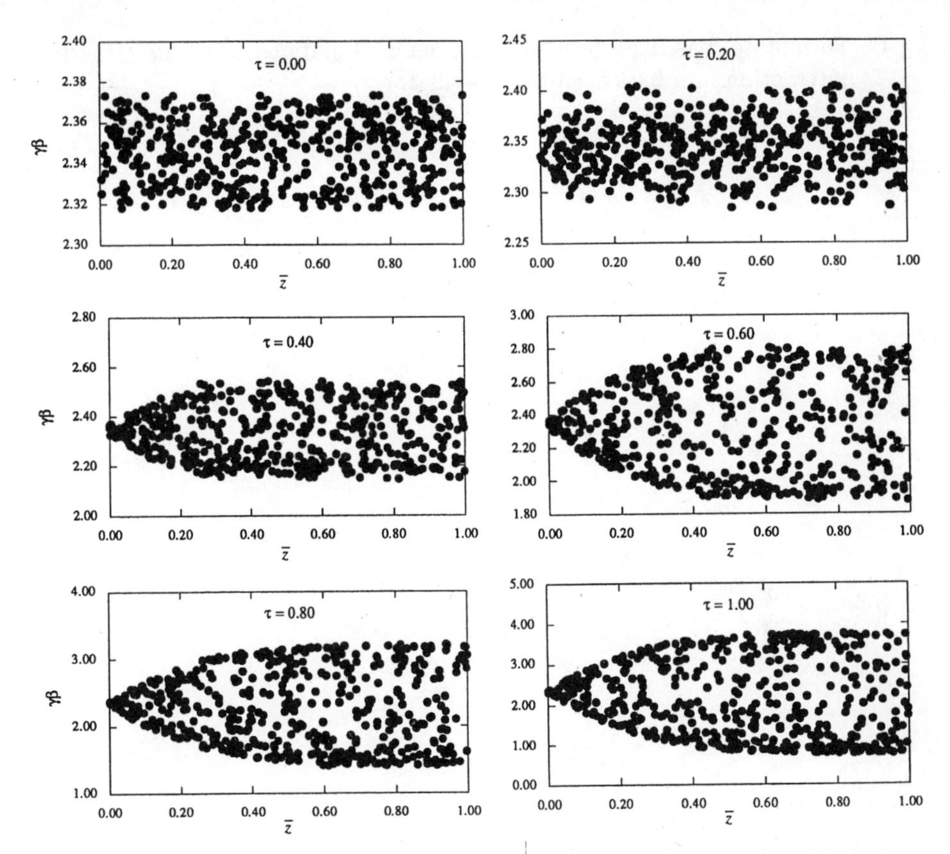

Fig. 4.21. Phase-space distribution in an oscillator at different instants of time

Exercises

4.1 Show that the modes in a partially dielectric loaded waveguide form a (complete) orthogonal set of functions.

4.2 Calculate the interaction impedance, as defined in (2.3.29), in a partially loaded waveguide for two cases: *(i)* pencil and *(ii)* annular beam. Express the coupling coefficient in terms of this impedance.

4.3 Develop the amplitude equation (as in Sect. 4.3) using in one case, the non-homogeneous wave equation for the magnetic vector potential and in the second case, wave equation for E_z.

4.4 Formulate the set of equations which determine the beam-wave interaction in the framework of the macro-particle approach of an annular beam in a slow-wave structure – Sect. 4.3.1.

4.5 Formulate the equations of the beam-wave interaction as in Sect. 4.4 assuming that the cross-section area of interaction may vary (adiabatically) in the longitudinal direction.

4.6 Formulate, based on Sect. 4.5, the set of equations which describe the interaction in a backward-wave oscillator.

5. Periodic Structures

One of the conditions for distributed beam-wave interaction to occur is phase velocity smaller than c. It is possible to satisfy this condition by loading the waveguide with a dielectric material, however the problem with dielectrics is that they are susceptible to breakdown at relatively low electric fields therefore it is difficult to generate high power microwave radiation with regular materials. Another difficulty is the electric charge which accumulates on the surface and must be drained for adequate operation. Consequently, in many microwave sources metallic structures with periodic boundary are preferred as the central component which facilitates the interaction. The periodic geometry can be conceived as a set of obstacles delaying the propagation of the wave due to the multi-reflection process and, as will be shown during our discussion on Floquet's theorem, an infinite spectrum of spatial harmonics develops. A few of these harmonics may propagate with a phase velocity larger or equal to c but the absolute majority have a smaller phase velocity.

This chapter presents various characteristics of periodic structures with emphasis on these aspects relevant to interaction with electrons. In particular, the interaction impedance, $Z_{\rm int}$, and the interaction dielectric coefficient, $\varepsilon_{\rm int}$, are calculated and analyzed since in the previous chapter we have shown that they play an important role in the collective beam-wave interaction. It will be assumed that only a single mode participates in the interaction and from the infinite spectrum of spatial harmonics of this single mode, only one harmonic interacts directly with the electrons. Since our treatment of periodic structures is limited to the objectives of the above, we refer the reader to Elachi (1976) for a broader review on periodic structures. Tutorial discussion of this subject can be found in a book by Brillouin (1953) and aspects associated with solid state physics are presented by Kittel (1956) or Ashcroft and Mermin (1976).

In the first section we present the basic theorem of periodic structures namely, Floquet's theorem. This is followed by an investigation of closed periodic structures in Sect. 5.2 and open structures in the third. Smith-Purcell effect is considered as a particular case of a Green's function calculation for an open structure and a simple scattering problem is also considered. The chapter concludes by presenting a simple transient solution in a periodic structure which is of importance in accelerators where wake fields left behind one bunch may affect trailing bunches.

5.1 The Floquet Theorem

A periodic function, $f(z)$, is a function whose value at a given point z is equal to its value at a point $z + L$ i.e.,

$$f(z) = f(z + L), \tag{5.1.1}$$

where L is the periodicity of the function. Any periodic function can be represented as a series of trigonometric functions $\exp(-j2\pi nz/L)$ and since this is an orthogonal and complete set of functions, it implies

$$f(z) = \sum_{n=-\infty}^{\infty} f_n e^{-j2\pi nz/L}. \tag{5.1.2}$$

The amplitudes f_n are determined by the value of the function $f(z)$ in a *single* cell. Specifically, we multiply (5.1.2) by $e^{j2\pi mz/L}$ and integrate over one cell i.e.,

$$\int_0^L dz f(z) e^{j2\pi mz/L} = \int_0^L dz e^{j2\pi mz/L} \sum_{n=-\infty}^{\infty} f_n e^{-j2\pi nz/L}. \tag{5.1.3}$$

Using the orthogonality of the trigonometric function we have

$$f_m = \frac{1}{L}\int_0^L dz f(z) e^{j2\pi mz/L}. \tag{5.1.4}$$

This presentation is called the Fourier series representation and it is valid for a *static* phenomenon in the sense that the value of $f(z)$ at the same relative location in two different cells is identical. It can not describe a propagation phenomenon, thus it can not represent a *dynamic* system. In the latter case the function $f(z)$ has to satisfy

$$f(z) = \xi f(z + L), \tag{5.1.5}$$

which means that the value of the function is proportional to the value of the function in the adjacent cell up to a constant, ξ, whose absolute value has to be unity otherwise at $z \to \pm\infty$ the function diverges or is zero as can be concluded from

$$f(z) = \xi^n f(z + nL), \tag{5.1.6}$$

where n is an arbitrary integer. Consequently, the coefficient ξ can be represented as a phase term of the form $\xi = \exp(j\psi)$ hence

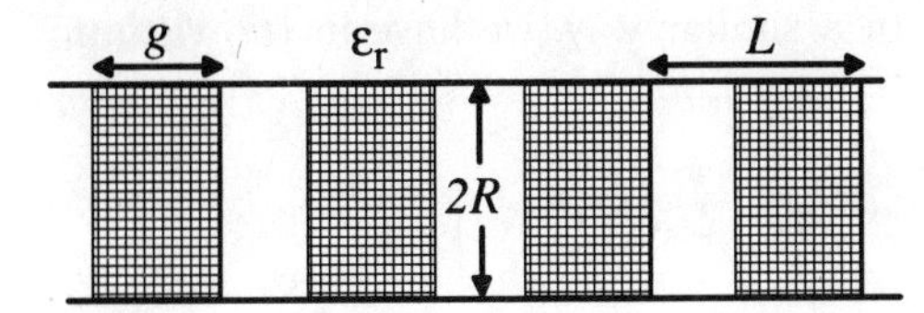

Fig. 5.1. Periodically loaded waveguide

$$f(z) = e^{j\psi} f(z+L)\,; \tag{5.1.7}$$

ψ is also referred to as the phase advance per cell. Without loss of generality one can redefine this phase to read $\psi = kL$. Since *a-priori* we do not know ψ, this definition does not change the information available. Nonetheless based on the Fourier series in (5.1.2) we can generalize the representation of a dynamic function in a periodic structure to

$$f(z) = \sum_{n=-\infty}^{\infty} f_n e^{-j2\pi nz/L} e^{-jkz}\,, \tag{5.1.8}$$

and realize that it satisfies

$$f(z) = e^{jkL} f(z+L)\,, \tag{5.1.9}$$

which is identical with the expression in (5.1.7). The last two expressions are different representations of the so-called *Floquet's Theorem.* Later we shall mainly use the form presented in (5.1.8), however in order to illustrate the use of Floquet's theorem in its latter representation, we investigate next the propagation of a TM wave in a periodically loaded waveguide.

Let us consider a waveguide of radius R which is loaded with dielectric layers: a representative cell ($0 \le z \le L$) consists of a region, $0 \le z \le g$, filled with a dielectric, ε_r, and the remainder is vacuum – see Fig. 5.1. We shall determine the dispersion relation of this structure and for this purpose we write the solution of the magnetic vector potential and electromagnetic field (steady state) in the dielectric ($0 \le z \le g$):

$$\begin{aligned}
A_z(r,z) &= \sum_{s=1}^{\infty} \mathrm{J}_0\left(p_s \frac{r}{R}\right)\left[A_s e^{-\Gamma_{d,s}z} + B_s e^{+\Gamma_{d,s}z}\right], \\
E_r(r,z) &= \frac{c^2}{j\omega\varepsilon_r} \sum_{s=1}^{\infty} \frac{p_s}{R}\Gamma_{d,s}\mathrm{J}_1\left(p_s \frac{r}{R}\right)\left[A_s e^{-\Gamma_{d,s}z} - B_s e^{+\Gamma_{d,s}z}\right], \\
E_z(r,z) &= -\frac{c^2}{j\omega\varepsilon_r} \sum_{s=1}^{\infty} \Gamma_{d,s}^2 \mathrm{J}_0\left(p_s \frac{r}{R}\right)\left[A_s e^{-\Gamma_{d,s}z} - B_s e^{+\Gamma_{d,s}z}\right], \\
H_\phi(r,z) &= \frac{1}{\mu_0} \sum_{s=1}^{\infty} \frac{p_s}{R}\mathrm{J}_1\left(p_s \frac{r}{R}\right)\left[A_s e^{-\Gamma_{d,s}z} + B_s e^{+\Gamma_{d,s}z}\right],
\end{aligned} \tag{5.1.10}$$

where $\Gamma_{d,s}^2 = (p_s/R)^2 - \varepsilon_r(\omega/c)^2$. In a similar way, we have in the vacuum $(g < z < L)$:

$$A_z(r,z) = \sum_{s=1}^{\infty} \mathrm{J}_0\left(p_s\frac{r}{R}\right)\left[C_s \mathrm{e}^{-\Gamma_s(z-g)} + D_s \mathrm{e}^{+\Gamma_s(z-g)}\right],$$

$$E_r(r,z) = \frac{c^2}{j\omega}\sum_{s=1}^{\infty} \frac{p_s}{R}\Gamma_s \mathrm{J}_1\left(p_s\frac{r}{R}\right)\left[C_s \mathrm{e}^{-\Gamma_s(z-g)} - D_s \mathrm{e}^{+\Gamma_s(z-g)}\right],$$

$$E_z(r,z) = -\frac{c^2}{j\omega}\sum_{s=1}^{\infty} \Gamma_s^2 \mathrm{J}_0\left(p_s\frac{r}{R}\right)\left[C_s \mathrm{e}^{-\Gamma_s(z-g)} - D_s \mathrm{e}^{+\Gamma_s(z-g)}\right],$$

$$H_\phi(r,z) = \frac{1}{\mu_0}\sum_{s=1}^{\infty} \frac{p_s}{R} \mathrm{J}_1\left(p_s\frac{r}{R}\right)\left[C_s \mathrm{e}^{-\Gamma_s(z-g)} + D_s \mathrm{e}^{+\Gamma_s(z-g)}\right], \tag{5.1.11}$$

with $\Gamma_s^2 = (p_s/R)^2 - (\omega/c)^2$. At this point we shall consider only the TM_{01} mode $(s = 1)$ thus the continuity of the radial electric field at $z = g$ implies

$$\frac{1}{\varepsilon_r}\Gamma_{d,1}\left[A_1 \mathrm{e}^{-\Gamma_{d,1}g} - B_1 \mathrm{e}^{+\Gamma_{d,1}g}\right] = \Gamma_1\left[C_1 - D_1\right], \tag{5.1.12}$$

and in a similar way the continuity of the azimuthal magnetic field reads

$$A_1 \mathrm{e}^{-\Gamma_{d,1}g} + B_1 \mathrm{e}^{+\Gamma_{d,1}g} = C_1 + D_1 . \tag{5.1.13}$$

Last two equations express the relation between the amplitudes of the field in the dielectric and vacuum.

In the dielectric filled region of next cell $(L \le z \le L + g)$ the field has a similar form as in (5.1.10) i.e.,

$$A_z(r,z) = \sum_{s=1}^{\infty} \mathrm{J}_0\left(p_s\frac{r}{R}\right)\left[A_s' \mathrm{e}^{-\Gamma_{d,s}(z-L)} + B_s' \mathrm{e}^{+\Gamma_{d,s}(z-L)}\right],$$

$$E_r(r,z) = \frac{c^2}{j\omega\varepsilon_r}\sum_{s=1}^{\infty} \frac{p_s}{R}\Gamma_{d,s} \mathrm{J}_1\left(p_s\frac{r}{R}\right)\left[A_s' \mathrm{e}^{-\Gamma_{d,s}(z-L)} - B_s' \mathrm{e}^{+\Gamma_{d,s}(z-L)}\right],$$

$$E_z(r,z) = -\frac{c^2}{j\omega\varepsilon_r}\sum_{s=1}^{\infty} \Gamma_{d,s}^2 \mathrm{J}_0\left(p_s\frac{r}{R}\right)\left[A_s' \mathrm{e}^{-\Gamma_{d,s}(z-L)} - B_s' \mathrm{e}^{+\Gamma_{d,s}(z-L)}\right],$$

$$H_\phi(r,z) = \frac{1}{\mu_0}\sum_{s=1}^{\infty} \frac{p_s}{R} \mathrm{J}_1\left(p_s\frac{r}{R}\right)\left[A_s' \mathrm{e}^{-\Gamma_{d,s}(z-L)} + B_s' \mathrm{e}^{+\Gamma_{d,s}(z-L)}\right]. \tag{5.1.14}$$

Accordingly, the boundary conditions at $z = L$ read

$$\frac{1}{\varepsilon_r}\Gamma_{d,1}\left[A_1' - B_1'\right] = \Gamma_1\left[C_1 \mathrm{e}^{-\Gamma_1(L-g)} - D_1 \mathrm{e}^{\Gamma_1(L-g)}\right], \tag{5.1.15}$$

and

$$A_1' + B_1' = C_1 \mathrm{e}^{-\Gamma_1(L-g)} + D_1 \mathrm{e}^{\Gamma_1(L-g)} \,. \tag{5.1.16}$$

The relation between the amplitudes of the wave in the second cell ($L < z < 2L$) and the first cell can be represented in a matrix form

$$\mathbf{a}' = \mathbf{T}\,\mathbf{a}\,, \tag{5.1.17}$$

where the components of $\mathbf{a}'$ are A_1' and B_1' and similarly the components of $\mathbf{a}$ are A_1 and B_1. According to Floquet's theorem (5.1.9) the two vectors are expected to be related by

$$\mathbf{a}' = \mathrm{e}^{-jkL}\mathbf{a}\,, \tag{5.1.18}$$

thus e^{-jkL} represents the eigen-values of the single cell transmission matrix $\mathbf{T}$:

$$\left|\mathbf{T} - \mathrm{e}^{-jkL}\mathbf{I}\right| = 0\,. \tag{5.1.19}$$

Explicitly this reads

$$\mathrm{e}^{-2jkL} - \mathrm{e}^{-jkL}(T_{11} + T_{22}) + T_{11}T_{22} - T_{12}T_{21} = 0\,. \tag{5.1.20}$$

For a passive system the determinant of the matrix $\mathbf{T}$ is unity, thus

$$\mathrm{e}^{-2jkL} - \mathrm{e}^{-jkL}(T_{11} + T_{22}) + 1 = 0\,. \tag{5.1.21}$$

The fact that the last term in this equation is unity indicates that if k is a solution of (5.1.21) $-k$ is also a solution. Consequently, we can write

$$\cos(kL) = \frac{1}{2}(T_{11} + T_{22})\,. \tag{5.1.22}$$

Note that this is an explicit expression for k as a function of the frequency and the other geometric parameters. In principle there are ranges of parameters where the right-hand side is larger than unity and there is no real k which satisfies this relation. If only the frequency is varied then this result indicates that there are frequencies for which the solution of the dispersion relation is real thus a wave can propagate, or the solution is imaginary and the amplitude of the wave is zero. The frequency range for which the wave is allowed to propagate is called the *passband*. Explicitly, the right-hand side of (5.1.22) reads

$$\frac{1}{2}(T_{11}+T_{22}) = \frac{(Z_1 + Z_2)^2}{4Z_1Z_2}\cosh(\psi+\chi) - \frac{(Z_1 - Z_2)^2}{4Z_1Z_2}\cosh(\psi-\chi)\,, \tag{5.1.23}$$

where $\psi = \Gamma_1(L-g)$, $\chi = \Gamma_{d,1}g$, the characteristic impedances are

$$Z_1 = \eta_0 \frac{c\Gamma_{d,1}}{j\omega\varepsilon_r},$$
$$Z_2 = \eta_0 \frac{c\Gamma_1}{j\omega}, \tag{5.1.24}$$

and $\eta_0 = 377\,\Omega$ is the impedance of the vacuum. Figure 5.2 illustrates the right-hand side of (5.1.22) as a function of the frequency ($\varepsilon_r = 10$, $R = 2\,\mathrm{cm}$, $L = 1\,\mathrm{cm}$ and $g = L/2$). The blocks at the bottom, illustrate the forbidden frequencies, namely at these frequencies TM waves can not propagate. In Fig. 5.3 the dispersion relation of the first three passbands is presented; these branches correspond only to the TM_{01} mode. Higher symmetric or asymmetric modes have additional contributions in this range of frequencies.

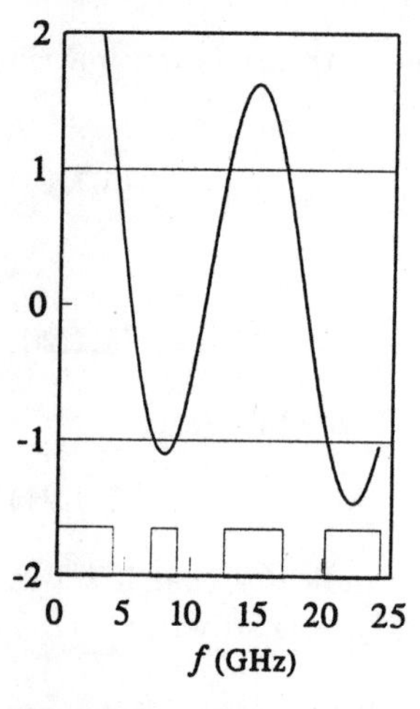

Fig. 5.2. Right-hand side of equation (5.1.22). For the frequencies marked at the bottom, no electromagnetic wave can propagate in the system

Remark 1. The expression in (5.1.22) is the dispersion relation of the periodic structure we introduced. From this simple example however we observe that *the dispersion relation of a periodic structure is itself periodic in k with a periodicity $2\pi/L$*. This is a general feature which can be deduced from (5.1.9). If the latter is satisfied for $k = k_0$ then (5.1.9) is satisfied also for $k = k_0 + 2\pi/L$ as shown next

$$\begin{aligned} f(z+L)\,\mathrm{e}^{j(k_0+2\pi/L)L} &= f(z+L)\,\mathrm{e}^{jk_0L}\,\mathrm{e}^{j2\pi}, \\ &= f(z+L)\,\mathrm{e}^{jk_0L} = f(z). \end{aligned} \tag{5.1.25}$$

Consequently, since the dispersion relation is periodic in k, it is sufficient to represent its variation with k in the range $-\pi/L \le k \le \pi/L$; this k domain is also called the first Brillouin zone.

Remark 2. Bearing in mind the last comment, we can re-examine the expression in (5.1.8) and realize that $f(z)$ is represented by a superposition of *spatial harmonics* $\exp(-jk_nz)$ where

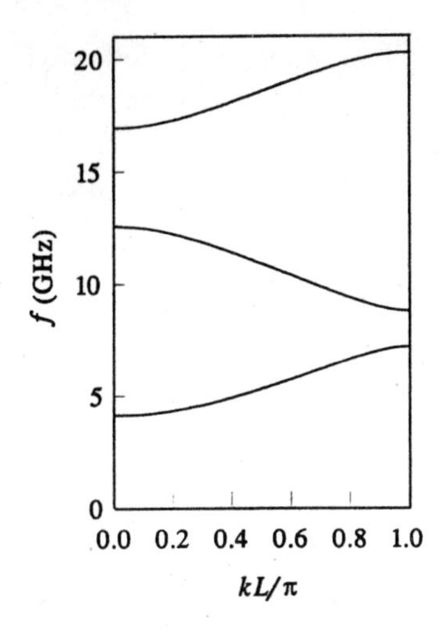

Fig. 5.3. A set of discrete points which represent the continuous dispersion relation corresponding to the same parameters as these for which Fig. 5.2 was plotted; the parameters $R =$ 2cm, $\varepsilon_{\rm r} = 10, L = 1$cm and $g = 0.5$cm

$$k_n = k + \frac{2\pi}{L} n\,, \tag{5.1.26}$$

which all correspond to the solution of the dispersion relation of the system. According to this definition the phase velocity of each harmonic is

$$v_{\mathrm{ph},n} = \frac{\omega}{ck_n}\,, \tag{5.1.27}$$

and for a high harmonic index, n, this velocity decreases as n^{-1}. Furthermore, all harmonics with negative index correspond to waves which propagate backwards. In addition, note that the zero harmonic ($n = 0$) has a positive group velocity for $\pi/L > k > 0$ and negative in the range $-\pi/L < k < 0$. This is a characteristic of *all* spatial harmonics. Since the group velocity is related to the energy velocity, one can conclude that although the the wave number of a particular space harmonic is positive, the power it carries may flow in the negative direction (if the group velocity is negative). This opens a whole new family of devices in which the power flows anti-parallel to the beam – the generic device is called backward-wave oscillator (BWO) and it was discussed in Sect. 4.2.4. Note also that at all π-points, i.e., $kL = \pi n$, the group velocity is zero.

5.2 Closed Periodic Structure

Based on what was shown in the previous section one can determine the dispersion relation of a TM_{01} mode which propagates in a corrugated waveguide [Brillouin (1948)]. Its periodicity is L, the inner radius is denoted by $R_{\rm int}$ and the external by $R_{\rm ext}$; the distance between two cavities (the drift region) is d – see Fig. 3.11. Using Floquet's Theorem (5.1.8) we can write for the magnetic potential in the inner cylinder ($0 < r < R_{\rm int}$) the following expression

$$A_z(r,z) = \sum_{n=-\infty}^{\infty} A_n e^{-jk_n z} I_0(\Gamma_n r)\,, \tag{5.2.1}$$

and accordingly the electromagnetic field components read

$$E_r(r,z) = \frac{c^2}{j\omega} \sum_{n=-\infty}^{\infty} (-jk_n \Gamma_n) A_n e^{-jk_n z} I_1(\Gamma_n r)\,,$$

$$E_z(r,z) = \frac{c^2}{j\omega} \sum_{n=-\infty}^{\infty} (-\Gamma_n^2) A_n e^{-jk_n z} I_0(\Gamma_n r)\,,$$

$$H_\phi(r,z) = -\frac{1}{\mu_0} \sum_{n=-\infty}^{\infty} \Gamma_n A_n e^{-jk_n z} I_1(\Gamma_n r)\,. \tag{5.2.2}$$

In these expressions,

$$\Gamma_n^2 = k_n^2 - \frac{\omega^2}{c^2}\,, \tag{5.2.3}$$

and $I_0(x), I_1(x)$ are the zero and first order modified Bessel functions of the first kind respectively. This choice of the radial functional variation is dictated by the condition of convergence of the electromagnetic field on axis.

In each individual groove the electromagnetic field can be derived from the following magnetic vector potential:

$$A_z^{(\sigma)}(r,z) = \sum_{\nu=0}^{\infty} B_\nu^{(\sigma)} \cos\left[q_\nu (z - z_\sigma - d)\right] t_{0,\nu}(r)\,, \tag{5.2.4}$$

where $q_\nu = \pi\nu/(L-d)$,

$$t_{0,\nu}(r) = I_0(\Lambda_\nu r) K_0(\Lambda_\nu R_{\rm ext}) - K_0(\Lambda_\nu r) I_0(\Lambda_\nu R_{\rm ext})\,, \tag{5.2.5}$$

and $\Lambda_\nu^2 = q_\nu^2 - (\omega/c)^2$. The electromagnetic field reads

$$E_r^{(\sigma)}(r,z) = \frac{c^2}{j\omega} \sum_{\nu=0}^{\infty} (-q_\nu) \Lambda_\nu B_\nu^{(\sigma)} \sin\Big[q_\nu (z - z_\sigma - d)\Big] t_{1,\nu}(r)\,,$$

$$E_z^{(\sigma)}(r,z) = \frac{c^2}{j\omega} \sum_{\nu=0}^{\infty} (-\Lambda_\nu^2) B_\nu^{(\sigma)} \cos\Big[q_\nu (z - z_\sigma - d)\Big] t_{0,\nu}(r)\,,$$

$$H_\phi^{(\sigma)}(r,z) = -\frac{1}{\mu_0} \sum_{\nu=0}^{\infty} \Lambda_\nu B_\nu^{(\sigma)} \cos\Big[q_\nu (z - z_\sigma - d)\Big] t_{1,\nu}(r)\,. \tag{5.2.6}$$

In these expressions $t_{1,\nu}(r)$ is the derivative of $t_{0,\nu}(r)$ defined by

$$t_{1,\nu}(r) = \mathrm{I}_1(\Lambda_\nu r)\mathrm{K}_0(\Lambda_\nu R_{\mathrm{ext}}) + \mathrm{K}_1(\Lambda_\nu r)\mathrm{I}_0(\Lambda_\nu R_{\mathrm{ext}})\,, \tag{5.2.7}$$

and except at $r = R_{\mathrm{int}}$ all the boundary conditions are satisfied; the index σ labels the "cavity".

5.2.1 Dispersion Relation

Our next step is to impose the continuity of the boundary conditions at the interface ($r = R_{\mathrm{int}}$). The continuity of the longitudinal component of the electric field $[E_z(r = R_{\mathrm{int}}, -\infty < z < \infty)]$ reads

$$\frac{c^2}{j\omega}\sum_{n=-\infty}^{\infty}(-\Gamma_n^2)A_n \mathrm{e}^{-jk_n z}\mathrm{I}_0(\Gamma_n R_{\mathrm{int}})$$
$$= \begin{cases} 0 & \text{for } z_\sigma < z < z_\sigma + d\,, \\ -\frac{c^2}{j\omega}\sum_{\nu=0}^{\infty}\Lambda_\nu^2 B_\nu^{(\sigma)}\cos\Big[q_\nu(z - z_\sigma - d)\Big]t_{0,\nu}(R_{\mathrm{int}}) & \text{for } z_\sigma + d < z < z_\sigma + L\,, \end{cases} \tag{5.2.8}$$

and the azimuthal magnetic field $[H_\phi(r = R_{\mathrm{int}}, z_\sigma + d < z < z_\sigma + L)]$ reads

$$-\frac{1}{\mu_0}\sum_{n=-\infty}^{\infty}\Gamma_n A_n \mathrm{e}^{-jk_n z}\mathrm{I}_1(\Gamma_n R_{\mathrm{int}})$$
$$= -\frac{1}{\mu_0}\sum_{\nu=0}^{\infty}\Lambda_\nu B_\nu^{(\sigma)}\cos\Big[q_\nu(z - z_\sigma - d)\Big]t_{1,\nu}(R_{\mathrm{int}})\,. \tag{5.2.9}$$

From these boundary conditions the dispersion relation of the structure can be developed. For this purpose we analyze the solution in the grooves having Floquet's theorem in mind. The latter implies that the longitudinal electric field in the σ's groove has to satisfy the following relation:

$$-\frac{c^2}{j\omega}\sum_{\nu=0}^{\infty}\Lambda_\nu^2 B_\nu^{(\sigma)}\cos\Big[q_\nu(z - z_\sigma - d)\Big]t_{0,\nu}(r)$$
$$= -\frac{c^2}{j\omega}\sum_{\nu=0}^{\infty}\Lambda_\nu^2 B_\nu^{(\sigma+1)}\mathrm{e}^{jkL}\cos\Big[q_\nu(z + L - z_{\sigma+1} - d)\Big]t_{0,\nu}(r)\,. \tag{5.2.10}$$

But by definition $z_{\sigma+1} - z_\sigma = L$ therefore, the last expression implies that

$$B_\nu^{(\sigma)} = B_\nu \mathrm{e}^{-jkz_\sigma}\,. \tag{5.2.11}$$

This result permits us to restrict the investigation to a single cell and without loss of generality we chose $z_{\sigma=0} = 0$ since if we know B_ν in one cell, the relation in (5.2.11) determines the value of this amplitude in all other cells. With this result in mind we multiply (5.2.8) by $\mathrm{e}^{jk_m z}$ and integrate over one cell; the result is

$$\sum_{n=-\infty}^{\infty} \Gamma_n^2 A_n \delta_{n,m} L \mathrm{I}_0(\Gamma_n R_{\mathrm{int}})$$
$$= \sum_{\nu=0}^{\infty} \Lambda_\nu^2 B_\nu t_{0,\nu}(R_{\mathrm{int}}) \int_d^L \mathrm{d}z \mathrm{e}^{jk_m z} \cos[q_\nu(z-d)] \,, \tag{5.2.12}$$

here $\delta_{n,m}$ is the Kroniker delta function which equals 1 if $n = m$ and zero otherwise. We also used the orthogonality of the Fourier spatial harmonics. We follow a similar procedure when imposing the continuity of the magnetic field with one difference, (5.2.9) is defined only in the groove aperture thus we shall utilize the orthogonality of the trigonometric function $\cos[q_\nu(z-d)]$. Accordingly, (5.2.9) is multiplied by $\cos[q_\mu(z-d)]$ and we integrate over $d < z < L$; the result is

$$\sum_{n=-\infty}^{\infty} \Gamma_n A_n \mathrm{I}_1(\Gamma_n R_{\mathrm{int}}) \int_d^L \mathrm{d}z \cos\left[q_\mu(z-d)\right] \mathrm{e}^{-jk_n z}$$
$$= \sum_{\nu=0}^{\infty} \Lambda_\nu B_\nu t_{1,\nu}(R_{\mathrm{int}})(L-d) g_\mu \delta_{\nu,\mu} \,, \tag{5.2.13}$$

where $g_0 = 1$ and $g_{\mu \neq 0} = 0.5$ otherwise. It is convenient to define the quantity

$$\mathcal{L}_{n,\nu}(k) = \frac{1}{L-d} \int_d^L \mathrm{d}z \cos\left[q_\nu(z-d)\right] \mathrm{e}^{jk_n z} \,, \tag{5.2.14}$$

which allows us to write (5.2.12) as

$$A_n = \frac{1}{\Gamma_n^2 \mathrm{I}_0(\Gamma_n R_{\mathrm{int}})} \frac{L-d}{L} \sum_{\nu=0}^{\infty} \Lambda_\nu^2 t_{0,\nu}(R_{\mathrm{int}}) \mathcal{L}_{n,\nu}(k) B_\nu \,, \tag{5.2.15}$$

and (5.2.13) as

$$B_\nu = \frac{1}{\Lambda_\nu t_{1,\nu}(R_{\mathrm{int}}) g_\nu} \sum_{n=-\infty}^{\infty} A_n \Gamma_n \mathrm{I}_1(\Gamma_n R_{\mathrm{int}}) \mathcal{L}_{n,\nu}^*(k) \,. \tag{5.2.16}$$

These are two equations for two unknown sets of amplitudes (A_n, B_ν) and the dispersion relation can be represented in two equivalent ways: One possibility is to substitute (5.2.16) in (5.2.15) and get

$$\sum_{m=-\infty}^{\infty} \left[\delta_{n,m} - \frac{L-d}{L} \frac{\Gamma_m \mathrm{I}_1(\Gamma_m R_{\mathrm{int}})}{\Gamma_n^2 \mathrm{I}_0(\Gamma_n R_{\mathrm{int}})} \sum_{\nu=0}^{\infty} \frac{t_{0,\nu}(R_{\mathrm{int}}) \Lambda_\nu}{t_{1,\nu}(R_{\mathrm{int}}) g_\nu} \mathcal{L}_{n,\nu} \mathcal{L}_{m,\nu}^* \right] A_m = 0 \,, \tag{5.2.17}$$

whereas the other possibility is to substitute (5.2.15) in (5.2.16) and obtain

$$\sum_{\mu=0}^{\infty} \left[\delta_{\nu,\mu} - \frac{L-d}{L} \frac{\Lambda_\mu^2 t_{0,\mu}(R_{\mathrm{int}})}{\Lambda_\nu t_{1,\nu}(R_{\mathrm{int}}) g_\nu} \sum_{n=-\infty}^{\infty} \frac{\mathrm{I}_1(\Gamma_n R_{\mathrm{int}})}{\Gamma_n \mathrm{I}_0(\Gamma_n R_{\mathrm{int}})} \mathcal{L}_{n,\nu}^* \mathcal{L}_{n,\mu} \right] B_\mu = 0 \,. \tag{5.2.18}$$

In both cases the dispersion relation is calculated from the requirement that the determinant of the matrix which multiplies the vector of amplitudes, is zero.

Although the two methods are equivalent, we found that the latter expression is by far more efficient for practical calculation because of the number of modes required to represent adequately the field in the groove compared to the number of spatial harmonics required to represent the field in the inner section. In the case of single mode operation we found that 1 to 3 modes are sufficient for description of the field in the grooves and about 40 spatial harmonics are generally used in the inner section. As indicated by these numbers it will be much easier to calculate the determinant of a 3×3 matrix rather than 40×40 one; we shall quantify this statement later. At this point we shall discuss the design of a disk-loaded structure assuming that the number modes in the grooves and harmonics in the inner space are sufficient.

Let assume that we want to determine the geometry of a disk-loaded structure which enables a wave at 10 GHz to be in resonance with electrons with $\beta = 0.9$ and the phase advance per cell is assumed to be $kL = 2\pi/3$. These two conditions determine the period of the structure. In our case $L =$ 9 mm. There are three additional geometric parameters to be determined: $R_{\rm ext}$, $R_{\rm int}$ and d. The last two have a dominant effect on the width of the passband and for the lowest mode, the passband increases with increasing $R_{\rm int}$ and decreases with increasing d. The passband, $\Delta\omega$, of a mode sets a limit on the maximum group velocity as can be seen bearing in mind that the half width of the first Brillouin zone is $\Delta k = \pi/L$. Consequently, $v_{\rm gr} = \Delta\omega/\Delta k <$ $\Delta\omega L/\pi$. A solution of the dispersion relation in (5.2.18) is illustrated in Fig. 5.4, the geometry chosen is: $R_{\rm int} = 8$ mm and $d = 2$ mm and from the condition of phase advance per cell of 120° at 10 GHz, we determined, using the dispersion relation, the value of the external radius to be $R_{\rm ext} =$ 13.96 mm.

In the remainder of this section we shall consider only a single mode in the groove. Therefore, before we conclude this subsection, we shall quantify the effect of higher modes. The first mode in the groove ($\nu = 0$) represents a TEM mode which propagates in the radial direction. Other modes ($\mathrm{TM}_{0,\nu>0}$) are either propagating or evanescent. The amplitudes of the magnetic and electric field (E_z) of the TEM mode are constant at the groove aperture thus the choice of using a single mode in the groove is equivalent to the average process at the boundary – approach usually adopted in the literature. Figure 5.5 illustrates the dependence of upper and lower cut-off frequency on the number of harmonics used in the calculation; the number of modes in the grooves is a parameter. For the geometry presented above, the number of harmonics required is 20 or larger; typically about 40 harmonics are being used. The effect of the $\nu = 1$ mode is negligible in this case as seen for both upper and lower cut-off frequencies. The effect of the higher mode introduces a correction on the order of 1% which for most practical purposes is sufficient.

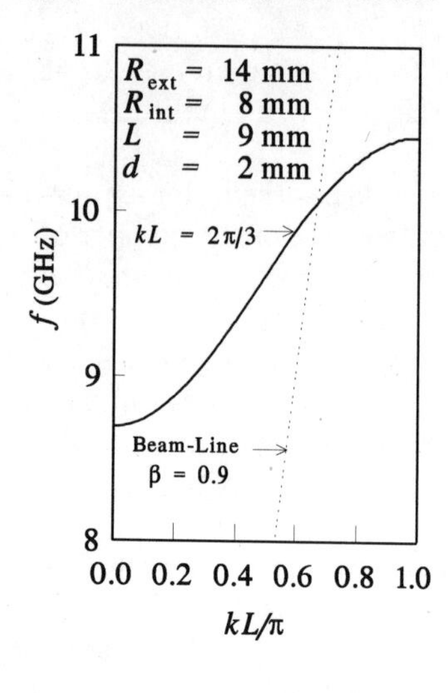

Fig. 5.4. A solution of the dispersion relation in (5.2.18). The geometry chosen corresponds to: $R_{\text{int}} = 8\,\text{mm}$, $L = 9\,\text{mm}$, $R_{\text{int}} = 13.96\,\text{mm}$ and $d = 2\,\text{mm}$

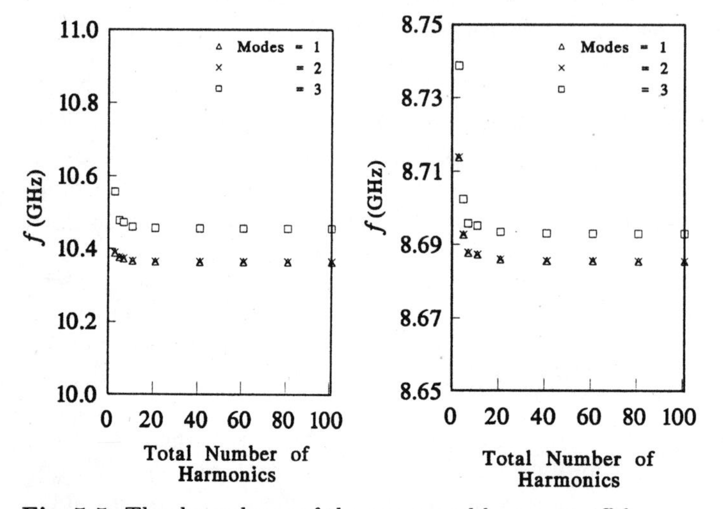

Fig. 5.5. The dependence of the upper and lower cut-off frequency on the number of harmonics used

5.2.2 Spatial Harmonics Coupling

Contrary to uniform dielectric structures, here each mode consists of a superposition of an infinite number of spatial harmonics. These harmonics are all coupled by the conditions imposed on the electromagnetic field by the geometry at $r = R_{\text{int}}$. We shall limit the investigation to the accuracy associated with a single mode taken in the groove, therefore according to (5.2.15), we have

$$A_n = -\frac{1}{\Gamma_n^2 \mathrm{I}_0(\Gamma_n R_{\text{int}})} \frac{\omega^2}{c^2} t_{0,0}(R_{\text{int}}) \mathcal{L}_{n,0}(k) B_0 \,, \tag{5.2.19}$$

and in this particular case

$$\mathcal{L}_{n,0}(k) = \frac{L-d}{L} \operatorname{sinc}\left[\frac{1}{2}k_n(L-d)\right] \exp\left[j\frac{1}{2}k_n(L+d)\right] . \tag{5.2.20}$$

Let us compare the first few spatial harmonics relative to the zero harmonic. For this purpose we take $f = 10\,\text{GHz}$, $v_0 = 0.9c$, $R_{\text{int}} = 8\,\text{mm}$, $L = 9\,\text{mm}$ and $d = 2\,\text{mm}$. The ratio of the first two amplitudes is

$$\begin{aligned} \left|\frac{A_{-1}}{A_0}\right| &= 8 \times 10^{-3} , \\ \left|\frac{A_1}{A_0}\right| &= 3 \times 10^{-6} , \\ \left|\frac{A_{-2}}{A_0}\right| &= 2 \times 10^{-6} , \\ \left|\frac{A_2}{A_0}\right| &= 1 \times 10^{-8} . \end{aligned} \tag{5.2.21}$$

This result indicates that *on axis*, the amplitude of the interacting harmonic is dominant. At the interface with the grooves ($r = R_{\text{int}}$) the ratio between the contribution of the zero and nth harmonic is much closer to unity and it can be checked that it reads

$$\frac{|E_{z,n}(r = R_{\text{int}})|}{|E_{z,0}(r = R_{\text{int}})|} = \frac{|\operatorname{sinc}(k_n(L-d)/2)|}{|\operatorname{sinc}(k_0(L-d)/2)|} , \tag{5.2.22}$$

which is a virtually unity. It indicates that there is a significant amount of energy in the high spatial harmonics which may cause breakdown due to the associated gradients on the metallic surface.

A more instructive picture is obtained by examining the average power flowing in one time and spatial period of the system:

$$P = 2\pi \int_0^{R_{\text{int}}} \mathrm{d}r\, r \frac{1}{L} \int_0^L \mathrm{d}z \left[\frac{1}{2} E_r(r,z) H_\phi^*(r,z)\right] . \tag{5.2.23}$$

According to the definition in (5.2.2) we have

$$P = \frac{\pi}{\eta_0} \sum_{n=-\infty}^{\infty} |cA_n|^2 \frac{ck_n}{\omega} \int_0^{\Gamma_n R_{\text{int}}} \mathrm{d}x\, x \mathrm{I}_1^2(x) ; \tag{5.2.24}$$

the integral can be calculated analytically [Abramowitz and Stegun (1968) p.484] and it reads

$$U(\xi) \equiv \int_0^{\xi} \mathrm{d}x\, x\mathrm{I}_1^2(x) = \xi \mathrm{I}_0(\xi)\mathrm{I}_1(\xi) + \frac{1}{2}\xi^2 \left[\mathrm{I}_1^2(\xi) - \mathrm{I}_0^2(\xi)\right] . \tag{5.2.25}$$

Based on these definitions we can calculate the average power carried by each harmonic as

$$P_n = \frac{\pi}{\eta_0} |cA_n|^2 \frac{ck_n}{\omega} U(\Gamma_n R_{\text{int}}) , \tag{5.2.26}$$

and the result is listed below

$$\frac{P_{-2}}{P_0} = -3 \times 10^{-3},$$
$$\frac{P_{-1}}{P_0} = -0.16,$$
$$\frac{P_1}{P_0} = 1 \times 10^{-4},$$
$$\frac{P_2}{P_0} = 3 \times 10^{-3}. \tag{5.2.27}$$

Although there is a total flow of power along (the positive) direction of the z axis, a substantial amount of power is actually flowing backwards. In this numerical example for all practical purposes we can consider only the lowest two harmonics and write the total power which flows, normalized to the power in the zero harmonic. Thus if the latter is unity then the power in the forward is $1-0.16 = 0.84$. This result indicates that if we have a finite length structure with finite reflections from the input end, then in this periodic structure we have an inherent feedback even if the output end is perfectly matched.

5.2.3 Interaction Parameters

Basically, even if an electron beam interacts only with a single mode the latter consists of an infinite number of harmonics and with this regard, we distinguish between direct and indirect interaction. By *direct interaction* we refer to the harmonic to which the electron transfers energy directly. For example, for a pencil beam (on axis) it is primarily the $n = 0$ harmonic which interacts with the beam, namely it has a phase velocity close to the velocity of the electrons. Because of the interaction we have shown in Chap. 4 that the wavenumber of the mode becomes complex, which in periodic structure, implies that the projection of the wavenumber in the first Brillouin zone (k) becomes complex. But this corresponds to all harmonics which finally implies that they all grow in space by the same relative amount such that locally the boundary conditions are satisfied. With this regard the beam indirectly interacts with all harmonics and this is referred to as *indirect interaction.* The condition for the beam to interact directly only with a single harmonic can be formulated in terms of the velocity spread of the beam and resonance condition: the latter reads in general $\omega/v_0 - k_n \simeq 0$ whose variation for a constant frequency reads $\omega|\Delta v|/v_0^2 = |\Delta k|$. Since two harmonics are separated by $\Delta k = 2\pi/L$, we conclude that the condition for single harmonic direct operation is that $(\omega L/c)(|\Delta\beta|/\beta^2) \ll 2\pi$. Subject to this condition, the beam-wave interaction is described primarily by a single parameter: the interaction impedance introduced in Sect. 2.3.3. For a pencil beam of radius R_b the effective field which acts on the electrons in a uniform periodic structure is

$$\left|E\right|^2 \equiv \frac{2}{R_\mathrm{b}^2}\int_0^{R_\mathrm{b}} \mathrm{d}r\, r \left|E_{z,n=0}(r,z)\right|^2 . \tag{5.2.28}$$

Using the explicit expression for E_z in (5.2.2) we find

$$\left|E\right|^2 \equiv \frac{c^4}{\omega^2}\left|A_0\right|^2 \Gamma_0^4 \frac{2}{R_\mathrm{b}^2}\int_0^{R_\mathrm{b}} \mathrm{d}r\, r\, \mathrm{I}_0^2(\Gamma_0 r)\, ; \tag{5.2.29}$$

the integral can be evaluated exactly [Abramowitz and Stegun (1968) p.484] and it reads

$$W_1(x) \equiv \int_0^x \mathrm{d}\xi\, \xi \mathrm{I}_0^2(\xi) = \frac{1}{2}x^2\left[\mathrm{I}_0^2(x) - \mathrm{I}_1^2(x)\right] . \tag{5.2.30}$$

With this expression the effective electric field reads

$$\left|E\right|^2 \equiv \frac{2c^4\Gamma_0^2}{\omega^2 R_\mathrm{b}^2}\left|A_0\right|^2 W_1(\Gamma_0 R_\mathrm{b})\, , \tag{5.2.31}$$

and finally we can determine the explicit expression for the interaction impedance in a periodic structure

$$\begin{aligned} Z_\mathrm{int} &\equiv \frac{1}{2}\frac{\left|E\right|^2 (\pi R_\mathrm{int}^2)}{P}\, , \\ &= \eta_0 \left(\frac{R_\mathrm{int}}{R_\mathrm{b}}\right)^2 \left(\frac{c\Gamma_0}{\omega}\right)^2 W_1(\Gamma_0 R_\mathrm{b}) \left[\sum_{n=-\infty}^{\infty} U(\Gamma_n R_\mathrm{int}) \left(\frac{ck_n}{\omega}\right) \frac{\left|A_n\right|^2}{\left|A_0\right|^2}\right]^{-1} . \end{aligned} \tag{5.2.32}$$

The ratio $\left|A_n/A_0\right|$ can be deduced from (5.2.19). Furthermore, the last expression reveals the effect of the beam radius on the interaction impedance. The latter can be formulated as

$$Z_\mathrm{int}(\Gamma_0 R_\mathrm{b}) = Z_\mathrm{int}(0)\left[\mathrm{I}_0^2(\Gamma_0 R_\mathrm{b}) - \mathrm{I}_1^2(\Gamma_0 R_\mathrm{b})\right] , \tag{5.2.33}$$

where

$$Z_\mathrm{int}(0) \equiv \eta_0 (\Gamma_0 R_\mathrm{int})^2 (\Gamma_0 \frac{c}{\omega})^2 \left[\sum_{n=-\infty}^{\infty} U(\Gamma_n R_\mathrm{int}) \left(\frac{ck_n}{\omega}\right) \frac{\left|A_n\right|^2}{\left|A_0\right|^2}\right]^{-1} . \tag{5.2.34}$$

The interaction impedance increases monotonically with the beam radius; this fact has been discussed also in Chap. 4 in the context of the interaction in a dielectrically loaded waveguide. Another aspect of the same phenomenon is illustrated in Fig. 5.6 where we present the interaction impedance as a function of the internal radius keeping L, d and the frequency ($f = 10\,\mathrm{GHz}$); the external radius is determined from the resonance condition and the phase advance per cell (which is chosen to be 120° for reasons which will be clarified in Sect. 8.1.1). In addition, the beam radius is taken to be $R_b = 3\,\text{mm}$. Note the rapid decrease of the interaction impedance with the increase in the internal radius of the structure. Again, this is a direct result of the exponential decay of the slow (evanescent) wave from the corrugated surface inwards.

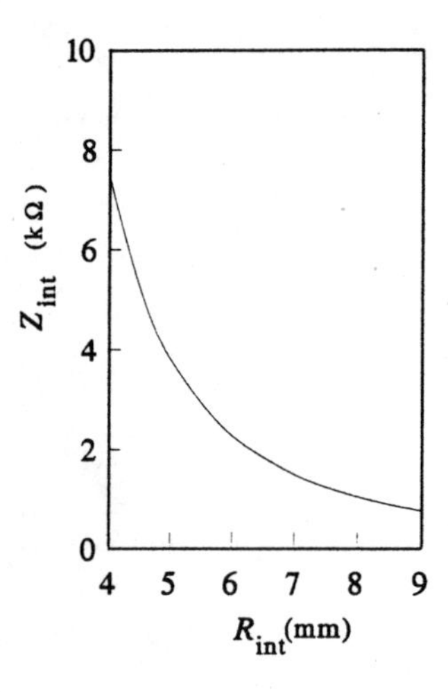

Fig. 5.6. Z_{int} as a function of the internal radius. The period L, the tooth width d and the frequency are kept constant. R_{ext} is determined from the resonance and phase advance conditions

The other parameter of interest is the interaction dielectric coefficient which is a measure of the total electromagnetic stored in one cell of the structure as defined in (2.3.31):

$$\varepsilon_{int} \equiv \frac{W_{em}}{\frac{1}{2}\varepsilon_0 \left|E\right|^2 \pi R_{int}^2} \,. \tag{5.2.35}$$

This parameter is important in the description of the operation of an oscillator. However, we have to bear in mind that according to (2.3.33) ε_{int} and Z_{int} are related. Let us now calculate this parameter. Firstly the electromagnetic energy stored in one groove, assuming one dominant mode, is given by

$$W_{em,groove} = \frac{1}{2}\varepsilon_0 (\pi R_{int}^2)(L-d)\omega^2 B_0^2 W_2 \,, \tag{5.2.36}$$

where

$$W_2 \equiv \int_1^{R_{ext}/R_{int}} dx\, x \left[T_0^2(\alpha x) + T_1^2(\alpha x)\right] , \tag{5.2.37}$$

$\alpha = \omega R_{int}/c$ and

$$T_0(\alpha x) \equiv \mathrm{J}_0(\alpha x)\mathrm{Y}_0\left(\alpha \frac{R_{ext}}{R_{int}}\right) - \mathrm{Y}_0(\alpha x)\mathrm{J}_0\left(\alpha \frac{R_{ext}}{R_{int}}\right) ,$$

$$T_1(\alpha x) \equiv \mathrm{J}_1(\alpha x)\mathrm{Y}_0\left(\alpha \frac{R_{ext}}{R_{int}}\right) - \mathrm{Y}_1(\alpha x)\mathrm{J}_0\left(\alpha \frac{R_{ext}}{R_{int}}\right) . \tag{5.2.38}$$

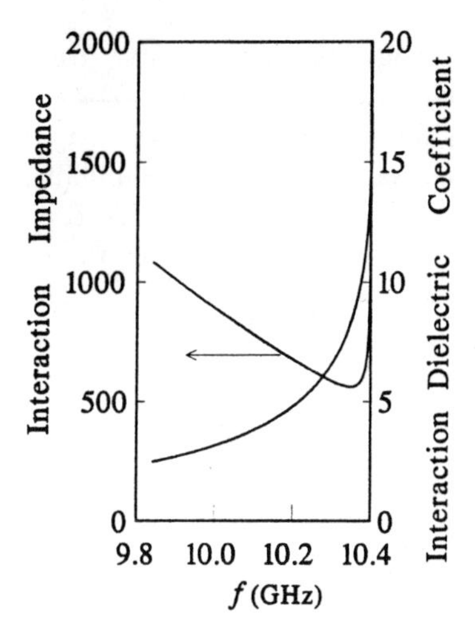

Fig. 5.7. ε_{int} and Z_{int} as a function of the frequency

The total averaged (in time and space) electromagnetic energy stored in one period of the inner cylinder ($0 \leq r \leq R_{\text{int}}$) is

$$W_{\text{em,cyl}} \equiv \frac{2\pi L}{4\mu_0} \sum_{n=-\infty}^{\infty} |A_n|^2 \left[U_n \left(1 + \frac{k_n^2 c^2}{\omega^2} \right) + \frac{\Gamma_n^2 c^2}{\omega^2} W_1(\Gamma_n R_{\text{int}}) \right] . \tag{5.2.39}$$

Consequently, the dielectric coefficient of the interaction is given by

$$\begin{aligned} \varepsilon_{\text{int}} =& \frac{1}{2} \frac{L}{L-d} \frac{(\Gamma_0 R_{\text{b}})^2 W_2}{W_1(\Gamma_0 R_{\text{b}})} \frac{\mathrm{I}_0^2(\Gamma R_{\text{int}})}{T_0^2(\alpha) \operatorname{sinc}^2 \left[\frac{1}{2} k(L-d) \right]} \\ &+ \frac{1}{2} \left(\frac{R_{\text{b}}}{R_{\text{int}}} \right)^2 \left(\frac{\omega}{c\Gamma_0} \right)^2 \frac{1}{W_1(\Gamma_0 R_{\text{b}})} \\ &\times \sum_{n=-\infty}^{\infty} \left[U_n \left(1 + \frac{k_n^2 c^2}{\omega^2} \right) + \frac{\Gamma_n^2 c^2}{\omega^2} W_1(\Gamma_n R_{\text{int}}) \right] \frac{|A_n|^2}{|A_0|^2} . \end{aligned} \tag{5.2.40}$$

Both the interaction impedance (Z_{int}) and its dielectric coefficient of the interaction (ε_{int}) are illustrated in Fig. 5.7 for these frequencies for which the phase velocity of the wave is smaller than c; the geometric parameters are: $R_{\text{int}} = 8\,\text{mm}$, $R_{\text{ext}} = 13.96\,\text{mm}$, $L = 9\,\text{mm}$, $d = 2\,\text{mm}$, $R_{\text{b}} = 3\,\text{mm}$ and the number of harmonics used is 13 ($-6 \leq n \leq 6$). Note that the interaction impedance has a *minimum* at a frequency which is higher than the frequency where the system was designed to operate ($f = 10\,\text{GHz}$). Close to the π-point ($kL = \pi$) the interaction impedance increases since the amount of power which can flow in the system diminishes. At the same time, the dielectric coefficient of the interaction increases, which means that the ratio of energy stored in the system to the electric field acting on the particles, increases. Consequently, in this frequency range the system will tend to oscillate.

From (5.2.40) we observe that ε_{int} has two contributions: the first from the energy stored in the grooves and the second represents the energy stored in the inner cylinder. Figure 5.8 illustrates again ε_{int} and the contribution of each region. It is readily seen that the effect of the groove is dominant at all frequencies of interest, emphasizing its cavity role. We conclude this subsection with a comparison of the group and energy velocity as illustrated in Fig. 5.9. Within the framework of our approximation the two are close but not identical. However, for most practical purposes the difference between the two is negligible.

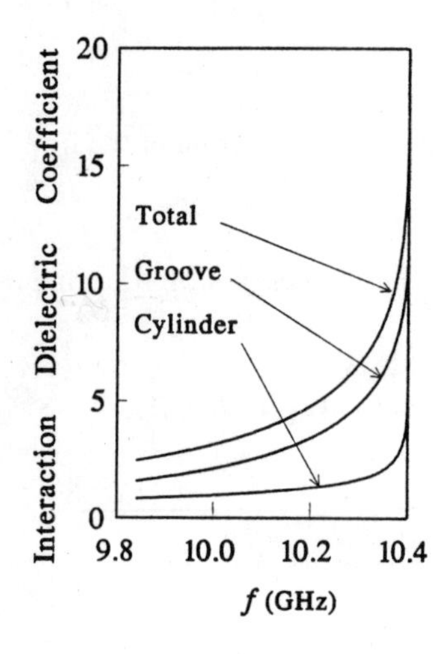

Fig. 5.8. The contribution of the cylinder and the groove to ε_{int}

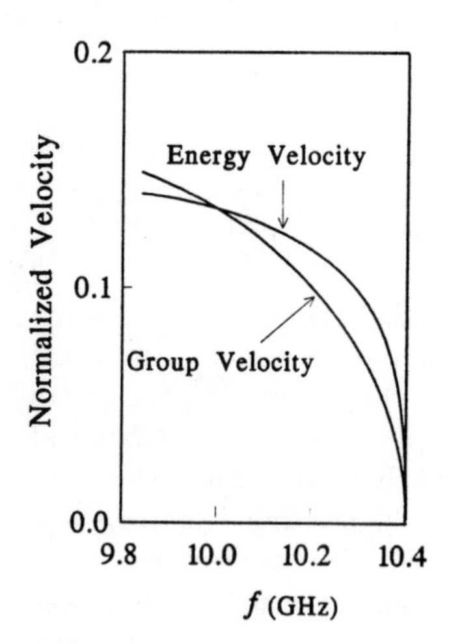

Fig. 5.9. Comparison of energy and group velocity as a function of the frequency

5.3 Open Periodic Structure

In this section an analysis similar to that in Sect. 5.2 is applied to an open periodic structure. As we shall see the number of modes which may develop in such a structure is small and therefore mode competition is minimized. This competition is a byproduct of the necessity to generate high power radiation which in the case of a single mode operation generates high gradients on the metallic surface. In order to avoid breakdown, it is necessary to increase the volume of the waveguide. Doing so, we allow more than one mode to coexist at the same frequency. Furthermore, the beam line intersects higher modes at frequencies higher than the operating one and these modes may deflect the electrons, as we shall see in Chap. 8.

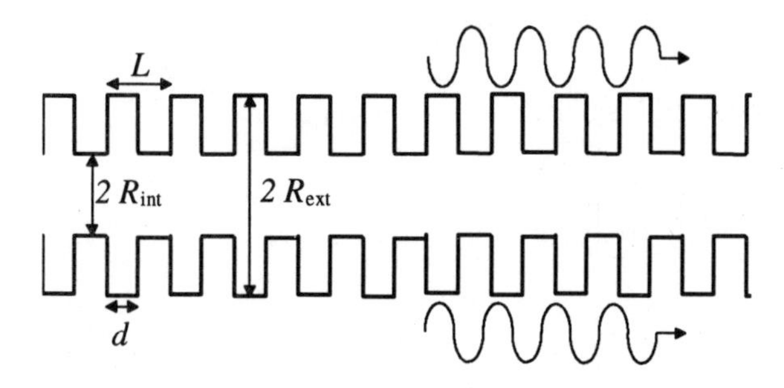

Fig. 5.10. Schematics of a cylindrical open periodic structure

We shall consider a system in which the wave propagates along the periodic structure which consists of a disk-loaded wire, as illustrated in Fig. 5.10 whose periodicity is L, the inner radius is denoted by $R_{\rm int}$, the external by $R_{\rm ext}$ and the distance between two cavities (the drift region) is d. Floquet's theorem as formulated in (5.1.8) allows us to write for the magnetic potential in the external region ($\infty > r \geq R_{\rm ext}$) the following expression

$$A_z(r,z) = \sum_{n=-\infty}^{\infty} A_n e^{-jk_n z} K_0(\Gamma_n r)\,, \tag{5.3.1}$$

and accordingly, the electromagnetic field components read

$$E_r(r,z) = \frac{c^2}{j\omega} \sum_{n=-\infty}^{\infty} (jk_n \Gamma_n) A_n e^{-jk_n z} K_1(\Gamma_n r)\,,$$

$$E_z(r,z) = \frac{c^2}{j\omega} \sum_{n=-\infty}^{\infty} (-\Gamma_n^2) A_n e^{-jk_n z} K_0(\Gamma_n r)\,,$$

$$H_\phi(r,z) = -\frac{1}{\mu_0} \sum_{n=-\infty}^{\infty} (-\Gamma_n) A_n e^{-jk_n z} K_1(\Gamma_n r)\,. \tag{5.3.2}$$

In these expressions $\mathrm{K}_0(x), \mathrm{K}_1(x)$ are the zero and first order modified Bessel functions of the second kind respectively and $\Gamma_n^2 = k_n^2 - (\omega/c)^2$. This choice of the radial functional variation is dictated by the condition of convergence of the electromagnetic field far away from the structure.

In each individual groove the electromagnetic field can be derived from the following magnetic vector potential:

$$A_z^{(\sigma)}(r,z) = \sum_{\nu=0}^{\infty} B_\nu^{(\sigma)} \cos\Big[q_\nu(z - z_\sigma - d)\Big] t_{0,\nu}(r)\,, \tag{5.3.3}$$

where $q_\nu = \pi\nu/(L-d)$,

$$t_{0,\nu}(r) = \mathrm{I}_0(\Lambda_\nu r)\mathrm{K}_0(\Lambda_\nu R_{\mathrm{int}}) - \mathrm{K}_0(\Lambda_\nu r)\mathrm{I}_0(\Lambda_\nu R_{\mathrm{int}})\,, \tag{5.3.4}$$

and $\Lambda_\nu^2 = q_\nu^2 - (\omega/c)^2$. The electromagnetic field reads

$$E_r^{(\sigma)}(r,z) = \frac{c^2}{j\omega}\sum_{\nu=0}^{\infty}(-q_\nu)\Lambda_\nu B_\nu^{(\sigma)} \sin\Big[q_\nu(z - z_\sigma - d)\Big] t_{1,\nu}(r)\,,$$

$$E_z^{(\sigma)}(r,z) = \frac{c^2}{j\omega}\sum_{\nu=0}^{\infty}(-\Lambda_\nu^2) B_\nu^{(\sigma)} \cos\Big[q_\nu(z - z_\sigma - d)\Big] t_{0,\nu}(r)\,,$$

$$H_\phi^{(\sigma)}(r,z) = -\frac{1}{\mu_0}\sum_{\nu=0}^{\infty}\Lambda_\nu B_\nu^{(\sigma)} \cos\Big[q_\nu(z - z_\sigma - d)\Big] t_{1,\nu}(r)\,. \tag{5.3.5}$$

The index σ labels the "cavity" and in these expressions we used

$$t_{1,\nu}(r) = \mathrm{I}_1(\Lambda_\nu r)\mathrm{K}_0(\Lambda_\nu R_{\mathrm{int}}) + \mathrm{K}_1(\Lambda_\nu r)\mathrm{I}_0(\Lambda_\nu R_{\mathrm{int}})\,. \tag{5.3.6}$$

The solution above satisfies all boundary conditions with the exception of $r = R_{\mathrm{ext}}$.

5.3.1 Dispersion Relation

Our next step is to impose the continuity of the boundary conditions at the interface ($r = R_{\mathrm{ext}}$). Continuity of the longitudinal component of the electric field implies $E_z(r = R_{\mathrm{ext}}, -\infty < z < \infty)$, reads

$$\frac{c^2}{j\omega}\sum_{n=-\infty}^{\infty}(-\Gamma_n^2)A_n \mathrm{e}^{-jk_n z}\mathrm{K}_0(\Gamma_n R_{\mathrm{ext}}) =$$

$$\begin{cases} 0 & \text{for } z_\sigma < z < z_\sigma + d\,, \\ -\frac{c^2}{j\omega}\sum_{\nu=0}^{\infty}\Lambda_\nu^2 B_\nu^{(\sigma)}\cos[q_\nu(z - z_\sigma - d)]t_{0,\nu}(R_{\mathrm{ext}}) & \text{for } z_\sigma + d < z < z_\sigma + L\,, \end{cases} \tag{5.3.7}$$

and the azimuthal magnetic field, $H_\phi(r = R_{\rm ext}, z_\sigma + d < z < z_\sigma + L)$, reads

$$\frac{1}{\mu_0}\sum_{n=-\infty}^{\infty} \Gamma_n A_n {\rm e}^{-jk_n z} {\rm K}_1(\Gamma_n R_{\rm ext}) = -\frac{1}{\mu_0}\sum_{\nu=0}^{\infty} \Lambda_\nu B_\nu^{(\sigma)} \cos\Big[q_\nu(z - z_\sigma - d)\Big] t_{1,\nu}(R_{\rm ext}) \,. \tag{5.3.8}$$

Following the same arguments as in Sect. 5.2.1 it can be shown that $B_\nu^{(\sigma)} = B_\nu\, {\rm e}^{-jkz_\sigma}$ and consequently we can limit the discussion to a single cell. We multiply (5.3.7) by ${\rm e}^{jk_m z}$ and integrate over one cell; the result is

$$\sum_{n=-\infty}^{\infty} \Gamma_n^2 A_n \delta_{n,m} L {\rm K}_0(\Gamma_n R_{\rm ext}) = \sum_{\nu=0}^{\infty} \Lambda_\nu^2 B_\nu t_{0,\nu}(R_{\rm ext}) \int_d^L {\rm d}z {\rm e}^{jk_m z} \cos[q_\nu(z - d)] \,. \tag{5.3.9}$$

We follow a similar procedure when imposing the continuity of the magnetic field; the difference in this case is that (5.3.8) is defined only in the groove's aperture thus we shall utilize the orthogonality of the trigonometric function $\cos[q_\nu(z - d)]$. Accordingly, (5.3.8) is multiplied by $\cos[q_\mu(z - d)]$ and we integrate over $d < z < L$; the result is

$$\sum_{n=-\infty}^{\infty} \Gamma_n A_n {\rm K}_1(\Gamma_n R_{\rm ext}) \int_d^L {\rm d}z \cos\Big[q_\mu(z - d)\Big] {\rm e}^{-jk_n z} = -\sum_{\nu=0}^{\infty} \Lambda_\nu B_\nu t_{1,\nu}(R_{\rm ext})(L - d) g_\mu \delta_{\nu,\mu} \,. \tag{5.3.10}$$

In this expression $g_0 = 1$ and $g_{n\neq 0} = 0.5$. It is convenient to define the quantity

$$\mathcal{L}_{n,\nu}(k) = \frac{1}{L - d}\int_d^L {\rm d}z \cos\Big[q_\nu(z - d)\Big]\, {\rm e}^{jk_n z} \,, \tag{5.3.11}$$

by whose means, (5.3.9) reads

$$A_n = \frac{1}{\Gamma_n^2 {\rm K}_0(\Gamma_n R_{\rm ext})} \frac{L - d}{L} \sum_{\nu=0}^{\infty} \Lambda_\nu^2 t_{0,\nu}(R_{\rm ext}) \mathcal{L}_{n,\nu}(k) B_\nu \,, \tag{5.3.12}$$

whereas (5.3.10)

$$B_\nu = -\frac{1}{\Lambda_\nu t_{1,\nu}(R_{\rm ext}) g_\nu} \sum_{n=-\infty}^{\infty} A_n \Gamma_n {\rm K}_1(\Gamma_n R_{\rm ext}) \mathcal{L}_{n,\nu}^*(k) \,. \tag{5.3.13}$$

These are two equations for two unknown sets of amplitudes (A_n, B_ν). The dispersion relation can be represented in two equivalent ways: One possibility

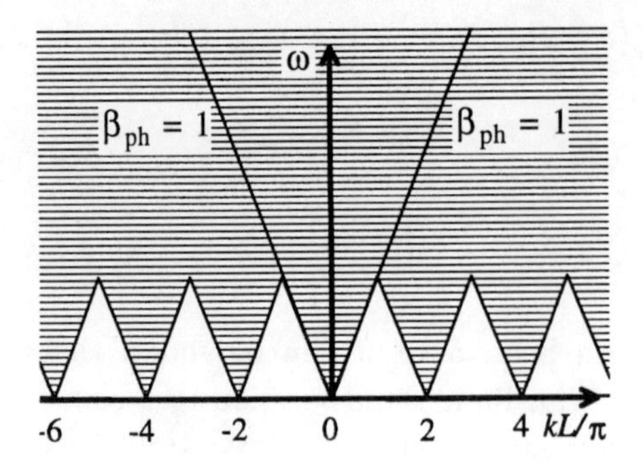

Fig. 5.11. Only the wavenumbers which are in the white triangles correspond to waves supported by an open structure

is to substitute (5.3.13) in (5.3.12) and obtain one equation for the amplitudes of the various harmonics

$$\sum_{m=-\infty}^{\infty}\left[\delta_{n,m}+\frac{L-d}{L}\frac{\Gamma_m \mathrm{K}_1(\Gamma_m R_{\mathrm{ext}})}{\Gamma_n^2 \mathrm{K}_0(\Gamma_n R_{\mathrm{ext}})}\sum_{\nu=0}^{\infty}\frac{t_{0,\nu}(R_{\mathrm{ext}})\Lambda_\nu}{t_{1,\nu}(R_{\mathrm{ext}})g_\nu}\mathcal{L}_{n,\nu}\mathcal{L}_{m,\nu}^*\right]A_m=0\,. \tag{5.3.14}$$

The other possibility is to substitute (5.3.12) in (5.3.13) and obtain one equation for the amplitudes of the various modes in the groove

$$\sum_{\mu=0}^{\infty}\left[\delta_{\nu,\mu}+\frac{L-d}{L}\frac{\Lambda_\mu^2 t_{0,\mu}(R_{\mathrm{ext}})}{\Lambda_\nu t_{1,\nu}(R_{\mathrm{ext}})g_\nu}\sum_{n=-\infty}^{\infty}\frac{\mathrm{K}_1(\Gamma_n R_{\mathrm{ext}})}{\Gamma_n \mathrm{K}_0(\Gamma_n R_{\mathrm{ext}})}\mathcal{L}_{n,\nu}^*\mathcal{L}_{n,\mu}\right]B_\mu=0\,. \tag{5.3.15}$$

In both cases the dispersion relation is calculated from the requirement that the determinant of the matrix which multiplies the vector of amplitudes, is zero. As in the closed structure the two methods are equivalent, but the last expression is by far more efficient for practical calculation.

There is one substantial difference between open and closed periodic structures. In the latter case, the radiation is guided by the waveguide and there is an infinite discrete spectrum of frequencies which can propagate along the system. In open structures, modes can propagate provided that the projection of the wavenumbers of *all harmonics* in the first Brillouin zone corresponds to waves whose phase velocity is smaller than c; in other words, no radiation propagates outwards (radially). Figure 5.11 illustrates the two regions of interest: in the shadowed region no solutions are permissible and in the remainder the solution is possible with an adequate choice of the geometric parameters. It is evident from this picture that waves at frequencies higher than

$$f \geq \frac{1}{2}\frac{c}{L}\,, \tag{5.3.16}$$

can not be supported by a disk-loaded wire, regardless of the geometrical details of the cavity. With this regard, an open structure forms a low pass

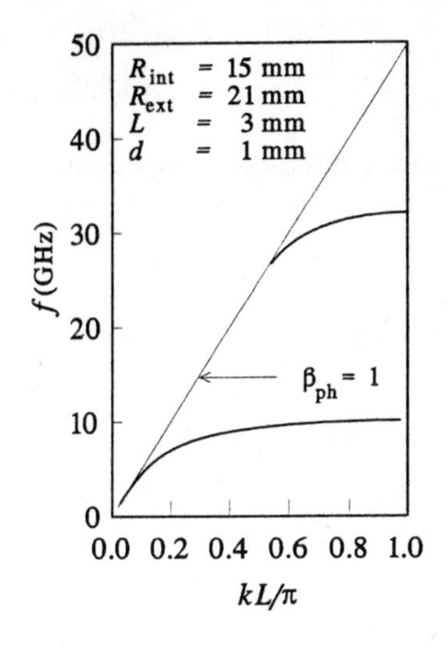

Fig. 5.12. Dispersion relation of the open periodic structure

filter. Figure 5.12 illustrates the dispersion relation of such a system for $L = 3\,\text{mm}$, $d = 1\,\text{mm}$, $R_{\text{int}} = 15\,\text{mm}$ and $R_{\text{ext}} = 21\,\text{mm}$. For comparison, in the same frequency range (0–50 GHz) there are 6 symmetric TM modes which can propagate in a closed system of the same geometry; obviously there are many others at higher frequencies.

5.3.2 Interaction Parameters

Provided that the electrons are interacting primarily with one harmonic (say $n = 0$) then we assume that the spatial component of the interaction in an amplifier is controlled by one parameter: the interaction impedance. For an annular beam of radius R_{b} and width Δ the effective field which acts on the electrons is

$$|E|^2 \equiv \frac{1}{R_{\text{b}}\Delta}\int_{R_{\text{b}}-\Delta/2}^{R_{\text{b}}+\Delta/2} \mathrm{d}r\, r\,\left|E_{z,n=0}(r,z)\right|^2 . \tag{5.3.17}$$

Using the explicit expression for E_z in (5.3.2) we find

$$|E|^2 \equiv \frac{c^4}{\omega^2}|A_0|^2 \Gamma_0^2 \frac{1}{R_{\text{b}}\Delta}\int_{\Gamma_0(R_{\text{b}}-\Delta/2)}^{\Gamma_0(R_{\text{b}}+\Delta/2)} \mathrm{d}\xi\,\xi\,\mathrm{K}_0^2(\xi)\,; \tag{5.3.18}$$

assuming that the variations of the wave across the beam section are negligible the integral reads

$$\int_{\Gamma_0(R_{\text{b}}-\Delta/2)}^{\Gamma_0(R_{\text{b}}+\Delta/2)} \mathrm{d}\xi\,\xi\,\mathrm{K}_0^2(\xi) \simeq \left(\Gamma_0^2 R_{\text{b}}\Delta\right)\mathrm{K}_0^2\left(\Gamma_0 R_{\text{b}}\right); \tag{5.3.19}$$

With this expression the effective electric field reads

$$\left|E\right|^2 \equiv \frac{c^4\Gamma_0^4}{\omega^2}\left|A_0\right|^2 \mathrm{K}_0^2(\Gamma_0 R_\mathrm{b}) \,. \tag{5.3.20}$$

In order to determine the explicit expression for the interaction impedance the total power which flows along the structure has to be determined. According to (5.3.2) it is given by

$$P = \frac{1}{2}(2\pi)\frac{1}{\eta_0}\sum_{n=-\infty}^{\infty}\left|cA_n\right|^2\left(\frac{ck_n}{\omega}\right)\int_{\Gamma_n R_\mathrm{ext}}^{\infty} \mathrm{d}\xi\, \xi\, \mathrm{K}_1^2(\xi) \,. \tag{5.3.21}$$

The last integral can be evaluated analytically [Abramowitz and Stegun (1968) p.484] and it reads

$$\begin{aligned} W_3(x) &\equiv \int_x^{\infty} \mathrm{d}\xi\, \xi\, \mathrm{K}_1^2(\xi) \\ &= x\mathrm{K}_0(x)\mathrm{K}_1(x) + \frac{1}{2}x^2\left[\mathrm{K}_0^2(x) - \mathrm{K}_1^2(x)\right] , \end{aligned} \tag{5.3.22}$$

hence

$$\begin{aligned} Z_\mathrm{int} \equiv& \frac{1}{2}\frac{\left|E\right|^2(\pi R_\mathrm{ext}^2)}{P} \\ =& \frac{1}{2}\eta_0\left[\frac{\mathrm{K}_0(\Gamma_0 R_\mathrm{b})}{\mathrm{K}_0(\Gamma_0 R_\mathrm{ext})}\right]^2 \\ &\times\left[\sum_{n=-\infty}^{\infty}\frac{W_3(\Gamma_n R_\mathrm{ext})}{(\Gamma_n R_\mathrm{ext})^2\mathrm{K}_0^2(\Gamma_n R_\mathrm{ext})}\left(\frac{\omega k_n}{c\Gamma_n^2}\right)\frac{\mathrm{sinc}^2[\frac{1}{2}k_n(L-d)]}{\mathrm{sinc}^2[\frac{1}{2}k(L-d)]}\right]^{-1} . \end{aligned} \tag{5.3.23}$$

The other parameter of interest in an oscillator is the interaction dielectric coefficient which is a measure of the total electromagnetic stored in the open structure and is defined by

$$\varepsilon_\mathrm{int} \equiv \frac{W_\mathrm{em}}{\frac{1}{2}\varepsilon_0\left|E\right|^2(\pi R_\mathrm{ext}^2)} \,. \tag{5.3.24}$$

Firstly, the electromagnetic energy stored in one groove, assuming one dominant mode, is given by

$$W_\mathrm{em,groove} = \frac{1}{2}\varepsilon_0(\pi R_\mathrm{int}^2)(L-d)\omega^2 B_0^2 R_\mathrm{int}^2 W_4 \,, \tag{5.3.25}$$

The total averaged (in time and space) electromagnetic energy stored in one period of the structure in its outer region ($r \geq R_\mathrm{ext}$) is

$$\begin{aligned} W_\mathrm{em,out} \equiv& \frac{2\pi L}{4\mu_0}\sum_{n=-\infty}^{\infty}\left|A_n\right|^2 \\ &\times\left[\left(1+\frac{k_n^2c^2}{\omega^2}\right)W_3(\Gamma_n R_\mathrm{ext}) + \frac{\Gamma_n^2c^2}{\omega^2}W_5(\Gamma_n R_\mathrm{ext})\right] , \end{aligned} \tag{5.3.26}$$

where

$$W_4 \equiv \frac{1}{R_{\rm int}^2}\int_{R_{\rm int}}^{R_{\rm ext}} {\rm d}r\, r \left\{ \left[{\rm J}_0\left(\frac{\omega}{c}r\right){\rm Y}_0\left(\frac{\omega}{c}R_{\rm int}\right) - {\rm Y}_0\left(\frac{\omega}{c}r\right){\rm J}_0\left(\frac{\omega}{c}R_{\rm int}\right)\right]^2 \right.$$

$$\left. + \left[{\rm J}_1\left(\frac{\omega}{c}r\right){\rm Y}_0\left(\frac{\omega}{c}R_{\rm int}\right) - {\rm Y}_0\left(\frac{\omega}{c}r\right){\rm J}_0\left(\frac{\omega}{c}R_{\rm int}\right)\right]^2 \right\},$$

$$W_5(x) \equiv \int_x^\infty {\rm d}\xi\, \xi\, {\rm K}_0^2(\xi),$$

$$= \frac{1}{2}x^2\left[{\rm K}_1^2(x) - {\rm K}_0^2(x)\right]. \tag{5.3.27}$$

With these definitions, the dielectric coefficient of the interaction is given by

$$\varepsilon_{\rm int} = \frac{L}{L-d}\left(\frac{R_{\rm int}}{R_{\rm ext}}\right)^2 \left[\frac{{\rm K}_0(\Gamma_0 R_{\rm ext})}{{\rm K}_0(\Gamma_0 R_{\rm b})}\right]^2 \frac{1}{{\rm sinc}^2\left[\frac{1}{2}k(L-d)\right]}\frac{W_4}{T_0^2(\alpha)}$$

$$+ \left[\frac{{\rm K}_0(\Gamma_0 R_{\rm ext})}{{\rm K}_0(\Gamma_0 R_{\rm b})}\right]^2 \frac{1}{{\rm sinc}^2\left[\frac{1}{2}k(L-d)\right]}$$

$$\times \sum_{n=-\infty}^{\infty}\left[\left(1+\frac{k_n^2c^2}{\omega^2}\right)W_3(\Gamma_n R_{\rm ext}) + \frac{\Gamma_n^2c^2}{\omega^2}W_5(\Gamma_n R_{\rm ext})\right]$$

$$\times \frac{{\rm sinc}^2[\frac{1}{2}k_n(L-d)]}{(\Gamma_n R_{\rm ext})^2(\Gamma_n c/\omega)^2}. \tag{5.3.28}$$

Note that in the open system the effect of the beam distance from the structure is represented by

$$Z_{\rm int} \propto \left[\frac{{\rm K}_0(\Gamma_0 R_{\rm b})}{{\rm K}_0(\Gamma_0 R_{\rm ext})}\right]^2, \tag{5.3.29}$$

which for large arguments of the modified Bessel function implies $Z_{\rm int} \propto \exp[-2\Gamma_0(R_{\rm b} - R_{\rm ext})]$ whereas the dielectric coefficient of the interaction,

$$\varepsilon_{\rm int} \propto \left[\frac{{\rm K}_0(\Gamma_0 R_{\rm ext})}{{\rm K}_0(\Gamma_0 R_{\rm b})}\right]^2, \tag{5.3.30}$$

is proportional to $\varepsilon_{\rm int} \propto \exp[2\Gamma_0(R_{\rm b} - R_{\rm ext})]$ for large arguments.

The two frames in Fig. 5.13 illustrate the interaction impedance of the two modes which are supported by the structure introduced above. In contrast to closed structure where the impedance has a minimum, in the open structure presented here, the impedance has a maximum as a function of the frequency. The peak of the lower (frequency) branch occurs at 9.575 GHz, the phase velocity is 0.58 c, the interaction impedance, for $R_{\rm b} = 25$ mm, is $Z_{\rm int} = 196\,\Omega$ and the coupling coefficient $K_0 = 29.2\,{\rm m}^{-1}$ (see (4.1.18) when the total current is 500 A. The peak at the upper branch occurs at 31.75 GHz and $Z_{\rm int} = 31.9\,\Omega$ corresponding to $K_0 = 9.3\,{\rm m}^{-1}$; the phase velocity in this case is 0.877 c.

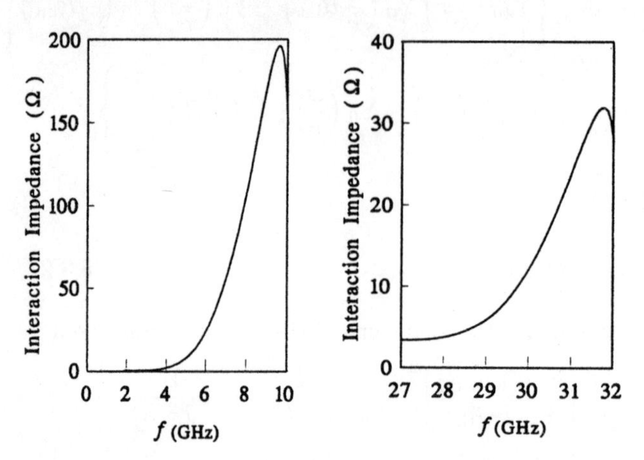

Fig. 5.13. The interaction impedance of the two modes presented in Fig. 5.12

We indicated previously that the advantage of an open periodic structure is that it supports the propagation of a small number of modes. Clearly it would be convenient to utilize this fact to generate radiation at high frequency. From the previous example, we observe that we could generate radiation at 32 GHz provided that we can suppress the lower frequency and no additional frequencies (of TM like modes) can develop in the system. In order to suppress the lower frequency we can take advantage of the fact that the interacting waves decay exponentially in the radial direction and a high frequency wave decays radially much rapidly than a low frequency one. It implies that in principle we can put an absorbing wall (say at $R = 30\,{\rm mm}$) which will virtually absorb all the energy from the low frequency wave but practically it does not affect the higher frequency mode since its amplitude is virtually zero at its location.

5.3.3 Green's Function: The Smith-Purcell Effect

When we investigated the electromagnetic field generated by a charged particle in its motion in the vicinity of a dielectric material it was shown that radiation can be generated if the velocity of the particle exceeds the phase velocity of the plane wave in the medium. A similar process may occur in a metallic periodic structure. Qualitatively the process is as follows: it was indicated in Sect. 2.2.4 that a point charge moving at a velocity v_0 generates a continuous spectrum of evanescent (non-radiating) waves; these waves impinge upon the grating whose periodicity is L. The incident wavenumber in the direction parallel to the motion of the particle is given by

$$k_z^{\rm inc} = \frac{\omega}{v_0}\,. \tag{5.3.31}$$

Although the great majority of the reflected waves are evanescent under certain circumstances, there might be a few which can propagate. An observer, located far away from the grating at an angle θ relative to the motion of the particle (z direction), measures the outcoming radiation. The projection on the z direction of the wavenumber as measured by this observer is

$$k_z^{\rm obs} = \frac{\omega}{c}\cos\theta\,. \tag{5.3.32}$$

The periodic structure couples between the wavenumbers in the z direction therefore the difference between the incident and observed (scattered) wavenumbers is attributed to the grating and it is an integer number ν of grating wavenumbers $2\pi\nu/L$,

$$k_z^{\rm inc} - k_z^{\rm obs} = \frac{2\pi\nu}{L}\,. \tag{5.3.33}$$

Substituting the previous two equations into the latter we obtain

$$\frac{\omega}{c} = \frac{2\pi}{L}\frac{\nu}{\beta^{-1} - \cos\theta}\,. \tag{5.3.34}$$

It indicates that for a given velocity and a given periodicity, different frequencies are emitted in different directions. The effect was first reported by Smith and Purcell (1953). Toraldo di Francia (1960) has formulated the problem in terms of the coupling between evanescent and propagating waves and Van den Berg (1973) has calculated the effect numerically. Salisbury (1970) observed a similar spectrum of radiation but in his experiment he found that there is a correlation between the radiation intensity and the current associated with electrons scattered by the grating. His interpretation is based on the oscillation of the electrons in the periodic potential induced by the scattered electrons. However, estimates of the acceleration associated with this process indicate that it can not account for the intensity of the observed radiation - Chang (1989). Recently, the Smith-Purcell effect was re-examined at much higher energies (3.6MeV) and in the angle range $56° - 150°$; the agreement between the dispersion relation and the experiment was excellent - see Doucas (1992). In this subsection we shall discuss in detail the dynamics of the Smith-Purcell effect as a particular case of Green's function formulation of the electromagnetic problem in an open periodic structure.

Consider a charged ring moving at a constant velocity v_0 along the same system as in Fig. 5.10. The beam has an azimuthal symmetry, forming an annular beam of radius $R_{\rm b}$ therefore the current density is given by

$$\begin{aligned} J_z(r,z,t) &= -ev_0\delta(z - v_0 t)\frac{1}{2\pi r}\delta(r - R_{\rm b})\,, \\ &\equiv \int_{-\infty}^{\infty} {\rm d}\omega\, J_z(r,z,\omega)\,{\rm e}^{j\omega t}\,; \end{aligned} \tag{5.3.35}$$

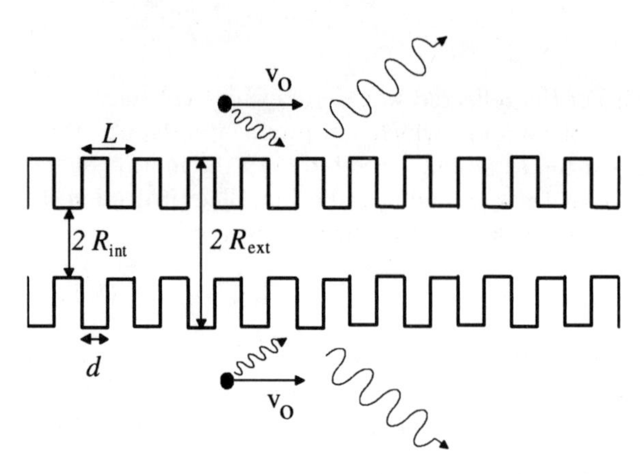

Fig. 5.14. Evanescent waves of a moving charged ring are scattered by an open periodic structure. Part of the scattered waves are propagating: Smith-Purcell radiation

this system is illustrated in Fig. 5.14.

This current density excites the longitudinal component of the magnetic vector potential. Its Fourier transform is governed by

$$\left[\frac{1}{r}\frac{\partial}{\partial r}r\frac{\partial}{\partial r}+\frac{\partial^2}{\partial z^2}+\frac{\omega^2}{c^2}\right]A_z(r,z,\omega) = -\mu_0 J_z(r,z,\omega)\,, \tag{5.3.36}$$

where the explicit expression for the Fourier transform of the current density is

$$J_z(r,z,\omega) = -e\frac{1}{2\pi\, r}\delta(r-R_\mathrm{b})\frac{1}{2\pi}\mathrm{e}^{-j(\omega/v_0)z}\,. \tag{5.3.37}$$

Green's function associated with the boundless problem is

$$G(r,z|r',z') = \int_{-\infty}^{\infty}\mathrm{d}q\, g(r|r';q)\,\mathrm{e}^{-jq(z-z')}\,, \tag{5.3.38}$$

where $g(r|r';q)$ is given by

$$g(r|r';q) = \frac{1}{(2\pi)^2}\begin{cases}\mathrm{I}_0(\Gamma r)\mathrm{K}_0(\Gamma r') & \text{for} \quad 0\le r\le r'<\infty\,,\\ \mathrm{K}_0(\Gamma r)\mathrm{I}_0(\Gamma r') & \text{for} \quad 0< r'\le r<\infty\,,\end{cases} \tag{5.3.39}$$

and $\Gamma^2 = q^2-(\omega/c)^2$.

With this function, Green's theorem [(2.4.6)] and the current density as given in (5.3.37), we can determine the magnetic vector potential

$$A_z(r,z,\omega) = 2\pi\mu_0 \int_{R_{\rm ext}}^{\infty} {\rm d}r' r' \int_{-\infty}^{\infty} {\rm d}z' G(r,z|r',z') J_z(r,z,\omega)$$
$$+ \int_{-\pi/L}^{\pi/L} {\rm d}k \sum_{n=-\infty}^{\infty} A_n(k) {\rm e}^{-jk_n z} {\rm K}_0(\Gamma_n r) \,. \tag{5.3.40}$$

Note that with this notation, the continuous variable q was replaced by $k + 2\pi n/L$ i.e.,

$$\int_{-\infty}^{\infty} {\rm d}q \rightarrow \int_{-\pi/L}^{\pi/L} {\rm d}k \sum_{n=-\infty}^{\infty} \,. \tag{5.3.41}$$

In one groove the solution, taking only the lowest mode, is given by

$$A_z(r,z,\omega) = DT_0\left(\frac{\omega}{c} r\right) , \tag{5.3.42}$$

where $T_0(\frac{\omega}{c}r) \equiv {\rm J}_0(\frac{\omega}{c}r){\rm Y}_0(\frac{\omega}{c}R_{\rm int}) - {\rm Y}_0(\frac{\omega}{c}r){\rm J}_0(\frac{\omega}{c}R_{\rm int})$. The next step is to substitute the explicit expression for the current density and integrate over the variables r', z':

$$A_z(r,z,\omega) = (-e)\mu_0 \int_{-\infty}^{\infty} {\rm d}q g(r|R_{\rm b})\, {\rm e}^{-jqz} \delta\left(q - \frac{\omega}{v_0}\right)$$
$$+ \int_{-\pi/L}^{\pi/L} {\rm d}k \sum_{n=-\infty}^{\infty} A_n(k) {\rm e}^{-jk_n z} {\rm K}_0(\Gamma_n r) \,. \tag{5.3.43}$$

The Dirac delta function in the first expression indicates that the spatial spectrum of waves is determined by the frequency and the velocity of the particle. Without loss of generality we can express this wavenumber in terms of the periodicity of the structure

$$\frac{\omega}{v_0} \equiv k_0 + \frac{2\pi M}{L} , \tag{5.3.44}$$

which means that at a frequency ω, the moving charge "emits" an evanescent wave which corresponds to the M'th harmonic and whose projection in the first Brillouin zone is k_0. For imposing the boundary conditions we can express the magnetic vector potential in the region $R_{\rm b} \geq r \geq R_{\rm ext}$ as

$$A_z(r,z,\omega) = \int_{-\pi/L}^{\pi/L} {\rm d}k \sum_{n=-\infty}^{\infty} {\rm e}^{-jk_n z} \Big[A_n(k) {\rm K}_0(\Gamma_n r) + B_n(k) {\rm I}_0(\Gamma_n r)\Big] , \tag{5.3.45}$$

where

$$B_n(k) = -e\mu_0 \frac{1}{(2\pi)^2} {\rm K}_0(\Gamma_n R_{\rm b}) \delta_{n,M} \delta(k - k_0) \,. \tag{5.3.46}$$

With this definition we can omit the notation of the integration over the first Brillouin zone and the two field components which are relevant to the boundary conditions at $r = R_{\rm ext}$ read

$$E_z(r,z,\omega) = \frac{c^2}{j\omega}\sum_{n=-\infty}^{\infty}(-\Gamma_n^2)e^{-jk_nz}\left[A_n(k)K_0(\Gamma_n r) + B_n(k)I_0(\Gamma_n r)\right],$$

$$H_\phi(r,z,\omega) = \frac{-1}{\mu_0}\sum_{n=-\infty}^{\infty}\Gamma_n e^{-jk_nz}\left[-A_n(k)K_1(\Gamma_n r) + B_n(k)I_1(\Gamma_n r)\right]. \tag{5.3.47}$$

Our next step is to impose the continuity of the boundary conditions at the interface ($r = R_{\rm ext}$). Continuity of the longitudinal component of the electric field ($E_z(r = R_{\rm ext}, -\infty < z < \infty)$) reads

$$\frac{c^2}{j\omega}\sum_{n=-\infty}^{\infty}(-\Gamma_n^2)e^{-jk_nz}\left[A_nK_0(\Delta_n) + B_nI_0(\Delta_n)\right]$$

$$= \begin{cases} 0 & \text{for } 0 < z < d, \\ -j\omega DT_0(\frac{\omega}{c}R_{\rm ext}) & \text{for } d < z < L, \end{cases} \tag{5.3.48}$$

and the continuity of the azimuthal magnetic field at the aperture of the groove

$$\frac{-1}{\mu_0}\sum_{n=-\infty}^{\infty}\Gamma_n e^{-jk_nz}\left[-A_nK_1(\Delta_n) + B_nI_1(\Delta_n)\right] = \frac{1}{\mu_0}\frac{\omega}{c}DT_1\left(\frac{\omega}{c}R_{\rm ext}\right); \tag{5.3.49}$$

here $\Delta_n = \Gamma_n R_{\rm ext}$ and $T_1(\frac{\omega}{c}r) \equiv J_1(\frac{\omega}{c}r)Y_0(\frac{\omega}{c}R_{\rm int}) - Y_1(\frac{\omega}{c}r)J_0(\frac{\omega}{c}R_{\rm int})$.

Following the same procedure as at the beginning of this section we obtain from (5.3.48)

$$A_nK_0(\Delta_n) + B_nI_0(\Delta_n) = -\frac{\omega^2}{c^2\Gamma_n^2}\frac{L-d}{L}T_0\left(\frac{\omega}{c}R_{\rm ext}\right)\mathcal{L}_n(k)D, \tag{5.3.50}$$

and from (5.3.49)

$$DT_1\left(\frac{\omega}{c}R_{\rm ext}\right) = -\sum_{n=-\infty}^{\infty}\frac{c\Gamma_n}{\omega}\left[-A_nK_1(\Delta_n) + B_nI_1(\Delta_n)\right]\mathcal{L}_n^*(k). \tag{5.3.51}$$

It is convenient to substitute (5.3.50) into the latter and determine the amplitude of the magnetic vector potential in the groove:

$$D(k) = \frac{e\mu_0}{(2\pi)^2}\frac{K_0\left(\frac{\omega}{c}R_b\frac{1}{\gamma\beta}\right)}{K_0\left(\frac{\omega}{c}R_{\rm ext}\frac{1}{\gamma\beta}\right)}\frac{\mathcal{L}_M^*(k)}{\frac{\omega}{c}R_{\rm ext}\mathcal{D}(\omega,k)}\delta(k-k_0), \tag{5.3.52}$$

where $\mathcal{D}(\omega,k)$ denotes the dispersion relation in this structure within the framework of the single mode approximation,

$$\begin{aligned}\mathcal{D}(\omega,k) =& T_1\left(\frac{\omega}{c}R_{\rm ext}\right)+\frac{L-d}{L}T_0\left(\frac{\omega}{c}R_{\rm ext}\right)\left(\frac{\omega}{c}R_{\rm ext}\right)\\ &\times\sum_{n=-\infty}^{\infty}\left|\mathcal{L}_n(k)\right|^2\frac{\mathrm{K}_1(\Delta_n)}{\Delta_n\mathrm{K}_0(\Delta_n)}\,. \end{aligned}\tag{5.3.53}$$

In principle, with the expression in (5.3.52) the electromagnetic problem in the frequency domain has been solved, since the amplitudes of all harmonics out of the grooves can be determined using (5.3.50). In the context of Smith-Purcell effect we are interested in those harmonics which may contribute to the radiation far away from the particle. Consequently, the B_n term (in (5.3.50)) can be ignored since it corresponds to an evanescent mode and the magnetic vector potential which may contribute to the radiation field is

$$A_z(r,z,\omega)=\frac{e\mu_0}{(2\pi)^2}\frac{L}{R_{\rm ext}}\sum_{n=-\infty}^{\infty}\mathcal{A}_n(\omega)\mathrm{K}_0(\Gamma_{n,0}r)\,\mathrm{e}^{-jk_{n,0}z}\,,\tag{5.3.54}$$

where $k_{n,0}=k_0+2\pi n/L$, $\Gamma_{n,0}^2\equiv k_{n,0}^2-(\omega/c)^2$ and

$$\begin{aligned}\mathcal{A}_n(\omega)\equiv&\frac{L-d}{L}T_0\left(\omega R_{\rm ext}/c\right)\frac{\mathrm{K}_0(\omega R_{\rm b}/\gamma v_0)}{\mathrm{K}_0(\omega R_{\rm ext}/\gamma v_0)}\\ &\times\left[\left(\frac{\omega}{c\Gamma_n}\right)^2\frac{\mathcal{L}_n(k)\mathcal{L}_M^*(k)}{\mathcal{D}(\omega,k)\mathrm{K}_0(\Delta_n)}\right]_{k=k_0}\,.\end{aligned}\tag{5.3.55}$$

This quantity determines the amplitude of the radiation field corresponding to the wavenumber, k_0, and the harmonics index, n. According to the $k-\omega$ diagram in Fig. 5.11 the former has two possibilities: (i) $\omega/v_0<\pi/L$ in which case only the surface modes are excited and no radiation can be emitted perpendicular to the grating. In other words, no radiation will occur transverse to the grating in the frequency range

$$0<\omega<\frac{\pi}{L}v_0\,,\tag{5.3.56}$$

and all radiation propagates parallel to the grating. (ii) For all the values of frequencies $\omega/v_0>\pi/L$ there can be radiation, whose wavenumber is

$$\frac{\omega}{c}\cos\alpha=k_0+\frac{2\pi}{L}N\,.\tag{5.3.57}$$

Therefore the exact expression for the frequency of the radiation generated is

$$\frac{\omega}{c}=\frac{2\pi}{L}\frac{M-N}{\beta^{-1}-\cos\alpha}\,.\tag{5.3.58}$$

Comparing this expression with (5.3.34) it is readily observed that the coupling due to the grating is $\nu = M - N$ and we expect that the angle α at which the wave is propagating will ultimately coincide with the angle at which the observer is located. This will be shown in a systematic way in the remainder of this subsection.

We next examine the power emitted at a given angle; for this purpose we investigate the propagation of the Nth harmonic as given by

$$A_z(r,z,\omega) = \frac{e\mu_0}{(2\pi)^2}\frac{L}{R_{\rm ext}}\mathcal{A}_N\left(-j\frac{\pi}{2}\right)H_0^{(2)}\left(\frac{\omega}{c}r\sin\alpha\right)\exp[-j(\omega/c)z\cos\alpha]\,. \tag{5.3.59}$$

An observer located at

$$\begin{aligned} r &= \varrho\sin\theta\,,\\ z &= \varrho\cos\theta\,, \end{aligned} \tag{5.3.60}$$

measures the outgoing radiation and assuming that it is located many wavelengths from the grating i.e., $\varrho \to \infty$ we can use the asymptotic form of Hankel function [Abramowitz and Stegun (1968) p.364],

$$\lim_{x\to\infty} H_0^{(2)}(x) \simeq \sqrt{\frac{2}{\pi x}}\,\mathrm{e}^{-j(x-\pi/4)}\,, \tag{5.3.61}$$

in order to simplify the magnetic vector potential in the time domain

$$\begin{aligned} A_z(\varrho,\theta,t) = &-\frac{e\mu_0}{(2\pi)^2}\frac{L}{R_{\rm ext}}\int_{-\infty}^{\infty}\mathrm{d}\omega\mathrm{e}^{j\omega t}\\ &\times\left[j\mathcal{A}_N(\omega)\sqrt{\frac{\pi}{2}\frac{1}{(\omega/c)\varrho\sin\alpha\,\sin\theta}}\mathrm{e}^{j\pi/4}\right]\\ &\times\exp[-j(\omega/c)\varrho\cos(\theta-\alpha)]\,. \end{aligned} \tag{5.3.62}$$

The frequency of the wave which propagates in a direction, α, varies according to (5.3.58) therefore

$$d\omega = -\frac{2\pi c}{L}(M-N)\frac{\sin\alpha}{(\beta^{-1}-\cos\alpha)^2}d\alpha \tag{5.3.63}$$

and

$$\begin{aligned} A_z(\varrho,\theta,t) = &\frac{2\pi c}{L}(M-N)\frac{e\mu_0}{(2\pi)^2}\frac{L}{R_{\rm ext}}\int_0^{\pi}\mathrm{d}\alpha\frac{\sin\alpha}{(\beta^{-1}-\cos\alpha)^2}\\ &\times\left[j\mathrm{e}^{j\pi/4}\mathcal{A}_N(\omega)\sqrt{\frac{\pi/2}{(\omega/c)\varrho\sin\alpha\sin\theta}}\right.\\ &\left.\times\exp\left(j\omega t - j\frac{\omega}{c}\varrho\cos(\alpha-\theta)\right)+\mathrm{c.c.}\right]. \end{aligned} \tag{5.3.64}$$

The last integral can be evaluated for $\varrho \to \infty$ using the stationary phase method. Assuming that the exponential function varies much faster than all other functions of the integrand, then the main contribution to the integral occurs from the stationary point at $\alpha = \theta$ hence

$$\begin{aligned} A_z(\varrho,\theta,t) =& \frac{e\mu_0}{(2\pi)^2}\frac{L}{R_{\text{ext}}}\frac{4\sqrt{\pi}c\int_0^\infty \mathrm{d}x\cos\left(\frac{1}{2}x^2\right)}{\varrho(\beta^{-1}-\cos\theta)} \\ & \times \left|\mathcal{A}_N(\omega)\right|\cos\left[\omega(t-\varrho/c)-\Phi_N\right], \end{aligned} \tag{5.3.65}$$

where $\mathcal{A}_N = \left|\mathcal{A}_N\right|\mathrm{e}^{-j\Phi_N}$; the Fresnel integral can be evaluated analytically and it reads

$$\int_0^\infty \mathrm{d}x\cos\left(\frac{1}{2}x^2\right) = \frac{1}{2}\sqrt{\pi}. \tag{5.3.66}$$

The angular frequency in these expressions is a function of the angle θ as determined by

$$\frac{\omega}{c} = \frac{2\pi}{L}\frac{M-N}{\beta^{-1}-\cos\theta}, \tag{5.3.67}$$

which is the exact expression to describe the Smith-Purcell effect. The far magnetic field can now be evaluated and it reads

$$\begin{aligned} H_\phi(\varrho,\theta,t) =& \frac{\omega}{c}\sin\theta\left[\frac{e}{2\pi}\frac{L}{R_{\text{ext}}}\frac{c}{\varrho(\beta^{-1}-\cos\theta)}\left|\mathcal{A}_N\right|\right] \\ & \times \sin\left[\omega(t-\varrho/c)-\Phi_N\right]. \end{aligned} \tag{5.3.68}$$

The radial (in spherical coordinates) component of the average Poynting vector is

$$S_\varrho(\varrho,\theta) = \frac{1}{2}\eta_0\frac{\omega^2}{c^2}\sin^2\theta\frac{e^2}{(2\pi)^2}\frac{L^2}{R_{\text{ext}}^2}\frac{c^2}{\varrho^2(\beta^{-1}-\cos\theta)^2}\left|\mathcal{A}_N[\omega(\theta)]\right|^2, \tag{5.3.69}$$

and the average power in an angular interval $\theta \to \theta + d\theta$ is given by $P(\theta) = 2\pi\varrho^2\sin(\theta)S_\varrho(\varrho,\theta)$ or explicitly

$$P(\theta) = \omega(\theta)\left[\frac{e^2}{4\pi\varepsilon_0}\frac{\omega(\theta)}{c}\frac{L^2}{R_{\text{ext}}^2}\right]\frac{\sin^3\theta}{(\beta^{-1}-\cos\theta)^2}\left|\mathcal{A}_N[\omega(\theta)]\right|^2. \tag{5.3.70}$$

For evaluation of the radiated power it is convenient to define the normalized power emitted as

$$\begin{aligned} \bar{P}(\theta) \equiv& P(\theta)\left[\frac{e^2c}{4\pi\varepsilon_0 R_{\text{ext}}^2}\right]^{-1}, \\ =& (2\pi)^2\frac{\sin^3\theta}{(\beta^{-1}-\cos\theta)^4}\left|\mathcal{A}_N[\omega(\theta)]\right|^2. \end{aligned} \tag{5.3.71}$$

Figure 5.15 consists of 4 frames: the top-left illustrates the frequency emitted by a 1 MeV charge at the various angles; the geometric parameters are $R_{\rm int} = 15\,{\rm mm}$, $R_{\rm ext} = 21\,{\rm mm}$, $R_{\rm b} = 22\,{\rm mm}$, $L = 3\,{\rm mm}$ and $d = 1\,{\rm mm}$. As anticipated from (5.3.56) no radiation is emitted at frequencies lower than 50 GHz. The top-right frame presents the variation of the power as a function of the angle θ. The lobes in the radiation pattern are the result of interference due to the geometry of the grooves. There are four terms which contribute to the lobes formation: *(i)* effect of the groove height represented by the term $T_0(\omega R_{\rm ext}/c)$, *(ii)* effect of the "incident" wave represented by $\mathcal{L}^*_M(k) \propto {\rm sinc}[\omega(L-d)/2c\beta]$, *(iii)* the effect of the scattered wave represented by $\mathcal{L}^*_N(k) \propto {\rm sinc}[\omega(L-d)\cos\theta/2c]$ and *(iv)* the dispersion term $\mathcal{D}[\omega, k_0(\omega)]$. The lower two frames represent the first and last terms; the other two terms are slowly varying functions for angles larger than 90°. The two strong lobes of the radiation pattern are primarily controlled by the dispersion term and the resonances associated with groove height. Note that the bottom-left frame indicates that there are frequencies and angles of zero emitted power.

5.3.4 Scattering from Cylindrical Gratings

Gratings play an important role in optics [Born and Wolf (1984)] and their analysis is reviewed in a book by Petit (1980). In most cases these are planar structures [DeSanto (1971,1972)]. However in this subsection we continue our approach of investigating cylindrical structures, namely for cases when the radius of curvature is not by orders of magnitude larger than the vacuum wavelength. The system under consideration is illustrated in Fig. 5.11 but in this subsection a single propagating wave impinges upon the structure.

In the context of microwave sources cylindrical gratings are important in multi-mode devices which come to solve the problem of the large gradients which develop on the metallic surface when ultra-high power is generated. For this purpose, the internal radius of the device is increased allowing several modes to propagate at the same frequency. Since the impedance of each mode is different it is practically impossible to tune the system using the conventional tuning methods. However, with proper design of the gratings it is possible to cascade two or more periodic structures [Bugaev 1990] such that at a certain frequency, neither of the modes emits radiation in a given direction, in this way isolating one stage from another. In order to demonstrate the concept briefly, let us consider a plane wave which is described by

$$A_z(x, z; \omega) = B\exp\left(-j\frac{\omega}{c}z\cos\theta_{\rm inc} + j\frac{\omega}{c}x\sin\theta_{\rm inc}\right), \tag{5.3.72}$$

and which impinges upon the corrugated cylinder. We can use the fact that

$$e^{\frac{1}{2}\zeta(\tau-1/\tau)} \equiv \sum_{\nu=-\infty}^{\infty} \tau^\nu {\rm J}_\nu(\zeta), \tag{5.3.73}$$

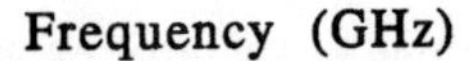

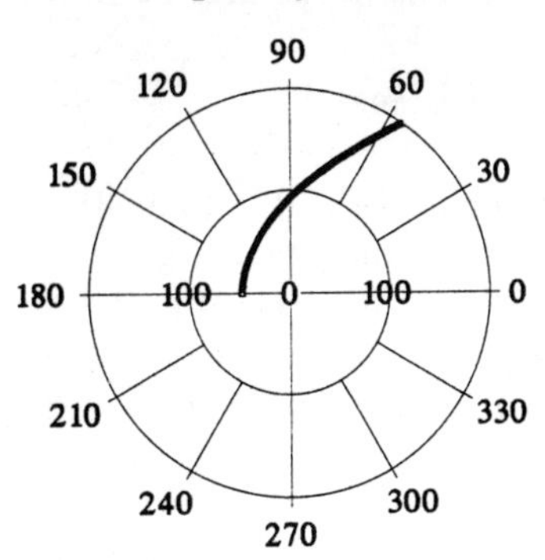

Normalized Power (dB)

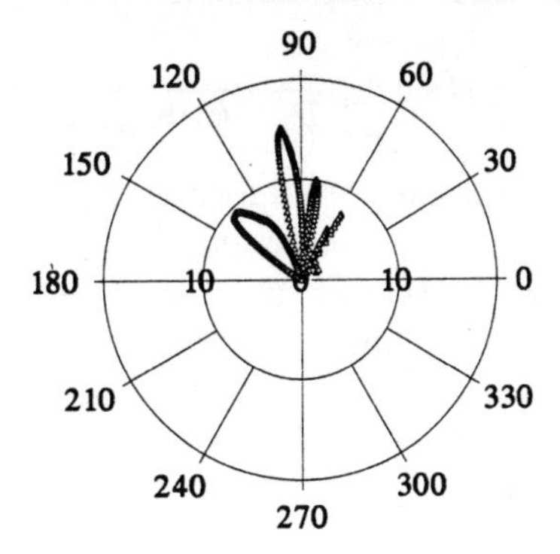

Groove Height Term

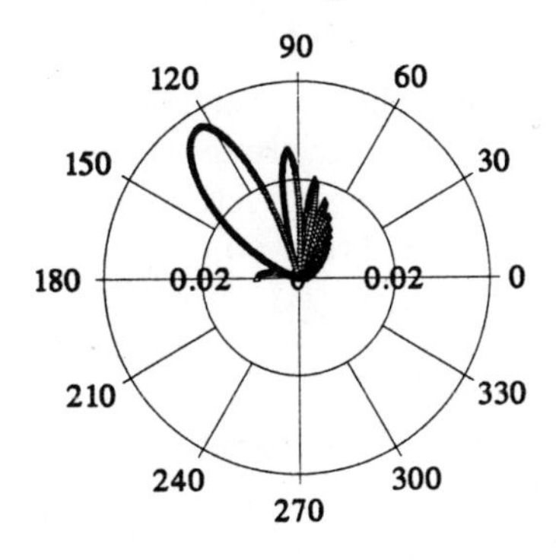

Dispersion Term

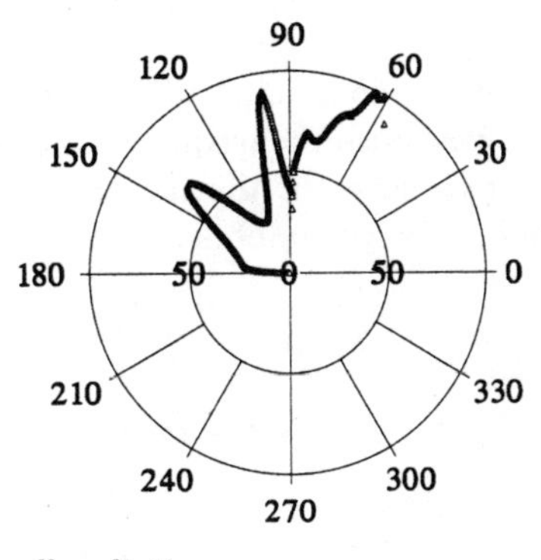

Fig. 5.15. Characteristics of Smith-Purcell radiation

in order to write the magnetic vector potential as

$$A_z(x, z; \omega) = B\mathrm{e}^{-j(\omega/c)\cos\theta_{\mathrm{inc}} z} \sum_{\nu=-\infty}^{\infty} \mathrm{e}^{j\nu(\phi+\pi/2)} \mathrm{J}_\nu\left(\frac{\omega}{c} r \sin\theta_{\mathrm{inc}}\right) , \quad (5.3.74)$$

where we assumed that $x = r\cos\phi$. The corrugated cylinder is azimuthally symmetric therefore we can treat each azimuthal harmonic independently. We shall consider for the moment only the contribution of the symmetric harmonic ($\nu = 0$) therefore one can now use the results from the previous subsection and only need to identify the incident wave with B_n of (5.3.45). For this purpose we use the fact that $\mathrm{I}_0(j\xi) = \mathrm{J}_0(\xi)$ to write

$$B_n = B\delta(k - k_0)\delta_{n,N} , \quad (5.3.75)$$

where

$$\frac{\omega}{c}\cos\theta_{\mathrm{inc}} = k_0 + \frac{2\pi N}{L} . \quad (5.3.76)$$

The amplitude of the magnetic vector potential in one of the grooves is

$$D = -jB\sin\theta_{\text{inc}}\frac{1}{\mathcal{D}(\omega,k_0)\text{K}_0(j(\omega/c)r\sin\theta_{\text{inc}})}\mathcal{L}_N^*(k_0)\,; \tag{5.3.77}$$

$\mathcal{D}(\omega,k)$ is the expression for the dispersion relation as defined in (5.3.53) and the modified Bessel function of the second kind with imaginary argument is related to the Hankel function of the second kind: $\text{K}_0(j\xi) = -\frac{1}{2}j\pi H_0^{(2)}(\xi)$. With the expression for D we can now determine the amplitude of the scattered waves using the relation in (5.3.50) but in contrast with the Smith-Purcell case, in this subsection, the B_n term does contribute to the radiation field therefore

$$\begin{aligned} A_n = &-\frac{B}{\text{K}_0(\Delta_n)} \\ &\times\left[\delta_{n,N}\text{I}_0(\Delta_n) - j\frac{\omega^2 R_{\text{ext}}^2}{c^2\Delta_n^2}\frac{L-d}{L}\frac{T_0(\frac{\omega}{c}R_{\text{ext}})\sin\theta_{\text{inc}}\mathcal{L}_n(k_0)\mathcal{L}_N^*(k_0)}{\mathcal{D}(\omega,k_0)\text{K}_0(j\frac{\omega}{c}r\sin\theta_{\text{inc}})}\right]\,; \end{aligned} \tag{5.3.78}$$

in all the expressions (Δ_n) where k does not occur explicitly it equals k_0.

In this case both the frequency and the angle of incidence are set therefore the scattered wave (in the far field) can propagate along directions which correspond to wavenumbers which satisfy $k^{\text{sca}} = k^{\text{inc}} \pm 2\pi\nu/L$ or

$$\frac{\omega}{c}\left[\cos\theta_{\text{sca}} - \cos\theta_{\text{inc}}\right] = \pm\frac{2\pi}{L}\nu\,. \tag{5.3.79}$$

In particular for $n \neq N$ the term $\mathcal{L}_n(k_0)$ can be shown to have zeros for

$$\frac{1}{2}\frac{\omega}{c}(L-d)\cos\theta_{\text{sca}} = \pi\,, \tag{5.3.80}$$

therefore no power is scattered in this direction. This means that regardless of what is the incident angle, at this particular (vacuum) wavelength if the geometry $(L-d)$ is chosen to satisfy (5.3.80) no power is reflected in this direction. This fact can be utilized to avoid reflections in multi-mode devices.

5.4 Transients

When several bunches of electrons are injected in a structure as is the case in an accelerator they not only interact with the electromagnetic field which was prepared for their acceleration, but they also generate a whole spectrum of waves at different frequencies. These form a so-called *wake field* which in turn can deteriorate the acceleration process. In order to visualize the process, imagine a pulse which consists of two bunches. When the first enters the periodic structure it generates a wake-field and if this is not "drained" fast enough then it may affect the interaction of the trailing bunch.

Propagation of a pulse in a disk-loaded waveguide should, in principle, account for all the modes and all the reflections from the disks. The difficulties in the analysis of transients generated by charged particles in periodic closed structures arise from the fact that *(i)* the frequency spectrum of a moving particle is infinite and *(ii)* although the spectrum of frequencies in a closed periodic structure is discrete, it spans to infinity. The analysis is somewhat simplified by the fact that in the transverse direction the (evanescent) wave decays exponentially ($\mathrm{e}^{-\omega r/c\gamma\beta}$) therefore the contribution of the high frequencies might be small – at least at low energies. The situation is different in open periodic structures where, as we already indicated, the spectrum is discrete and finite. Therefore, potentially less energy is induced in the system. In this context Smith-Purcell effect can be regarded as a transient generated by a moving particle.

In order to illustrate the effect of the periodicity on the propagation of a wave packet we shall consider at $t = 0$ the same wavepacket $a(z)$, in vacuum and in a periodic structure. The propagation in vacuum will be represented by a dispersion relation $k^2 = \omega^2/c^2$, therefore a scalar wave function $\Psi(z,t)$ is given by

$$\Psi(z,t) = \int_{-\infty}^{\infty} \mathrm{d}k\psi(k)\mathrm{e}^{-jkz}\frac{1}{2}\left[\mathrm{e}^{jkct} + \mathrm{e}^{-jkct}\right] . \tag{5.4.1}$$

Since at $t = 0$ this function equals $a(z)$, the amplitudes $\psi(k)$ can be readily determined using the inverse Fourier transform hence

$$\psi(k) = \frac{1}{2\pi}\int_{-\infty}^{\infty} \mathrm{d}za(z)\mathrm{e}^{jkz} . \tag{5.4.2}$$

Substituting back into (5.4.1) we find that

$$\Psi(z,t) = \frac{1}{2}\left[a(z-ct) + a(z+ct)\right] , \tag{5.4.3}$$

which basically indicates that the pulse moves at the speed of light in both directions and asymptotically, it preserves its shape.

In a periodic structure the description of the wavepacket is complicated by the dispersion relation which in its lowest order approximation (e.g., first TM symmetric mode in a waveguide) can be expressed as

$$\omega(k) = \bar{\omega} - \delta\omega\cos(kL) , \tag{5.4.4}$$

where $\bar{\omega} = (\omega_0 + \omega_\pi)/2$ is the average frequency between the low ($kL = 0$) cut-off denoted by ω_0 and the high ($kL = \pi$) cut-off denoted by ω_π. The quantity $\delta\omega = (\omega_\pi - \omega_0)/2$ is half the passband width and L is the period of the structure. Contrary to the previous case k here denotes the wavenumber in the first Brillouin zone. In the framework of this approximation we can use Floquet's representation to write

$$\Psi(z,t) = \mathrm{Re}\left[\sum_{n=-\infty}^{\infty}\int_{-\pi/L}^{\pi/L} \mathrm{d}k\psi_n(k)\mathrm{e}^{j\omega(k)t-jk_n z}\right], \tag{5.4.5}$$

where $k_n = k + 2\pi n/L$. The amplitudes $\psi_n(k)$ are determined by the value of the function at $t = 0$ hence

$$\psi_n(k) = \frac{1}{2\pi}\int_{-\infty}^{\infty} \mathrm{d}z a(z)\mathrm{e}^{jk_n z}. \tag{5.4.6}$$

Substituting back into (5.4.5) we have

$$\Psi(z,t) = \mathrm{Re}\left[\frac{1}{2\pi}\int_{-\infty}^{\infty} \mathrm{d}\zeta a(\zeta)\sum_{n=-\infty}^{\infty}\int_{-\pi/L}^{\pi/L} \mathrm{d}k\mathrm{e}^{jt[\bar{\omega}-\delta\omega\cos(kL)]-jk_n(z-\zeta)}\right]. \tag{5.4.7}$$

At this point we can take advantage of (5.3.73) and simplify the last equation to read

$$\Psi(z,t) = \mathrm{Re}\left[\frac{1}{2\pi}\int_{-\infty}^{\infty} \mathrm{d}\zeta a(\zeta)\sum_{n=-\infty}^{\infty}\int_{-\pi/L}^{\pi/L} \mathrm{d}k\mathrm{e}^{j\bar{\omega}t} \times \sum_{\nu=-\infty}^{\infty} \mathrm{J}_\nu(\delta\omega t)\mathrm{e}^{j(kL-\pi/2)\nu}\mathrm{e}^{-jk_n(z-\zeta)}\right], \tag{5.4.8}$$

which after the evaluation of the integrals and summation (over n) reads

$$\Psi(z,t) = \sum_{\nu=-\infty}^{\infty} a(z-\nu L)\mathrm{J}_\nu(\delta\omega t)\cos(\bar{\omega}t - \pi\nu/2). \tag{5.4.9}$$

Figure 5.16 illustrates the propagation of two wavepackets in vacuum (dashed line) and in a periodic structure. The latter is characterized by $\bar{\omega} = 2\pi \times 10\,\mathrm{GHz}$, $\delta\omega = \bar{\omega}/30$ and a spatial periodicity of $L = 1\,\mathrm{cm}$. At $t = 0$ the distribution is a Gaussian, $a(z) = \exp[-(z/L)^2]$. In each one of the frames $\Psi(z,t)$ was plotted at a different time as a function of z. Characteristic to all the frames is the relatively large peak following the front of the pulse.

It is evident that although the front of the pulse propagates at the speed of light (as in vacuum) the main pulse propagates slower. In fact, a substantial fraction of the energy remains at the origin even a long time after $t = 0$. For the parameters used, the amplitude of the signal at the origin ($z = 0$) is dominated by the zero order Bessel function i.e., $\mathrm{J}_0(\delta\omega t)$ therefore the energy is drained on a time scale which is determined by the asymptotic behavior of the Bessel function namely $\propto 1/\sqrt{\delta\omega t}$. Clearly the wider the passband the faster the energy is drained from the origin.

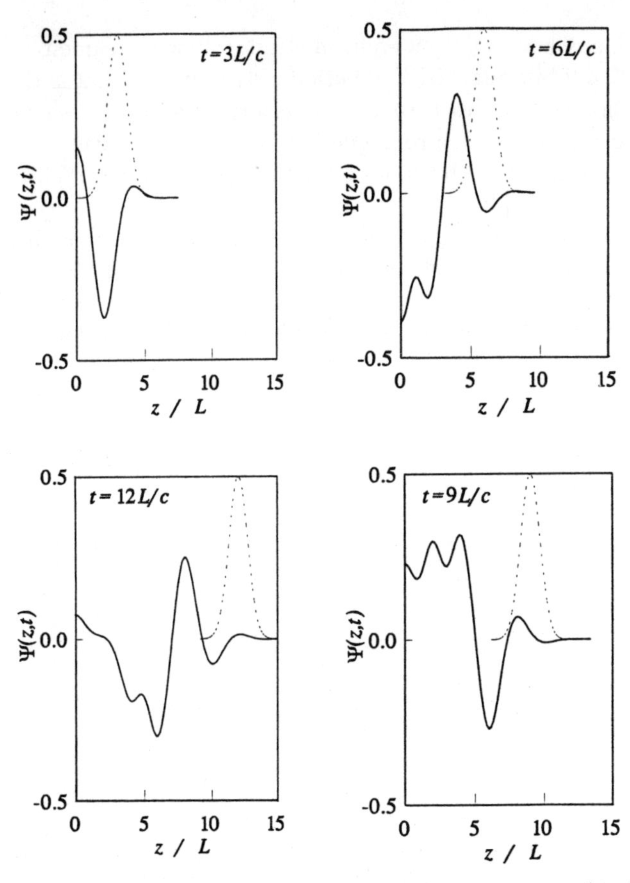

Fig. 5.16. Propagation of the wavepacket in vacuum (dashed line) and in the periodic structure at four instants

Exercises

5.1 Based on the solution for $A_z(r, z)$ in Sect. 5.1 determine the Floquet representation of the magnetic vector potential (TM_{01}). In other words write

$$A_z(r, z) = \mathrm{J}_0(p_1 \frac{r}{R}) \sum_n a_n(k)\, \mathrm{e}^{-jk_n z}$$

and determine $a_n(k)$.

5.2 Find all the waves which can propagate between $f = 0$ to $20\,\mathrm{GHz}$, including asymmetric modes for the system described in Sect. 5.1. Repeat this exercise for the branches of the TE modes.

5.3 Calculate Green's function for the closed periodic structure in Sect. 5.2. As a particular case consider a train of particles moving with a velocity

v_0. Calculate the Cerenkov radiation emitted (Hint: follow the same procedure as in Sect. 2.4.2 but adapted to a periodic structure). What is the role of the dispersion relation in this case? It is suggested that a single mode be used to describe the electromagnetic mode in the groove.

5.4 Analyze the coupling of spatial harmonics for the system in Sect. 5.3 in a similar way as in Sect. 5.2.2.

5.5 In the context of Sect. 5.3.3, calculate the force which acts on the moving charge. Based on this expression, determine the total power emitted.

5.6 Repeat the calculation of the propagation of a transient in a periodic structure (Sect. 5.4) but this time for a TEM-like mode. [Hint: take $\omega = \omega_\pi \frac{1}{2}(1 - \cos kL)$.]

6. Quasi-periodic Structures

Periodic structures play an important role in the interaction of electrons with waves since they support harmonic whose phase velocity is smaller than c and with an adequate design, this can be set equal to the average velocity of the electrons. As the electrons interact with the wave and lose energy, they slip out of phase and consequently, the interaction is degraded. In order to avoid this situation the phase velocity of the wave has to be adjusted and the geometry change associated with this process should be designed for minimum reflections, otherwise the system oscillates.

In a periodic structure, at a given frequency and single mode operation, the electromagnetic wave is characterized by a single wavenumber k and quantities like phase velocity, group velocity and interaction impedance are well defined. In principle, if the structure is no longer periodic the field can not be represented by a single wavenumber except if the variations are adiabatic in which case these characteristics are assumed to be determined by the geometry of the *local* cell. Adiabatic perturbations in the geometry may improve the efficiency from a few percent level in uniform structures to the 30% level. But one can not expect to achieve 60–80% efficiency by slow variation of the structure, in particular, bearing in mind that in contrast to accelerators where these changes occur over many wavelengths, in traveling-wave output structures these changes should occur in one or, at the most, two wavelengths.

Non-adiabatic change of geometry dictates a wide spatial spectrum in which case the formulation of the interaction in terms of a single wave with a varying amplitude and phase is inadequate. In fact, the electromagnetic field can not be expressed in a simple (analytic) form if substantial geometric variations occur from one cell to another. To be more specific: in a uniform or weakly tapered disk-loaded waveguide, the beam-wave interaction is analyzed assuming that the general functional form of the electromagnetic wave is known i.e., $A(z)\cos[\omega t - kz - \phi(z)]$ and as indicated in Chap. 4 the beam affects the amplitude $A(z)$ and the phase, $\phi(z)$. Furthermore, it is assumed that the variation due to the interaction is small on the scale of one wavelength of the radiation. Both assumptions are not acceptable in the case of a structure designed for high efficiency interaction. In order to emphasize even further this difficulty, we recall that a non-adiabatic local perturbation of geometry affects *global* electromagnetic characteristics, this is to say that a change in a given cell affects the interaction impedance or the group velocity several cells before and after the point where the geometry was altered.

In order to overcome these difficulties, we present in this chapter, an analytical technique which has been developed in order to design and analyze quasi-periodic metallic structures of the type discussed in Chap. 5. The method relies on a model which consists of a cylindrical waveguide to which a number of pill-box cavities and radial arms, are attached. In principle the number of cavities and arms is arbitrary. The boundary condition problem is formulated in terms of the amplitudes of the electromagnetic field in the cavities and arms. The elements of the matrix which relates these amplitudes with the source term are analytic functions and no a-priori knowledge of the functional behavior of the electromagnetic field is necessary. In Sect. 6.1 we examine the homogeneous electromagnetic characteristic of quasi-periodic structures. The technique is further developed to include Green's function formulation in Sect. 6.2 followed by the investigation of space-charge waves (Sect. 6.3) within the framework of the linear hydrodynamic approximation for the beam dynamics. In Sect. 6.4 the method is further generalized to include effects of large deviations from the initial average velocity of the electrons by formulating the beam-wave interaction in the framework of the macro-particle dynamics. Additional aspects of beam-wave interaction in quasi-periodic structures will be discussed in Chap. 7 in the context of a free-electron laser with a tapered wiggler.

The study presented in this chapter was triggered by experimental work performed at Cornell University. In the introduction to Chap. 4 we indicated that power levels in excess of 200 MW were generated in a 50MHz bandwidth. The 200 MW generated with this structure were accompanied by gradients larger than 200 MV/m and no rf breakdown was observed experimentally. However, for any further increase in the power levels it is necessary to increase the volume of the last two or three cells in order to minimize the electric field on the metallic surface. The system becomes then quasi-periodic. In order to envision the process in a clearer way let us assume that 80% efficiency is required from our source. If the initial beam is not highly relativistic, which is the case in most systems, such an efficiency implies a dramatic change in the geometry of the structure over a short distance. Specifically, for a 500 keV beam, the initial velocity is $v_0 \sim 0.86c$ and 80% efficiency would imply a phase velocity of $0.55c$ at the output. This corresponds to a 36% change in the phase velocity and a similar change will be required in the geometry which is by no means an adiabatic change when it occurs in one period of the wave.

Based on our experience the main problems of an extraction section based on a quasi-periodic traveling-wave structure are: *(i)* minimize the reflections primarily at the output end of the structure in order to maintain a clean spectrum and to avoid oscillations and *(ii)* taper the output section to avoid breakdown and *(iii)* compensate for the decrease in the velocity of the electrons. The technique presented in this chapter helps us to optimize these conflicting requirements.

6.1 Homogeneous Solution

The model used to analyze a quasi-periodic structure consists of a set of radial arms and pill-box cavities attached to a cylindrical waveguide. Their number and order is arbitrary. However for this presentation we shall consider a situation in which the input arm is the first cell (subscript 1) and the output arm is the last (subscript N) – as illustrated in Fig. 6.1. Each aperture, whether it corresponds to a cavity or an arm, has a width denoted by d_n where n is the index ascribed to each unit ($n = 1, 2 \cdots N$); N is the total number of cells and arms. The height, width and separation of each cavity can be arbitrary. Only the internal radius ($R_{\rm int}$) has to be the same throughout the device. The height of each cavity is determined by its external radius denoted by $R_{\rm ext,n}$. A cylindrical coordinates system is used: its origin is chosen in the center of the first aperture. Furthermore, the system is azimuthally symmetric and so is the electromagnetic excitation. Consequently, throughout this chapter we shall consider only symmetric transverse magnetic (TM) modes. Specifically, we shall examine the transmission and reflection characteristics.

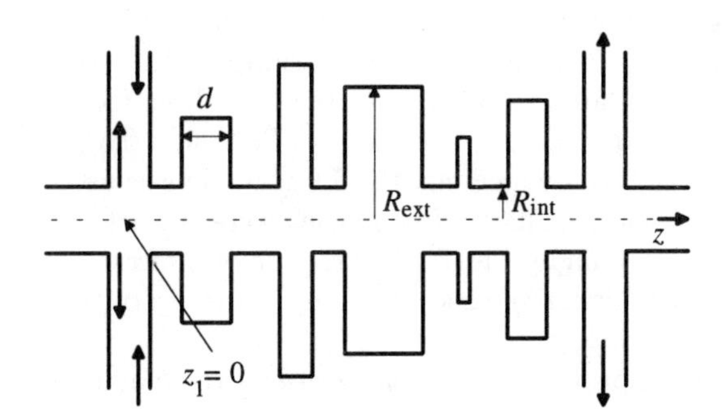

Fig. 6.1. Schematic of the model used for investigation of quasi-periodic structures

One way to analyze the electromagnetic characteristics of such a structure is by mode decomposition and formulating the boundary condition problem in terms of a transmission matrix from each discontinuity – see Sect. 2.5.2. This method is addressed in literature [Mittra and Lee (1971) or Lewin (1975)], but its performance is poor whenever more than one discontinuity is involved. This is due to the large and small numbers evolving from the evanescent modes associated with each discontinuity and their advance from one discontinuity to another.

6.1.1 Definition of the Model

Contrary to a periodic structure, where the field in the inner cylinder ($0 < r < R_{\text{int}}$) can be represented by Floquet series, in this system we have to consider the entire spatial spectrum of waves, therefore the magnetic vector potential reads

$$A_z(r, z; \omega) = \int_{-\infty}^{\infty} \mathrm{d}k A(k) \mathrm{I}_0(\Gamma r) \mathrm{e}^{-jkz}, \tag{6.1.1}$$

where $\Gamma^2 = k^2 - \omega^2/c^2$ and $\mathrm{I}_0(x)$ is the zero order modified Bessel function of the first kind. All the transients are assumed to be zero or in other words the system has reached a steady state regime thus a phasor notation, $\mathrm{e}^{j\omega t}$, is adopted. In the arms or grooves the electromagnetic field should be represented by a superposition of modes which satisfy the boundary conditions on the metallic walls. In principle an infinite number of such modes is required. However, as long as the vacuum wavelength is about 5 times larger than the groove or arm width, the first mode [transverse electric and magnetic (TEM)] is sufficient for most practical purposes. This assumption is by no means critical for the present analysis and the calculation is similar when a larger number of modes is required, however we use it since it makes the presentation simpler. In order to quantify this statement let us give a simple example of a periodic disk-loaded structure: consider the case when $R_{\text{ext}} = 15.9$ mm, $R_{\text{int}} = 9.0$ mm, the period of the system is 10.0 mm and the disk is 5mm wide. For this geometry it is required that the phase advance per cell will be 120° at 9 GHz. With 39 spatial Floquet harmonics, the lower cutoff frequency ($kL = 0$) was calculated to be 8.206 GHz using three modes (TEM, TM_{01} and TM_{02}) in the grooves, with two modes (TEM and TM_{01}) the cutoff was 8.192 GHz and 8.192 GHz when only the TEM mode was used. For the higher cutoff ($kL = \pi$) the calculated frequencies were 9.270 GHz, 9.229 GHz and 9.229 GHz correspondingly. Thus in the regime of interest the typical error associated with the higher modes omission in the grooves is expected to be of the order of 1% or less.

Within the framework of this approximation we can write for the magnetic vector potential in the input arm,

$$A_z(r, z; \omega) = A_{\text{in}} \mathrm{H}_0^{(1)}\left(\frac{\omega}{c} r\right) + D_1 \mathrm{H}_0^{(2)}\left(\frac{\omega}{c} r\right), \tag{6.1.2}$$

where $\mathrm{H}_0^{(1)}(x)$ and $\mathrm{H}_0^{(2)}(x)$ are the zero order Hankel function of the first and second kind respectively; A_{in} represents the amplitude of the incoming wave and D_1 is the amplitude of the reflected wave which is yet to be determined. In the nth ($1 < n < N$) groove we have

$$A_z^n(r, z; \omega) = D_n T_{0,n}\left(\frac{\omega}{c} r\right), \tag{6.1.3}$$

where D_n is the amplitude of the magnetic vector potential, $T_{0,n}(\frac{\omega}{c}r) = \mathrm{J}_0(\frac{\omega}{c}r)\,\mathrm{Y}_0(\frac{\omega}{c}R_{\mathrm{ext,n}}) - \mathrm{Y}_0(\frac{\omega}{c}r)\mathrm{J}_0(\frac{\omega}{c}R_{\mathrm{ext,n}})$; later we shall also use the function $T_{1,n}(\frac{\omega}{c}r) = \mathrm{J}_1(\frac{\omega}{c}r)\,\mathrm{Y}_0(\frac{\omega}{c}R_{\mathrm{ext,n}}) - \mathrm{Y}_1(\frac{\omega}{c}r)\mathrm{J}_0(\frac{\omega}{c}R_{\mathrm{ext,n}})$. Finally in the output arm,

$$A_z(r,z;\omega) = D_N \mathrm{H}_0^{(2)}\left(\frac{\omega}{c}r\right), \tag{6.1.4}$$

represents a cylindrical outgoing wave.

In order to determine the various amplitudes we next impose the boundary conditions in a way which is similar to the case of a periodic structure. But we no longer consider a single cell to characterize the entire system, instead we examine each individual region. From the condition of continuity of the longitudinal electric field we can conclude that

$$A(k) = -\frac{1}{2\pi}\frac{\alpha^2}{\Delta^2 \mathrm{I}_0(\Delta)}\left[A_{\mathrm{in}}\mathrm{H}_0^{(1)}(\alpha)d_1\mathcal{L}_1(k) + \sum_{n=1}^{N} D_n\psi_{0,n}d_n\mathcal{L}_n(k)\right], \tag{6.1.5}$$

where $\alpha = \omega R_{\mathrm{int}}/c$ is the normalized angular frequency, $\Delta = \Gamma R_{\mathrm{int}}$ is the normalized wavenumber in the radial direction and

$$\mathcal{L}_n(k) = \frac{1}{d_n}\int_{z_n-d_n/2}^{z_n+d_n/2} \mathrm{d}z e^{jkz}; \tag{6.1.6}$$

z_n is the location of the center of the nth groove or arm and in the first cell its value is zero ($z_1 = 0$). The function

$$\psi_{\nu,n} = \begin{cases} H_\nu^{(2)}(\alpha) & n=1 \quad \text{or} \quad n=N, \\ T_{\nu,n}(\alpha) & n\neq 1 \quad \text{or} \quad n\neq N, \end{cases} \tag{6.1.7}$$

is a generalized function defined in the aperture of either the grooves or the arms and $\nu = 0,1$.

Imposing the continuity of the tangential magnetic field on each aperture we find

$$A_{\mathrm{in}}\mathrm{H}_1^{(1)}(\alpha)\delta_{n,1} + D_n\psi_{1,n} = -\frac{1}{\alpha}\int_{-\infty}^{\infty} \mathrm{d}kA(k)\Delta \mathrm{I}_1(\Delta)\mathcal{L}_n^*(k). \tag{6.1.8}$$

It is now convenient to substitute (6.1.5) in (6.1.8) in order to represent the entire electromagnetic problem in terms of the amplitudes of the mode in the grooves and arms i.e.,

$$\sum_{m=1}^{N} \tau_{n,m}D_m = S_n, \tag{6.1.9}$$

where

$$\tau_{n,m} = \psi_{1,n}\delta_{n,m} - \psi_{0,m}\chi_{n,m}\,,$$
$$S_n = -\mathrm{H}_1^{(1)}(\alpha)\delta_{n,1}A_{\mathrm{in}} + \mathrm{H}_0^{(1)}(\alpha)\chi_{n,1}A_{\mathrm{in}}\,, \tag{6.1.10}$$

and

$$\chi_{n,m} = \frac{d_m\alpha}{2\pi}\int_{-\infty}^{\infty} \mathrm{d}k \frac{\mathrm{I}_1(\Delta)}{\Delta \mathrm{I}_0(\Delta)}\mathcal{L}_n^*(k)\mathcal{L}_m(k)\,. \tag{6.1.11}$$

In principle, with the matrix τ established, the electromagnetic problem is solved.

6.1.2 Evaluation of Green's Function

Our next step is to simplify the expression for the matrix τ and for this purpose we evaluate the integral which defines the matrix χ in terms of analytic functions. For this purpose we use Cauchy's residue theorem. First we substitute the explicit expressions for $\mathcal{L}_n(k)$ from (6.1.6); the result is

$$\begin{aligned}\chi_{n,m} =& \frac{d_m\alpha}{2\pi}\frac{1}{d_m}\int_{z_m-d_m/2}^{z_m+d_m/2} \mathrm{d}x_1 \frac{1}{d_n}\int_{z_n-d_n/2}^{z_n+d_n/2} \mathrm{d}x_2 \\ &\times \int_{-\infty}^{\infty} \mathrm{d}k \frac{\mathrm{I}_1(\Delta)}{\Delta \mathrm{I}_0(\Delta)} \mathrm{e}^{jk(x_1-x_2)}\,. \end{aligned}\tag{6.1.12}$$

If we now examine the integrand we observe that there are an infinite set of poles which correspond to $\mathrm{I}_0(\Delta) = 0$ since the modified Bessel function and the regular one $[\mathrm{J}_0(x)]$ are related thus we realize that the condition above is satisfied for $k^2 = (\omega/c)^2 - (p_s/R_{\mathrm{int}})^2$; here p_s are all the zeros of the zero order Bessel function of the first kind i.e., $\mathrm{J}_0(p_s) \equiv 0$. According to Cauchy's theorem the contribution to the integral will come from the poles of the integrand thus the last integral in (6.1.12) reads

$$\frac{1}{2\pi}\int_{-\infty}^{\infty} \mathrm{d}k \frac{\mathrm{I}_1(\Delta)}{\Delta \mathrm{I}_0(\Delta)} \mathrm{e}^{jk(x_1-x_2)} = \frac{1}{\pi R_{\mathrm{int}}^2}\sum_{s=1}^{\infty}\int_{-\infty}^{\infty} \mathrm{d}k \frac{\mathrm{e}^{jk(x_1-x_2)}}{k^2+\Gamma_s^2}\,, \tag{6.1.13}$$

where $\Gamma_s^2 = (p_s/R_{\mathrm{int}})^2 - (\omega/c)^2$. The last integral corresponds to Green's function for a uniform waveguide and is easily evaluated as $G(x_1|x_2) = \frac{\pi}{\Gamma_s}\mathrm{e}^{-\Gamma_s|x_1-x_2|}$. This result permits us to express the matrix χ in terms of analytic functions since the integration over x_1 and x_2 in (6.1.13) can be performed explicitly; the result reads

$$\chi_{n,m} = \frac{\alpha}{R_{\mathrm{int}}^2}\sum_{s=1}^{\infty}\begin{cases} \frac{2}{\Gamma_s^2}\left[1 - \exp(-\Gamma_s d_n/2)\,\mathrm{sinhc}(\Gamma_s d_n/2)\right] & n = m\,, \\ (d_m/\Gamma_s)\exp[-\Gamma_s|z_n - z_m|]\,\mathrm{sinhc}(\Gamma_s d_n/2) & \\ \quad\times\,\mathrm{sinhc}(\Gamma_s d_m/2) & n \neq m. \end{cases} \tag{6.1.14}$$

In this expression $\mathrm{sinhc}(x) = \sinh(x)/x$. The electromagnetic problem has now been simplified to inversion of a matrix whose components are analytic functions.

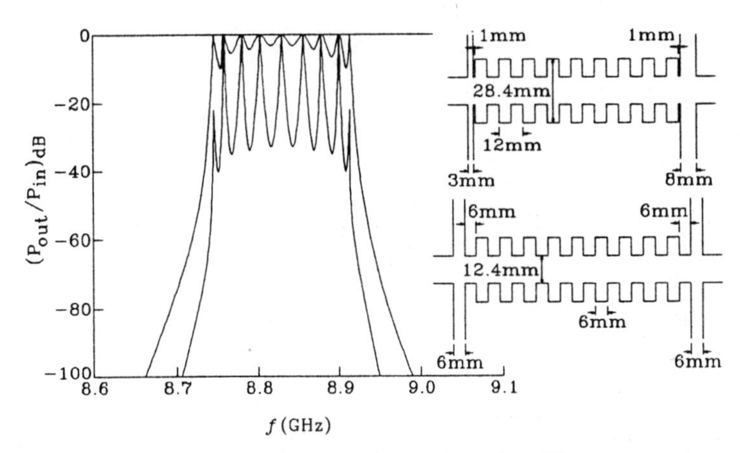

Fig. 6.2. Power transmitted for two geometries. The upper geometry corresponds to the upper curve and the distance between the groove and the arm is 1 mm. In the lower geometry and corresponding curve this distance is 6 mm

6.1.3 Transmission and Reflection

In order to test the method we used a set of identical cells. We were able to calculate the pass-band in the transmission coefficient and it fits very well that calculated using the dispersion relation of an infinite periodic structure. The following example illustrates the potential of this method: our first goal is to determine what should be the location of the arms to feed power adequately into a 9 cell structure ($R_{\mathrm{ext}} = 14.2\,\mathrm{mm}$, $R_{\mathrm{int}} = 6.2\,\mathrm{mm}$, $L = 12\,\mathrm{mm}$ and $d = 6\,\mathrm{mm}$). Figure 6.2 illustrates the geometry of the narrow band structure with 9 cavities and two arms. In the first case the arms are 6mm from the adjacent cells (see lower system) and we observe that the *average* transmission coefficient, as illustrated in the lower curves, is about −20dB. Thus the bandwidth is much narrower than that of a practical source and to this extent the fact that the peaks reach the 0 dB level becomes irrelevant to any experimental consideration. For this reason we prefer to consider here the average transmission coefficient over a range of frequencies. As the length of the waveguide between the arm and adjacent cell was shortened to 1mm (both at the output and input), the transmission coefficient increases dramatically to an average value of −3 dB.

Let us now assume that we have matched the system for a given frequency, i.e., the transmission coefficient in dB, defined by $10\log(|D_N|^2 d_N/|A_{\mathrm{in}}|^2\, d_1)$, is zero. It is known that in a narrow pass-band structure high gradients may develop in the (high power) interaction process – in particular in the

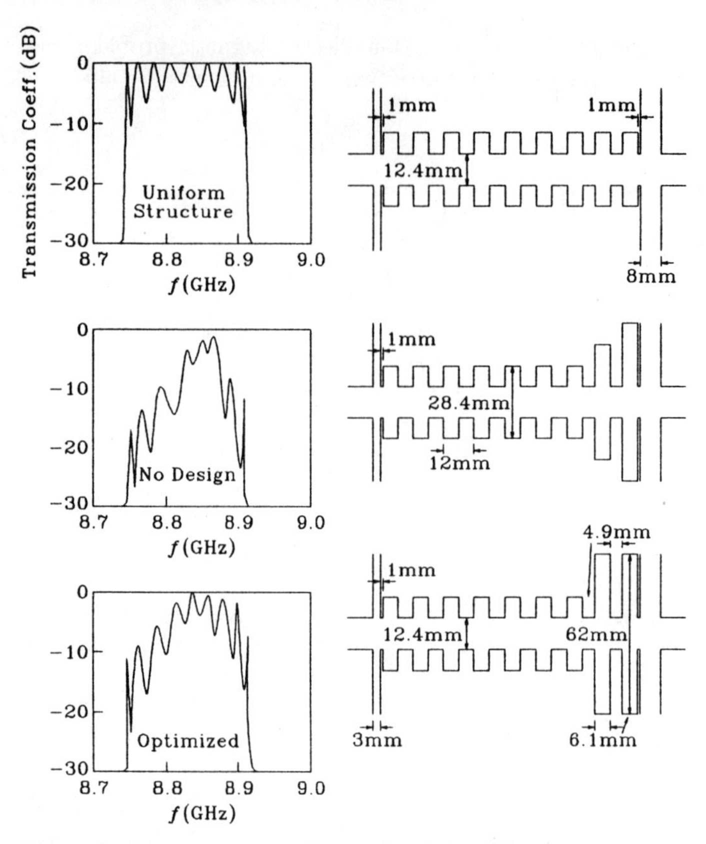

Fig. 6.3. Transmission coefficient for three different geometries

last couple of cells. In order to avoid rf breakdown the volume in which the electromagnetic energy is stored has to be increased, thus reducing in the process the energy density, and consequently reducing the field. Initially a *linear* tapering of the external radius of the last three cells was tried. In the process the width of these cells and their separation was varied in a wide range of parameters to bring the transmission coefficient to 0 dB at given frequency and the best we could achieve was −3 dB which is not acceptable (see Fig. 6.3). At this stage we returned to the initial geometry but doubled the external radius of the last two cells. These cavities have two (rather than one) resonant frequencies, one of which is close to that of a cavity in the uniform structure. After some fine tuning we obtained the transmission which is optimized to the required frequency – as indicated in the lower frame of Fig. 6.3.

This example emphasizes the dual way we can examine a quasi-periodic structure: *(i)* as a traveling-wave structure or *(ii)* as a set of coupled cavities. It is the latter which is of great importance in the design of extraction regions since, as we indicated above, quantities like phase or group velocity have practically no meaning when the geometry of the structure *varies rapidly* in space. Such variation is a direct result of the broad spectrum of wavenumbers compared to a single wavenumber in a regular periodic structure.

6.2 Non-homogeneous Solution

The homogeneous solution presented above assumes that the source of the electromagnetic field is far away from the structure and the electromagnetic energy is guided by the arm into the system. In this section we shall consider the case when the source is in the structure. By virtue of linearity of Maxwell equations we can assume that the remote sources are zero and we calculate only the contribution of the inner source. A general solution is obviously a superposition of the two solutions.

6.2.1 Green's Function

When a current distribution is present in the structure we have to solve the non-homogeneous wave equation

$$\left[\nabla^2 + \frac{\omega^2}{c^2}\right] A_z(r,z,\omega) = -\mu_0 J_z(r,z,\omega)\,, \tag{6.2.1}$$

and we shall proceed by calculating Green's function of the system. For this purpose, consider, instead of the general source of the above, a simple one, namely a narrow ring located at $z = z'$ and $r = r'$ which is a source to a field G which satisfies

$$\left[\nabla^2 + \frac{\omega^2}{c^2}\right] G(r,z|r',z') = -\frac{1}{2\pi r}\delta(r-r')\delta(z-z')\,, \tag{6.2.2}$$

and when subject to the same boundary conditions as A_z it is exactly Green's function of the system. In the absence of any boundaries this function is given by (see Sect. 2.4.1)

$$G(r,z|r',z') = \int_{-\infty}^{\infty} \mathrm{d}k e^{-jk(z-z')} g_k(r|r')\,, \tag{6.2.3}$$

where

$$g_k(r|r') = \frac{1}{(2\pi)^2}\begin{cases} \mathrm{I}_0(\Gamma r)\mathrm{K}_0(\Gamma r') & \text{for} \quad 0 \le r \le r'\,, \\ \mathrm{K}_0(\Gamma r)\mathrm{I}_0(\Gamma r') & \text{for} \quad r' \le r < \infty\,. \end{cases} \tag{6.2.4}$$

Accordingly, the solution for the vector magnetic potential reads

$$A_z(r,z,\omega) = 2\pi\mu_0 \int_0^{R_b} dr'r' \int_{-\infty}^{\infty} dz' G(r,z|r',z') J_z(r',z',\omega) + \int_{-\infty}^{\infty} dk A(k) e^{-jkz} I_0(\Gamma r) . \quad (6.2.5)$$

The second term is the solution of the homogeneous equation which does not diverge on axis and is a direct result of the presence of the metallic surface; R_b is the radius of the source. In the region outside the beam ($r \geq R_b$), this expression can also be written as

$$A_z(r > R_b, z, \omega) = \int_{-\infty}^{\infty} dk \, [B(k) K_0(\Gamma r) + A(k) I_0(\Gamma r)] \, e^{-jkz} , \quad (6.2.6)$$

where

$$B(k) = \frac{\mu_0}{2\pi} \int_0^{R_b} dr'r' I_0(\Gamma r') \int_{-\infty}^{\infty} dz' e^{jkz'} J_z(r',z',\omega) , \quad (6.2.7)$$

is the spatial Fourier transform of the current density. For the boundary condition problem the relevant components of the electromagnetic field are

$$E_z(r > R_b, z, \omega) = \frac{c^2}{j\omega} \int_{-\infty}^{\infty} dk (-\Gamma^2) \, [B(k) K_0(\Gamma r) + A(k) I_0(\Gamma r)] \, e^{-jkz} ,$$

$$H_\phi(r > R_b, z, \omega) = -\frac{1}{\mu_0} \int_{-\infty}^{\infty} dk (\Gamma) \, [-B(k) K_1(\Gamma r) + A(k) I_1(\Gamma r)] \, e^{-jkz} . \quad (6.2.8)$$

In the grooves and arms the solution is identical with (6.1.2–4) except that $A_{in} \equiv 0$. When imposing the continuity of the longitudinal component of the electric field, we obtain

$$-2\pi \frac{\Delta^2}{\alpha^2} \Big[B(k) K_0(\Delta) + A(k) I_0(\Delta) \Big] = D_1 H_0^{(2)}(\alpha) d_1 \mathcal{L}_1(k) + \sum_{n=2}^{N-1} D_n T_{0,n}(\alpha) d_n \mathcal{L}_n(k) + D_N H_0^{(2)}(\alpha) d_N \mathcal{L}_N(k) , \quad (6.2.9)$$

and the continuity of the azimuthal magnetic field provides us with an additional set of equations similar to (6.1.8):

$$\alpha D_n \psi_{1,n} = \int_{-\infty}^{\infty} dk \Delta \, [B(k) K_1(\Delta) - A(k) I_1(\Delta)] \, \mathcal{L}_n^*(k) . \quad (6.2.10)$$

In these two equations $\mathcal{L}_n(k)$ was defined in (6.1.6). Based on these two equations we can determine the amplitudes D_n in the arms and grooves by substituting (6.2.9) in (6.2.10). The result is similar to the homogeneous case:

$$\sum_{m=1}^{N} \tau_{n,m} D_m = S_n \,, \tag{6.2.11}$$

where

$$S_n = \frac{1}{\alpha}\int_{-\infty}^{\infty} \mathrm{d}k \frac{1}{\mathrm{I}_0(\Delta)} B(k)\mathcal{L}_n^*(k) \,. \tag{6.2.12}$$

In this expression we used the property of the modified Bessel functions: $\mathrm{I}_0(x)\mathrm{K}_1(x)+\mathrm{I}_1(x)\mathrm{K}_0(x) = 1/x$. Expression (6.2.11) indicates that if we know the source term S_n we can determine all the amplitudes D_n using the inverse of exactly the same matrix τ we defined in the previous section. Therefore, we next direct our efforts to simplify the expression for the source term S_n.

Based on the definition of $B(k)$ in (6.2.7) we can write

$$S_n = \frac{\mu_0}{\alpha}\int_0^{R_b} \mathrm{d}r' r' \int_{-\infty}^{\infty} \mathrm{d}z' J_z(r',z',\omega)\sigma_n(r',z') \,, \tag{6.2.13}$$

where

$$\sigma_n(r',z') = \frac{1}{2\pi}\int_{-\infty}^{\infty} \mathrm{d}k \frac{\mathrm{I}_0(\Gamma r')}{\mathrm{I}_0(\Delta)} \mathcal{L}_n^*(k) \mathrm{e}^{jkz'} \,. \tag{6.2.14}$$

Thus in order to simplify the source term S_n one has first to simplify the function $\sigma_n(r',z')$. We substitute the explicit expression for $\mathcal{L}_n(k)$ and then evaluate the integral on k based on Cauchy's residue theorem; the result is

$$\sigma(r',z') = \frac{1}{2\pi}\frac{1}{d_n}\int_{z_n-d_n/2}^{z_n+d_n/2} \mathrm{d}\xi \sum_{s=1}^{\infty} \frac{2p_s \mathrm{J}_0(p_s r'/R_{\mathrm{int}})}{R_{\mathrm{int}}^2 \mathrm{J}_1(p_s)} \frac{\pi}{\Gamma_s} \mathrm{e}^{-\Gamma_s|\xi - z'|} \,. \tag{6.2.15}$$

Our next step is to perform the integration over ξ thus

$$S_n = \frac{\mu_0}{\alpha}\int_0^{R_b} \mathrm{d}r' r' \int_{-\infty}^{\infty} \mathrm{d}z' J_z(r',z',\omega) \sum_{s=1}^{\infty} \frac{p_s \mathrm{J}_0(p_s r'/R_{\mathrm{int}})}{\Gamma_s R_{\mathrm{int}}^2 \mathrm{J}_1(p_s)} \sigma_{s,n}(z') \,, \tag{6.2.16}$$

where

$$\sigma_{s,n}(z') \equiv \frac{1}{d_n}\int_{z_n-d_n/2}^{z_n+d_n/2} \mathrm{d}\xi \mathrm{e}^{-\Gamma_s|\xi - z'|} \tag{6.2.17}$$

$$= \begin{cases} \mathrm{e}^{-\Gamma_s|z'-z_n|}\,\mathrm{sinhc}(\Gamma_s d_n/2) & \text{for } |z'-z_n| > d_n/2 \\ 2\left[1-\mathrm{e}^{-\Gamma_s d_n/2}\cosh(\Gamma_s(z'-z_n))\right]/\Gamma_s d_n & \text{for } |z'-z_n| < d_n/2 \,. \end{cases}$$

Formally this concludes the formulation of the boundary condition problem in (6.2.11) and in the remainder we present two examples.

6.2.2 Stationary Dipole

In order to simplify the analysis further we shall now make the following assumptions: (*i*) the current density varies very slowly in the transverse direction such that it can be considered constant and (*ii*) we shall examine a Dirac delta function current distribution in the longitudinal direction, such that the field due to any other current distribution can be represented as a superposition of such point sources i.e.,

$$J_z(r,z) = \frac{I}{\pi R_{\rm b}^2}\Delta_z \delta(z-z_c)h(R_{\rm b}-r)\,; \tag{6.2.18}$$

$h(x)$ is the regular step function. In this expression I is the current of the particular dipole, Δ_z is its characteristic length and z_c is its location in the z direction. This is a stationary (motionless) dipole which oscillates at an angular frequency ω. With these assumptions and bearing in mind that

$$\int_0^{R_{\rm b}} {\rm d}r\, r {\rm J}_0(p_s r/R_{\rm int}) = R_{\rm b} R_{\rm int}\frac{1}{p_s}{\rm J}_1\left(p_s \frac{R_{\rm b}}{R_{\rm int}}\right), \tag{6.2.19}$$

the source term in (6.2.11) is given by

$$S_n = \frac{\mu_0 I \Delta_z}{\pi a R_{\rm b}}\sum_{s=1}^{\infty}\frac{{\rm J}_1(p_s R_{\rm b}/R_{\rm int})}{{\rm J}_1(p_s)}\frac{1}{\sqrt{p_s^2-\alpha^2}}\sigma_{s,n}(z_c)\,. \tag{6.2.20}$$

In order to present the radiation emitted by such a stationary dipole it is convenient to normalize both the source term and the amplitude with the term $a \equiv \mu_0 I \Delta_z/(\pi R_{\rm b})$ hence $\bar{S}_n = S_n/a$ and $\bar{D}_n = D_n/a$. The average power which flows through the νth arm is $P_\nu = 2\omega d_\nu |D_\nu|^2/\mu_0$. Accordingly, the average normalized power flowing through each one of the arms in the structure is given by

$$\bar{P}_\nu \equiv P_\nu \left[\frac{1}{2}\eta_0 I^2 \frac{\Delta_z^2}{\pi R_{\rm b}^2}\right]^{-2} = \frac{4}{\pi}\frac{d_\nu}{R_{\rm int}}\alpha\left|\bar{D}_\nu(\alpha)\right|^2; \tag{6.2.21}$$

here, the index ν indicates the input or output arms only i.e., $\nu = 1$ represents the input arm and $\nu = N$, the output. First to be examined was the effect of the arm location on the radiation emitted by a single dipole and for this purpose two quantities are defined: the total emitted power, $(P_{\rm tot})_{\rm dB} \equiv 10\log(\bar{P}_1 + \bar{P}_N)$, and the ratio between the power emitted in the output arm and input arm i.e., $(P_N/P_1)_{\rm dB} \equiv 10\log(\bar{P}_N/\bar{P}_1)$. The geometry considered next is somewhat different than in the previous section: $R_{\rm ext} = 17.3\,$mm, $R_{\rm int} = 9\,$mm, $L = 10.4\,$mm and $d = 1.4\,$mm. This choice of parameters was determined by the need to increase the internal radius of the structure while at the same time maintain the group velocity relatively low. The phase advance per cell was chosen to be 120° at 9 GHz which is the resonant frequency with a 1 MeV electron.

Figure 6.4 illustrates the power emitted by a dipole oscillating at 9 GHz as its location is varied along the structure, for two different geometries. The upper frame represents the case we showed previously to be the optimal from the point of view of feeding the system; namely, minimum distance between the arms and adjacent cavities ($g_{\rm in} = g_{\rm out} = 1\,{\rm mm}$). In the lower frame the separation of the input arm is $g_{\rm in} = g_1 = 5\,{\rm mm}$. There are several features which should be emphasized. Firstly, for the upper frame there is a clear pattern of larger emission when the dipole is in the cavity compared to the case when it oscillates in the drift region between two cavities. Secondly, comparing the power in the output arm with that in the input arm for the upper case we observe that both are of the same order of magnitude. Thirdly, breaking the symmetry of the system ($g_{\rm in} = 5\,{\rm mm}$), causes a preferred direction of emission towards the output (since the input is "blocked") as indicated in the lower frame. Note that although the dipole current is the same, the peak power is larger. In addition, the clear pattern of maximum power obtained when the dipole is in the cavity (see Fig. 6.4, upper frame) is not as clear in the case shown in the lower frame of Fig. 6.4.

Another case of interest is to examine the effect of the length of the drift regions between two adjacent cavities. We increased the distance between the third and the fourth cavity from 9 mm to $g_4 = 20\,{\rm mm}$. The effect is illustrated in the upper frame of Fig. 6.5, and the lower frame shows the case when $g_3 = 20\,{\rm mm}$. After examining the previous case the results are intuitive: in the first part of the structure the emission is primarily towards the input arm whereas in the second part, practically all the radiation is emitted through the output arm. It should be mentioned that since the current density is imposed, the emitted power is a direct measure of the longitudinal component of the electric field in the structure. As such, we observe that the main difference between the upper and lower frame is the field pattern – directly associated with the change in the geometry. We shall return later to this geometry since it can simulate the operation of a two stage traveling-wave structure or a klystron with a traveling-wave output.

6.2.3 Distributed Current Density

In a uniform section of a traveling-wave amplifier the modulation amplitude grows exponentially. In this subsection we shall calculate the electromagnetic field generated when imposing a current density which is similar to that which develops in the interaction process in a traveling-wave amplifier. The current density is given by

$$J_z(r,z;\omega) = J_0 {\rm e}^{-jKz} h(R_{\rm b} - r)\,; \tag{6.2.22}$$

K is a complex wavenumber which represents the phase advance and the amplitude variation. According to (6.2.13) the source term is given by

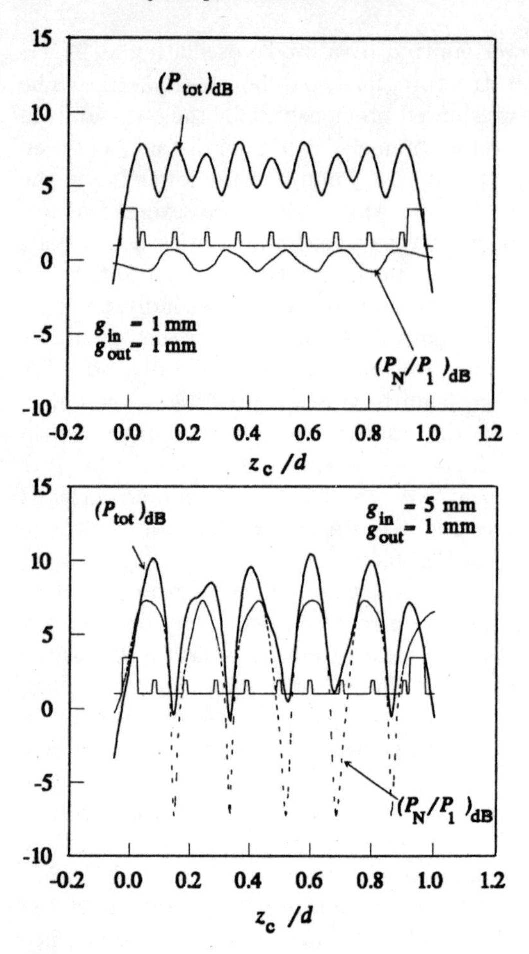

Fig. 6.4. Power emitted by the dipole as a function of its location in two different geometries which differ only by the distance of the first cavity from the input arm

$$S_n = \frac{J_0\mu_0}{\alpha}\frac{R_{\rm b}}{R_{\rm int}}\sum_{s=1}^{\infty}\frac{R_{\rm int}^2}{\sqrt{p_s^2-\alpha^2}}\frac{{\rm J}_1(p_s R_{\rm b}/R_{\rm int})}{{\rm J}_1(p_s)}$$
$$\times\frac{1}{R_{\rm int}}\int_{-\infty}^{\infty}{\rm d}z{\rm e}^{-jKz}\int_{z_n-d_n/2}^{z_n+d_n/2}\frac{{\rm d}z'}{d_n}{\rm e}^{-\Gamma_s|z-z'|}\,. \qquad (6.2.23)$$

Changing the order of integration we have

$$\frac{1}{R_{\rm int}}\int_{-\infty}^{\infty}{\rm d}z{\rm e}^{-jKz}\cdots=\frac{1}{d_n}\int_{z_n-d_n/2}^{z_n+d_n/2}{\rm d}z'\frac{1}{R_{\rm int}}\int_{-\infty}^{\infty}{\rm d}z{\rm e}^{-jKz}{\rm e}^{-\Gamma_s|z-z'|}\,, \qquad (6.2.24)$$

thus

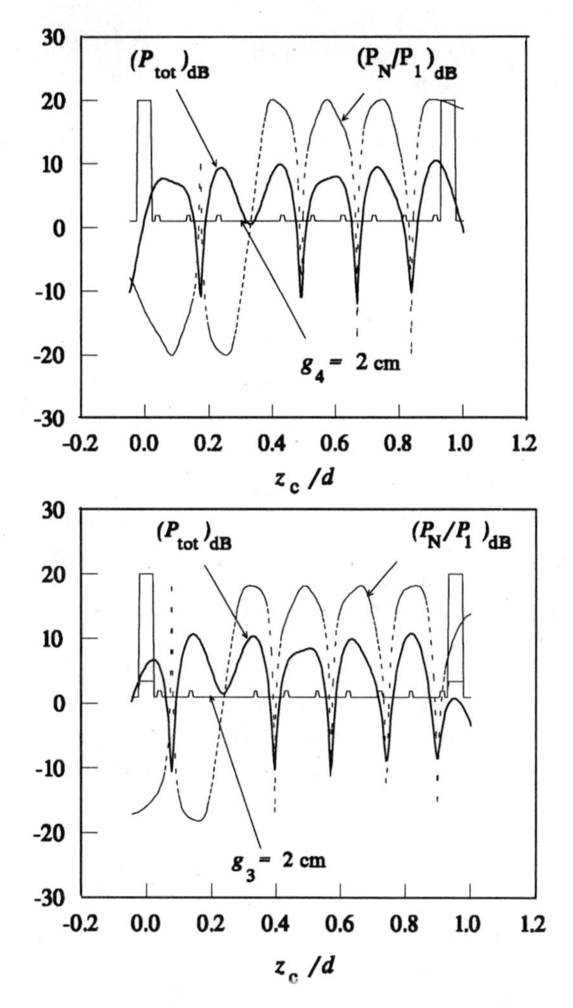

Fig. 6.5. Power emitted by the dipole as a function of its location in two different geometries which differ in the distance between the third and fourth groove (top) and second and third groove (bottom)

$$S_n = 2\frac{J_0 R_b^2 \mu_0}{\alpha}\frac{R_{\rm int}}{R_b}\sum_{s=1}^{\infty}\frac{{\rm J}_1(p_s R_b/R_{\rm int})}{{\rm J}_1(p_s)}\frac{{\rm sinc}(Kd_n/2)}{p_s^2-\alpha^2+(KR_{\rm int})^2}{\rm e}^{-jKz_n}\,. \tag{6.2.25}$$

According to the coefficient in this expression we define the normalization factor $a \equiv 2J_0\mu_0 R_b R_{\rm int}$ which entails that $\bar{D}_n = D_n/a$ and $\bar{S}_n = S_n/a$. With these quantities the normalized average emitted power reads

$$\bar{P}_\nu = \frac{P_\nu}{\frac{1}{2}\eta_0(J_0\pi R_b^2)^2} = \frac{16}{\pi^2}\frac{d_\nu}{R_b}\frac{R_{\rm int}}{R_b}\alpha|\bar{D}_\nu(\alpha)|^2\,. \tag{6.2.26}$$

In order to describe phenomenologically the saturation effect we can consider a current density function which has the form

$$J_z(r,z;\omega) = J_0\left(1 - \frac{z}{d_{\rm sat}}\right)e^{-jKz}h(R_{\rm b} - r)\,; \tag{6.2.27}$$

in which case

$$\begin{aligned} S_n =& 2\frac{J_0 R_{\rm b}^2 \mu_0}{\alpha}\frac{R_{\rm int}}{R_{\rm b}} \\ &\times \sum_{s=1}^{\infty} \frac{{\rm J}_1(p_s R_{\rm b}/R_{\rm int})}{{\rm J}_1(p_s)}\left[1 + j\frac{1}{d_{\rm sat}}\frac{\rm d}{{\rm d}K}\right]\frac{{\rm sinc}(Kd_n/2)}{p_s^2 - \alpha^2 + (KR_{\rm int})^2}e^{-jKz_n}\,. \end{aligned} \tag{6.2.28}$$

Next we examine quantitatively the radiation emitted at 9 GHz by the current distribution in (6.2.27) for the following parameters: $d_{\rm sat} = 1.3 d_{\rm tot}$, $K = 0.5K_0(1 + j\sqrt{3}) + \omega/c\beta$, $\beta = 0.94$, $K_0 = 80\,{\rm m}^{-1}$ and $d_{\rm tot}$ is the total length of the system. The total power emitted by this current distribution when in a uniform structure is $P_{\rm tot} = 54.9$ dB (see definition in previous section), and most of this power is emitted forward due to the spatial phase correlation and the varying amplitude. The asymmetry associated with the current distribution is $P_N/P_1 = 8.6$ dB. As in the first section, we now increase the volume of the last cavity by increasing the width of the cell $d_{10} = 5$ mm. Its separation from the previous cavity remains the same ($g_9 = 9$ mm). Figure 6.6 illustrates the total power and the arms power ratio as the external radius of the 10th cavity is varied. It is evident from this figure the resonant character of the structure.

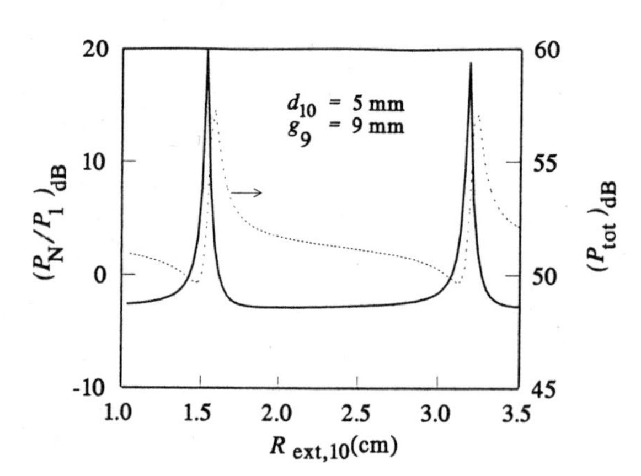

Fig. 6.6. Power emitted by the dipole as a function of the external radius of the last cavity

6.3 Beam-Wave Interaction: Hydrodynamic Approximation

In the previous subsection the current density was *imposed* and the effect of saturation was included phenomenologically. At this point we shall extend our investigation to a self-consistent solution of the current density and the electromagnetic field in the framework of the (linear) hydrodynamic approximation. This will permit us to investigate the propagation of space-charge waves in quasi-periodic structures.

6.3.1 Definition of the Model

In the framework of this model, the beam is considered to be an active linear medium which satisfies

$$J_z(r,k;\omega) = -j\omega\varepsilon_0 \frac{\omega_{\mathrm{p}}^2}{(\omega - kv_0)^2} E_z(r,k;\omega)\,, \tag{6.3.1}$$

namely, it is considered to be a fluid. The relativistic plasma frequency is defined as

$$\omega_{\mathrm{p}}^2 = \frac{\eta_0 eI}{mc^2} \frac{c^2}{\pi R_{\mathrm{b}}^2} \frac{1}{\beta\gamma^3}\,. \tag{6.3.2}$$

With (6.3.1) and the definition of the longitudinal electric field in terms of A_z [i.e., $E_z = -j\omega A_z + j(ck)^2 A_z/\omega$], the non-homogeneous wave equation for the magnetic vector potential

$$\left[\nabla^2 + \frac{\omega^2}{c^2}\right] A_z(r,z,\omega) = -\mu_0 J_z(r,z,\omega)\,, \tag{6.3.3}$$

becomes homogeneous and its solution (for a pencil beam) reads

$$A_z(r,z;\omega) = \int_{-\infty}^{\infty} \mathrm{d}k A(k) \mathrm{e}^{-jkz} \mathrm{I}_0(\Lambda r)\,, \tag{6.3.4}$$

with

$$\Lambda^2 = \Gamma^2 \left[1 - \frac{\omega_{\mathrm{p}}^2}{(\omega - kv_0)^2}\right]\,. \tag{6.3.5}$$

The longitudinal electric field and the azimuthal magnetic field read

$$E_z(r,z;\omega) = -\frac{c^2}{j\omega} \int_{-\infty}^{\infty} \mathrm{d}k \Gamma^2 A(k) \mathrm{e}^{-jkz} \mathrm{I}_0(\Lambda r)\,,$$

$$H_\phi(r,z;\omega) = -\frac{1}{\mu_0} \int_{-\infty}^{\infty} \mathrm{d}k \Lambda A(k) \mathrm{e}^{-jkz} \mathrm{I}_1(\Lambda r)\,. \tag{6.3.6}$$

In the gap between the beam and the metallic surface ($R_\mathrm{b} < r < R_\mathrm{int}$) the solution of the magnetic vector potential reads

$$A_z(r,z,\omega) = \int_{-\infty}^{\infty} \mathrm{d}k \, [B(k)\mathrm{I}_0(\Gamma r) + C(k)\mathrm{K}_0(\Gamma r)] \, \mathrm{e}^{-jkz} \,, \tag{6.3.7}$$

and correspondingly, the field components relevant for the boundary condition problem are

$$E_z(r,z,\omega) = -\frac{c^2}{j\omega}\int_{-\infty}^{\infty} \mathrm{d}k\Gamma^2 \, [B(k)\mathrm{I}_0(\Gamma r) + C(k)\mathrm{K}_0(\Gamma r)] \, \mathrm{e}^{-jkz} \,,$$
$$H_\phi(r,z,\omega) = -\frac{1}{\mu_0}\int_{-\infty}^{\infty} \mathrm{d}k\Gamma \, [B(k)\mathrm{I}_1(\Gamma r) - C(k)\mathrm{K}_1(\Gamma r)] \, \mathrm{e}^{-jkz} \,. \tag{6.3.8}$$

Continuity of these two components at $r = R_\mathrm{b}$ implies

$$\frac{1}{\Gamma}\frac{B(k)\mathrm{I}_0(b_\mathrm{v}) + C(k)\mathrm{K}_0(b_\mathrm{v})}{B(k)\mathrm{I}_1(b_\mathrm{v}) - C(k)\mathrm{K}_1(b_\mathrm{v})} = \frac{1}{\Lambda}\frac{\mathrm{I}_0(b_\mathrm{b})}{\mathrm{I}_1(b_\mathrm{b})} \,, \tag{6.3.9}$$

where $b_\mathrm{b} = \Lambda R_\mathrm{b}$ and $b_\mathrm{v} = \Gamma R_\mathrm{b}$ (subscript v stands for vacuum and subscript b for beam). This expression determines the relation between $B(k)$ and $C(k)$:

$$\varrho(k) \equiv \frac{C(k)}{B(k)} = \frac{b_\mathrm{v}\mathrm{I}_0(b_\mathrm{b})\mathrm{I}_1(b_\mathrm{v}) - b_\mathrm{b}\mathrm{I}_1(b_\mathrm{b})\mathrm{I}_0(b_\mathrm{v})}{b_\mathrm{v}\mathrm{I}_0(b_\mathrm{b})\mathrm{K}_1(b_\mathrm{v}) + b_\mathrm{b}\mathrm{I}_1(b_\mathrm{b})\mathrm{K}_0(b_\mathrm{v})} \,. \tag{6.3.10}$$

It is now convenient to extract $B(k)$ from the brackets of (6.3.7–8) and define the radial functions:

$$\bar{I}_0(k,r) \equiv \mathrm{I}_0(\Gamma r) + \varrho(k)\mathrm{K}_0(\Gamma r) \,,$$
$$\bar{I}_1(k,r) \equiv \mathrm{I}_1(\Gamma r) - \varrho(k)\mathrm{K}_1(\Gamma r) \,. \tag{6.3.11}$$

These can be considered generalizations of the modified Bessel functions we used in the homogeneous case therefore the magnetic vector potential and the field components relevant to the boundary condition problem are given by

$$A_z(r,z,\omega) = \int_{-\infty}^{\infty} \mathrm{d}kB(k)\bar{I}_0(k,r)\mathrm{e}^{-jkz} \,,$$
$$E_z(r,z,\omega) = -\frac{c^2}{j\omega}\int_{-\infty}^{\infty} \mathrm{d}k\Gamma^2 B(k)\bar{I}_0(k,r)\mathrm{e}^{-jkz} \,,$$
$$H_\phi(r,z,\omega) = -\frac{1}{\mu_0}\int_{-\infty}^{\infty} \mathrm{d}k\Gamma B(k)\bar{I}_1(k,r)\mathrm{e}^{-jkz} \,. \tag{6.3.12}$$

In the grooves and arms the functional form of the solution is identical with that established in Sect. 6.1. Therefore the formulation now is similar to the case when no beam is present and we can use the formal result we presented in Sect. 6.1 namely

$$\sum_{m=1}^{N} \tau_{n,m} D_m = S_n , \tag{6.3.13}$$

where

$$\tau_{n,m} = \psi_{1,n}\delta_{n,m} - \psi_{0,m}\chi_{n,m} , \tag{6.3.14}$$

$$S_n = -\mathrm{H}_1^{(1)}(\alpha)\delta_{n,1}A_{\mathrm{in}} + \mathrm{H}_0^{(1)}(\alpha)\chi_{n,1}A_{\mathrm{in}} , \tag{6.3.15}$$

and

$$\chi_{n,m} = \frac{d_m\alpha}{2\pi}\int_{-\infty}^{\infty} \mathrm{d}k \frac{\bar{I}_1(k,R_{\mathrm{int}})}{\Delta\bar{I}_0(k,R_{\mathrm{int}})}\mathcal{L}_n^*(k)\mathcal{L}_m(k) . \tag{6.3.16}$$

The only difference is that the modified Bessel functions (I_0 and I_1) were replaced by the generalized counterparts $\bar{I}_0$ and $\bar{I}_1$ defined in (6.3.11).

6.3.2 Evaluation of Green's Function

As in the first section we express the elements of the matrix χ in terms of analytic functions. Our first step is to substitute the explicit expressions for $\mathcal{L}_n(k)$; the result is

$$\begin{aligned}\chi_{n,m} =& \frac{d_m\alpha}{2\pi}\frac{1}{d_m}\int_{z_m-d_m/2}^{z_m+d_m/2} \mathrm{d}x_1 \frac{1}{d_n}\int_{z_n-d_n/2}^{z_n+d_n/2} \mathrm{d}x_2 \\ &\times \int_{-\infty}^{\infty} \mathrm{d}k \frac{\bar{I}_1(k,R_{\mathrm{int}})}{\Delta\bar{I}_0(k,R_{\mathrm{int}})} \mathrm{e}^{jk(x_1-x_2)} .\end{aligned} \tag{6.3.17}$$

It is convenient to define the following Green's function

$$G(x_1|x_2) = \frac{d_m}{2\pi}\int_{-\infty}^{\infty} \mathrm{d}k \frac{\alpha}{\Delta}\frac{\bar{I}_1(k,R_{\mathrm{int}})}{\bar{I}_0(k,R_{\mathrm{int}})} \mathrm{e}^{jk(x_1-x_2)} , \tag{6.3.18}$$

hence

$$\chi_{n,m} = \frac{1}{d_m}\int_{z_m-d_m/2}^{z_m+d_m/2} dx_1 \frac{1}{d_n}\int_{z_n-d_n/2}^{z_n+d_n/2} dx_2 G(x_1|x_2) . \tag{6.3.19}$$

If in the previous sections we evaluated Green's function G using a "simple" set of poles which were the zeros of $\mathrm{I}_0(\Delta)$, in this case we have to examine the poles of

$$\tilde{G} = \frac{\alpha}{\Delta}\frac{\bar{I}_1(k,R_{\mathrm{int}})}{\bar{I}_0(k,R_{\mathrm{int}})} . \tag{6.3.20}$$

For the sake of simplicity we consider the case when the beam fills the entire waveguide i.e., $R_{\mathrm{b}} = R_{\mathrm{int}}$, in which case

$$\tilde{G} = \frac{\alpha}{\Delta}\frac{b}{\Delta}\frac{\mathrm{I}_1(b)}{\mathrm{I}_0(b)}, \tag{6.3.21}$$

where $b = \Lambda R_{\text{int}}$. The poles of this expression correspond to the zeros of the dispersion relation of a waveguide filled with a beam. Since it was shown in Chap. 3 that in a cylindrical waveguide the electromagnetic modes and the space-charge modes are essentially "decoupled", we shall determine the poles accordingly. In other words the expression for $\tilde{G}$ is a superposition of the electromagnetic and space-charge modes

$$\tilde{G} = \tilde{G}_{\text{EM}} + \tilde{G}_{\text{SC}}. \tag{6.3.22}$$

The contribution of the electromagnetic modes is determined by ignoring the presence of the beam ($\omega_{\text{p}} = 0$) and it is practically identical with what was found in Sects. 6.1–2. Using (6.1.14) we have

$$\chi_{n,m}^{(\text{EM})} = \frac{\alpha}{R_{\text{int}}^2}\sum_{s=1}^{\infty}\begin{cases}\frac{2}{\Gamma_s^2}\left[1 - \mathrm{e}^{-\Gamma_s d_n/2}\,\mathrm{sinhc}(\Gamma_s d_n/2)\right] & n = m, \\ \frac{d_m}{\Gamma_s}\mathrm{e}^{-\Gamma_s|z_n - z_m|}\,\mathrm{sinhc}(\Gamma_s d_n/2)\,\mathrm{sinhc}(\Gamma_s d_m/2) & \text{otherwise}.\end{cases} \tag{6.3.23}$$

Next the contribution of the space-charge waves will be evaluated. As in the empty case, we consider the pole around $b = jp_s$, namely

$$b_s \simeq \sqrt{\left(\frac{\omega^2}{c^2}R_{\text{int}}^2\frac{1}{\beta^2} - \frac{\omega^2}{c^2}R_{\text{int}}^2\right)\left[1 - \frac{\omega_{\text{p}}^2}{(\omega - k_s v_0)^2}\right]} \simeq jp_s; \tag{6.3.24}$$

accordingly, Green's function for the space-charge waves reads

$$\tilde{G}_{\text{SC}} = -\sum_{s=1}^{\infty}\frac{K_{b,s}^2}{(k - \omega/v_0)^2 - K_{p,s}^2}, \tag{6.3.25}$$

where

$$K_{b,s}^2 = \frac{2\xi_s^2\gamma^2\beta^2}{\alpha(1 + \xi_s^2)}K_{p,s}^2, \tag{6.3.26}$$

$$K_{p,s}^2 = \frac{\omega_{\text{p}}^2}{v_0^2}\frac{1}{1 + \xi_s^2}, \tag{6.3.27}$$

and $\xi_s = p_s\gamma\beta/\alpha$. The next step is to evaluate Green's function

$$G_{\text{SC}}(x_1|x_2) = -\sum_{s=1}^{\infty}\frac{K_{b,s}^2 d_m}{2\pi}\int_{-\infty}^{\infty}\mathrm{d}k\frac{\mathrm{e}^{jk(x_1 - x_2)}}{(k - \omega/v_0)^2 - K_{p,s}^2}. \tag{6.3.28}$$

After adequate change of variables the latter reads

$$G_{\rm SC}(x_1|x_2) = -\sum_{s=1}^{\infty} \frac{K_{b,s}^2 d_m}{2\pi} e^{j\omega(x_1-x_2)/v_0} \int_{-\infty}^{\infty} dk \frac{e^{jk(x_1-x_2)}}{k^2 - K_{p,s}^2} . \tag{6.3.29}$$

The last integral is identical (except for the fact that the pole is real) to that in (6.1.13); however, for its evaluation we have to be more careful since contrary to the electromagnetic waves, the space-charge modes propagate only along the beam. Therefore, we may expect the integral in (6.3.29) to be identically zero for $x_1 > x_2$ otherwise the solution would indicate a wave propagating against the beam. In order to solve the integral it is convenient to follow the same approach as in Sect. 6.1. The function

$$g_s(x_1|x_2) \equiv \int_{-\infty}^{\infty} dk \frac{e^{jk(x_1-x_2)}}{k^2 - K_{p,s}^2} , \tag{6.3.30}$$

is defined and it can be shown to satisfy

$$\left[\frac{d^2}{dx_1^2} + K_{p,s}^2\right] g(_s x_1|x_2) = -2\pi\delta(x_1 - x_2) . \tag{6.3.31}$$

In the case of the space-charge waves we know that there exist two waves thus the solution reads

$$g_s(x_1|x_2) = \begin{cases} A e^{-jK_{p,s}(x_1-x_2)} + B e^{jK_{p,s}(x_1-x_2)} & \text{for } x_1 < x_2 , \\ 0 & \text{for } x_1 > x_2 . \end{cases} \tag{6.3.32}$$

These solutions are continuous at $x_1 = x_2$ whereas their derivatives as determined by integrating (6.3.31) are discontinuous at the same location. This ultimately implies that

$$g_s(x_1|x_2) = -\frac{2\pi}{K_{p,s}} \sin[K_{p,s}(x_1 - x_2)] h(x_2 - x_1) . \tag{6.3.33}$$

The contribution of the space-charge waves to the χ matrix can now be formulated as

$$\chi_{n,m}^{(\rm SC)} = \sum_{s=1}^{\infty} \int_{z_m-d_m/2}^{z_m+d_m/2} \frac{dx_1}{d_m} \int_{z_n-d_n/2}^{z_n+d_n/2} \frac{dx_2}{d_n} \frac{K_b^2 d_m}{K_{p,s}} e^{j\omega(x_1-x_2)/v_0} \times \sin[K_{p,s}(x_1 - x_2)] h(x_2 - x_1) . \tag{6.3.34}$$

In the evaluation of these integrals we take advantage of the fact that for $x_1 > x_2$ the integrand is zero which means that if $m > n$ then

$$\chi_{n,m}^{(\rm SC)} \equiv 0 . \tag{6.3.35}$$

Diagonal terms ($n = m$) of the matrix are given by

$$\chi_{n,n}^{(\mathrm{SC})} = -\sum_{s=1}^{\infty} \frac{K_{b,s}^2 d_n}{4K_{p,s}} \left[\frac{1}{\xi_{s,n,+}} \left(1 - \mathrm{e}^{-j\xi_{s,n,+}} \operatorname{sinc}(\xi_{s,n,+})\right) - \frac{1}{\xi_{s,n,-}} \left(1 - \mathrm{e}^{-j\xi_{s,n,-}} \operatorname{sinc}(\xi_{s,n,-})\right) \right], \tag{6.3.36}$$

where $\xi_{s,n,\pm} = \frac{1}{2} d_n (\omega/v_0 \pm K_{p,s})$ and the off-diagonal non-zero terms ($n > m$) are

$$\chi_{n,m}^{(\mathrm{SC})} = -\sum_{s=1}^{\infty} \frac{jK_{b,s}^2 d_m}{2K_{p,s}} \left[\mathrm{e}^{-j(z_n - z_m)(\omega/v_0 + K_{p,s})} \operatorname{sinc}(\xi_{s,n,+}) \operatorname{sinc}(\xi_{s,m,+}) \right] + \sum_{s=1}^{\infty} \frac{jK_{b,s}^2 d_m}{2K_{p,s}} \left[\mathrm{e}^{-j(z_n - z_m)(\omega/v_0 - K_{p,s})} \operatorname{sinc}(\xi_{s,n,-}) \operatorname{sinc}(\xi_{s,m,-}) \right]. \tag{6.3.37}$$

Finally, the χ matrix is the superposition of the electromagnetic term $\chi^{(\mathrm{EM})}$ defined in (6.3.23) and the space-charge term $\chi^{(\mathrm{SC})}$ defined in (6.3.35–37), i.e.,

$$\chi_{n,m} = \chi_{n,m}^{(\mathrm{EM})} + \chi_{n,m}^{(\mathrm{SC})}. \tag{6.3.38}$$

The sinc function in the χ matrix implies that this method is a hybrid of local beam-gap interaction as in a klystron where the cavities are electromagnetically decoupled and distributed beam-wave interaction as in a traveling-wave structure where the cavities are electromagnetically coupled. In the examples presented below, the finite size of the beam is accounted for by introducing the filling factor

$$F_{\mathrm{f}} = \left(\frac{R_{\mathrm{int}}}{R_{\mathrm{b}}} \right)^2 \frac{\mathrm{I}_1^2(\alpha R_{\mathrm{b}}/R_{\mathrm{int}}\gamma\beta)}{\mathrm{I}_1^2(\alpha/\gamma\beta)}, \tag{6.3.39}$$

which represents the actual overlap of the beam with the wave and it multiplies the plasma frequency term ($\omega_{\mathrm{p}}^2 \to \omega_{\mathrm{p}}^2 \times F_{\mathrm{f}}$).

6.3.3 Transmission and Reflection

In the remainder of this section we present a few results from simulations which use this method. The system used is identical with the one in Sect. 6.2.3. A 1MV, 1kA beam is propagating through the structure and the wave injected in the system is assumed to be of sufficiently low power such that the system operates in the linear regime. A frequency scan of a symmetric ($g_{\mathrm{in}} = g_{\mathrm{out}} = 1\,\mathrm{mm}$) system is illustrated in Fig. 6.7. The gain, defined as the power at the output divided by the power injected in the input arm, has a maximum at 9.0 GHz as designed. Another peak is close to the π-point and it occurs at 9.06 GHz. Note that the gain is relatively low – about 16 dB – and this is also the ratio (in dB) between the power in the output arm compared

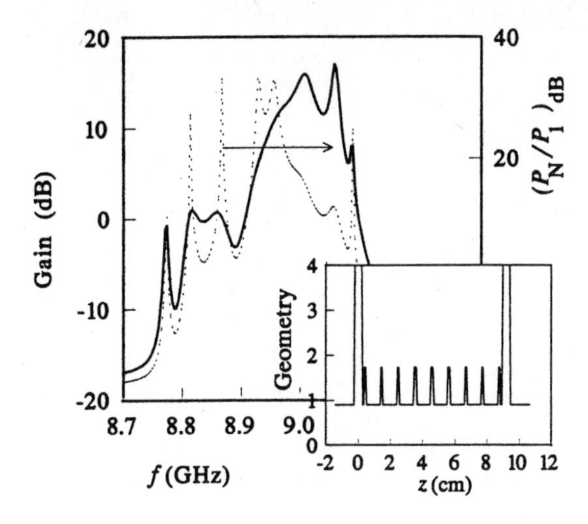

Fig. 6.7. Gain and power in a uniform structure with 1 mm separation between each arm and their adjacent cavities

to the input arm. For comparison, a Pierce like analysis predicts a growth rate of the order of 2.5 dB/cm which in a 10 cm long structure corresponds to a net gain of approximately 15 dB.

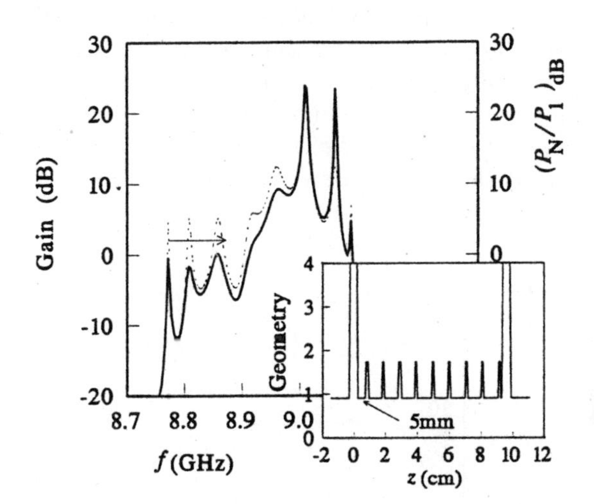

Fig. 6.8. Gain and power in a uniform structure with 5 mm separation between the input arm and its adjacent cavity

The next stage is to break the symmetry of the system by increasing the distance between the input arm and the first cavity. By doing so both the gain and the arms power ratio jumped to 24 dB (see Fig. 6.8). By increasing the distance between the second and the third cavity to 20mm (see Fig. 6.9) we were able to obtain a similar gain at the required frequency and minimize somewhat the effect of the higher frequency peak (9.06 GHz). If instead we changed the distance between the third and the fourth cavity, the gain dropped below 20 dB. With the former result in mind, we increased the ra-

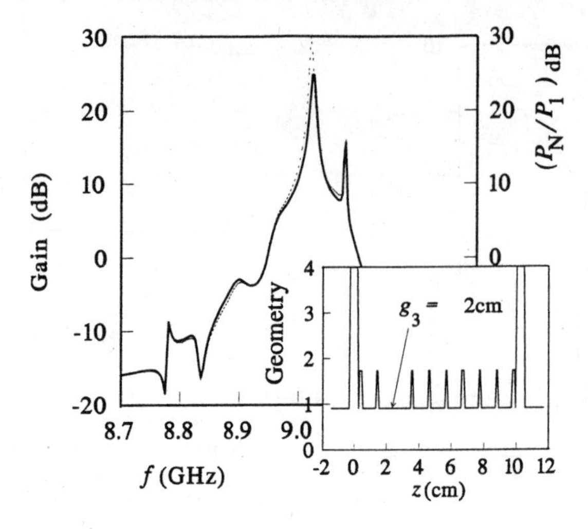

Fig. 6.9. Gain and power in a structure with 2 cm drift region between the second and third groove

dius of the last cavity to $R_{\text{ext},10} = 32$ mm and its width to $d_{10} = 5$ mm. The gain as a function of the separation from the 9th cavity was found to have an optimum for $g_9 = 11$ mm.

Next we considered the separation between the second and the third cavity and varied it in order to obtain maximum gain. Acccrding to the classical klystron theory we would expect the maximum to occur around $\lambda_p/4$ which in our case is roughly 4.6 cm. Figure 6.10 illustrates the gain and the arms power ratio for $g_3 = 50$ mm. In this case the gain approaches 40 dB. Finally, for the same geometry we calculated the gain as a function of the (normalized) average velocity of the electrons at the input. Figure 6.11 presents this gain and we observe that the gain may actually exceed the 40 dB level.

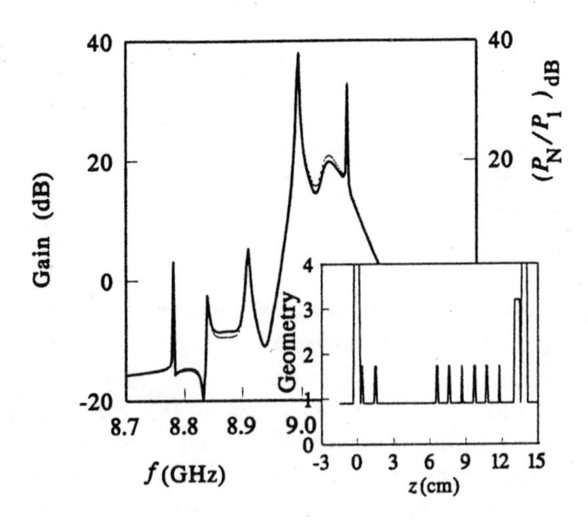

Fig. 6.10. Gain and power in a structure with $g_3 = 50$ mm, $g_9 = 11$ mm, $d_{10} = 5$ mm and $R_{\text{ext},10} = 32$ mm

Before we conclude this section we wish to emphasize the important steps and the main differences of the method presented here. When no beam is injected, the poles (which determine Green's function) correspond to the electromagnetic modes in a cylindrical waveguide. In the presence of the beam, there is an additional set of poles which correspond to the space-charge waves "family". Contrary to the electromagnetic modes, which can propagate in both directions, the space-charge waves propagate only along the beam (forwards). This fact has been addressed in the evaluation of the integrals. It is also important to point out that all poles are *real* (both electromagnetic and space-charge) as they all correspond to the eigen-modes in a cylindrical waveguide. Consequently the gain in the system is a result of the coupling between all these modes introduced by the cavities and arms – as in a klystron. This is different from the regular approach of beam-wave interaction in traveling-wave structures where the analysis relies strongly on the periodicity of the structure and the poles (eigen-wavenumbers) are *complex* – see Chap. 4.

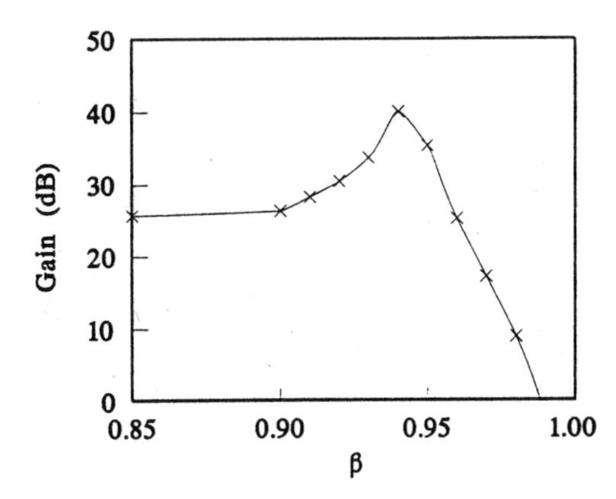

Fig. 6.11. Gain as a function of the beam velocity. The geometric parameters are identical with these in Fig. 6.10

6.4 Macro-Particle Approach

The investigation of beam-wave interaction in quasi-periodic structures was motivated by the large geometry variations required to obtain high efficiency. The latter in turn implies substantial variation in the kinetic energy of individual electrons from the ensemble average value. Consequently, the beam-wave interaction is investigated here in the framework of the macro-particle approach. Another issue addressed in this section is how one can design and analyze quasi-periodic structures when quantities like phase velocity, group velocity and interaction impedance are not well defined since there is an entire (spatial) spectrum of waves that the electrons interact with. To be more specific, in a periodic structure, for a given frequency, there is a single interacting wave (harmonic), and the interaction impedance is well defined.

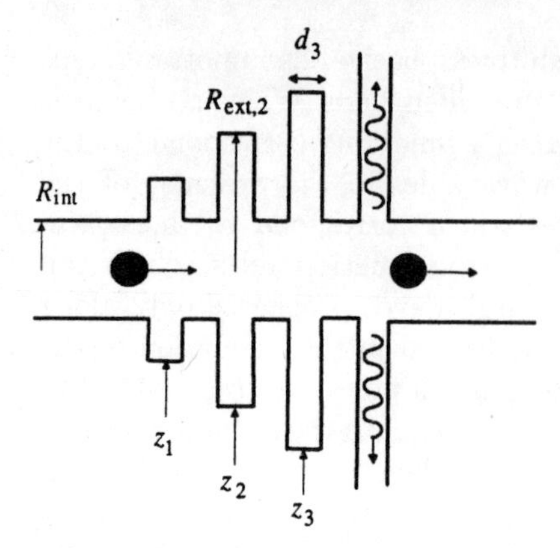

Fig. 6.12. A set of coupled pill-box cavities and an output arm model a quasi-periodic output structure

The question addressed here regards the analog in the case when substantial geometry variations occur.

It is shown that the interaction is controlled by a *matrix interaction impedance*, which is a generalization of the scalar interaction impedance concept, introduced for periodic structures. Its definition is possible after defining a set of functions which are characteristic to each aperture. The number of functions is determined by the number of apertures and number of modes which represent, at the required accuracy, the electromagnetic field in the grooves and arms. Each function has its peak at a different aperture but they are not necessarily orthogonal. The matrix interaction impedance is closely related to Green's function of the system in the representation of this set of functions. After we establish the basic formalism we illustrate the design and analysis of a high efficiency (70%) traveling-wave section, including space-charge effects.

6.4.1 Definition of the Model

A schematic of the system is presented in Fig. 6.12. It consists of a cylindrical waveguide of radius $R_{\rm int}$ to which an arbitrary number of pill-box cavities and one output arm, are attached; all the geometric definitions are like those in Sect. 6.1 with only one difference, there is only a single (output) arm. The system is driven by a modulated beam which is guided by a very strong ("infinite") magnetic field thus the motion of the electrons is confined to the z direction. Consequently, in the inner cylinder ($0 < r < R_{\rm int}$) the only non-zero component of the current density is in this direction i.e., $\mathbf{J}(r, z; t) = J_z(r, z; t)\mathbf{1}_z$ and it is given by

$$J_z(r, z; t) = -e \sum_i v_i(t)\delta[z - z_i(t)]\frac{1}{2\pi r}\delta[r - r_i(t)]\,. \tag{6.4.1}$$

In this expression $r_i(t)$ and $z_i(t)$ is the location of the ith particle at time t and subject to the assumptions above $r_i(t) = r_i(0)$.

The operation of the system as an amplifier, dictates a single frequency operation, thus the time dependence of all electromagnetic field components is assumed to be sinusoidal ($e^{j\omega t}$); this tacitly implies that all the transients associated with the front of the beam have decayed and for a particular phase-space distribution of electrons, the system can reach steady state. According to the assumptions above, the time Fourier transform of the current density is

$$J_z(r,z;\omega) = \frac{1}{T}\int_0^T dt e^{-j\omega t} J_z(r,z;t)\,; \tag{6.4.2}$$

$T = 2\pi/\omega$ is the period of the wave. Last expression can be simplified if no electrons are reflected. For this purpose we denote by $\tau_i(z)$ the time it takes the ith particle to reach the point z in the interaction region and by $v_i(z)$ the velocity of the ith particle at z; the two are related through

$$\tau_i(z) = \tau_i(0) + \int_0^z d\zeta \frac{1}{v_i(\zeta)}\,; \tag{6.4.3}$$

$\tau_i(0)$ is the time the ith particle reaches the $z = 0$ point chosen to be in the center of the first aperture. Using these definitions the integral in (6.4.2) can be evaluated analytically and the result is

$$J_z(r,z;\omega) = \frac{-e}{2\pi r T}\sum_i^{N_0} e^{-j\omega\tau_i(z)}\delta[r - r_i(0)]\,. \tag{6.4.4}$$

The summation is over all electrons (N_0) present in one time period of the wave and $I = eN_0/T$ is the average current. It is convenient to average over the transverse direction thus by denoting the beam radius by R_b and assuming that the electrons are uniformly distributed on the beam cross-section we find that

$$J(z) \equiv \frac{2}{R_b^2}\int_0^{R_b} dr\, r\, J_z(r,z;\omega) = -\frac{I}{\pi R_b^2}\langle e^{-j\omega\tau_i(z)}\rangle\,, \tag{6.4.5}$$

where $\langle\cdots\rangle \equiv N_0^{-1}\sum_i^{N_0}\cdots$. Finally, subject to the previous assumptions, the current density distribution reads

$$J_z(r,z;\omega) = -\frac{I}{\pi R_b^2}\langle e^{-j\omega\tau_i(z)}\rangle h(R_b - r)\,, \tag{6.4.6}$$

where $h(x)$ is the Heavyside step function, and in what follows, $\langle e^{-j\omega\tau_i(z)}\rangle$ is referred to as the normalized current density.

The longitudinal electric field averaged over the beam cross-section i.e.,

$$E(z) = \frac{2}{R_b^2}\int_0^{R_b} dr\, r\, E_z(r,z;\omega)\,, \tag{6.4.7}$$

determines the dynamics of the particles via the single particle equation of motion which in our case coincides with the single particle energy conservation

$$\frac{\mathrm{d}}{\mathrm{d}z}\gamma_i(z) = -\frac{1}{2}\frac{e}{mc^2}\left[E(z)\mathrm{e}^{j\omega\tau_i(z)} + c.c.\right]; \tag{6.4.8}$$

m is the rest mass of the electron. In the next subsection we shall determine the relation between the longitudinal electric field (averaged over the beam cross-section, (6.4.7)) and the current density (6.4.6).

6.4.2 Evaluation of Green's Function

A magnetic vector potential excited by the current distribution introduced above, satisfies

$$\left[\nabla^2 + \frac{\omega^2}{c^2}\right] A_z(r, z; \omega) = -\mu_0 J_z(r, z; \omega), \tag{6.4.9}$$

in the cylindrical waveguide and

$$\left[\nabla^2 + \frac{\omega^2}{c^2}\right] A_z(r, z; \omega) = 0, \tag{6.4.10}$$

in the grooves and output arm; in both cases the Lorentz gauge was tacitly assumed. The solution of the magnetic vector potential in the first region ($0 < r < R_{\rm int}$) reads

$$\begin{aligned} A_z(r, z; \omega) =& 2\pi\mu_0 \int_0^{R_{\rm b}} \mathrm{d}r'r' \int_{-\infty}^{\infty} \mathrm{d}z' G_\omega(r, z|r', z') J_z(r', z'; \omega) \\ &+ \int_{-\infty}^{\infty} \mathrm{d}k A(k)\mathrm{e}^{-jkz}\mathrm{I}_0(\Gamma r). \end{aligned} \tag{6.4.11}$$

where $\Gamma^2 = k^2 - (\omega/c)^2$, $G_\omega(r, z|r', z')$ is the vacuum Green's function:

$$G_\omega(r, z|r', z') = \int_{-\infty}^{\infty} \mathrm{d}k \mathrm{e}^{-jk(z-z')} g_{\omega,k}(r|r'), \tag{6.4.12}$$

and

$$g_{\omega,k}(r|r') = \frac{1}{(2\pi)^2}\begin{cases} \mathrm{I}_0(\Gamma r)\mathrm{K}_0(\Gamma r') & \text{for} \quad 0 \le r < r', \\ \mathrm{K}_0(\Gamma r)\mathrm{I}_0(\Gamma r') & \text{for} \quad r' \le r < \infty. \end{cases} \tag{6.4.13}$$

$\mathrm{I}_0(x)$ and $\mathrm{K}_0(x)$ are the zero order modified Bessel function of the first and second kind correspondingly. Due to the azimuthal symmetry of the current distribution and the metallic structure, only symmetric transverse magnetic (TM) modes have been considered.

In the grooves the electromagnetic field should be represented by a superposition of modes which satisfy the boundary conditions on the metallic walls. However, for the same reasons presented in Sect. 6.1, it is sufficient to consider only the first mode in the grooves. Within the framework of this approximation we can write

$$A_z^n(r,z;\omega) = D_n T_{0,n}\left(\frac{\omega}{c}r\right), \tag{6.4.14}$$

for the magnetic vector potential in the grooves, where D_n is the amplitude of the magnetic vector potential and $T_{0,n}(\frac{\omega}{c}r)$ was defined in the context of (6.1.3). In the output arm, the magnetic vector potential reads

$$A_z(r,z;\omega) = D_N \mathrm{H}_0^{(2)}\left(\frac{\omega}{c}r\right), \tag{6.4.15}$$

and $\mathrm{H}_0^{(2)}(x)$ is the zero order Hankel function of the second kind. This functional form is dictated by the boundary conditions which in this case assume no reflected wave along the output arm.

In order to determine the various amplitudes we next impose the boundary conditions following the same procedure as in the previous sections. From the condition of continuity of the longitudinal electric field we can conclude that

$$A(k)\mathrm{I}_0(\Delta) + B(k)\mathrm{K}_0(\Delta) = -\frac{1}{2\pi}\frac{\alpha^2}{\Delta^2}\sum_{n=1}^{N} D_n \psi_{0,n} d_n \mathcal{L}_n(k), \tag{6.4.16}$$

where $\alpha = \omega R_{\rm int}/c$ is the normalized angular frequency, $\Delta = \Gamma R_{\rm int}$ is the normalized wave number in the radial direction, $\mathcal{L}_n(k)$ was defined in (6.1.6) and is the normalized spatial Fourier transform of the first mode amplitude (whose amplitude is constant) in the domain of the nth aperture. The function

$$\psi_{\nu,n} = \begin{cases} H_\nu^{(2)}(\alpha) & n = N, \\ T_{\nu,n}(\alpha) & n \neq N, \end{cases} \tag{6.4.17}$$

determined at the internal radius and nth aperture; $\nu(=0,1)$ is the order of the function. In addition, z_n is the location of the center of the nth groove or arm and d_n is the corresponding width.

Imposing the continuity of the tangential magnetic field at each aperture (grooves and arm) we find

$$D_n \psi_{1,n} = -\frac{1}{\alpha}\int_{-\infty}^{\infty} \mathrm{d}k \left[A(k)\mathrm{I}_1(\Delta) - B(k)\mathrm{K}_1(\Delta)\right] \Delta \mathcal{L}_n^*(k). \tag{6.4.18}$$

In these expressions

$$B(k) = \frac{\mu_0}{2\pi}\int_0^{R_{\rm b}} \mathrm{d}r' r' \mathrm{I}_0(\Gamma r') \int_{-\infty}^{\infty} \mathrm{d}z' \mathrm{e}^{jkz'} J_z(r',z',\omega), \tag{6.4.19}$$

is the spatial Fourier transform of the current density averaged over the transverse direction with a weighting function which is proportional to the longitudinal electric field.

It is now convenient to substitute (6.4.16) in (6.4.18) in order to represent the entire electromagnetic problem in terms of the amplitudes of the mode in the grooves and output arm i.e.,

$$\sum_{m=1}^{N} \tau_{n,m} D_m = S_n \,. \tag{6.4.20}$$

The source term

$$S_n = -\frac{\mu_0 I}{2\pi\alpha} a_n \,, \tag{6.4.21}$$

is proportional to the average current and the Fourier transform of the normalized current density:

$$a_n = \frac{1}{R_{\text{int}}} \int_{-\infty}^{\infty} \mathrm{d}z f_n(z) \langle \mathrm{e}^{-j\omega\tau_i(z)} \rangle \,. \tag{6.4.22}$$

The Fourier transform is with respect to a function

$$f_n(z) = \sum_{s=1}^{\infty} \frac{p_s F_s}{\Delta_s \mathrm{J}_1(p_s)} \sigma_{s,n}(z) \,, \tag{6.4.23}$$

which is associated with the nth aperture. In particular, if all the modes in the inner cylinder (index s) are below cutoff, this function peaks in the center of the aperture; p_s are the zeros of the zero order Bessel function of the first kind i.e., $\mathrm{J}_0(p_s) = 0$. The function $f_n(z)$ is the product of two components,

$$\sigma_{s,n}(z) \equiv \frac{1}{d_n} \int_{z_n - d_n/2}^{z_n + d_n/2} \mathrm{d}\xi \mathrm{e}^{-\Gamma_s |\xi - z|} \tag{6.4.24}$$

$$= \begin{cases} \mathrm{e}^{-\Gamma_s |z - z_n|} \operatorname{sinhc}(\Gamma_s d_n/2) & \text{for } |z - z_n| > d_n/2 \\ 2\left[1 - \mathrm{e}^{-\Gamma_s d_n/2} \cosh(\Gamma_s (z - z_n))\right] / \Gamma_s d_n & \text{for } |z - z_n| < d_n/2 \,, \end{cases}$$

is the projection of Green's function ($s-$mode) on the nth aperture; $\Delta_s^2 = p_s^2 - \alpha^2$. The other component is the discrete spectrum filling factor, $F_s \equiv 2\mathrm{J}_1(p_s R_\mathrm{b}/R_\mathrm{int})/(p_s R_\mathrm{b}/R_\mathrm{int})$. To determine the amplitudes in (6.4.20) one has to multiply the source term by the inverse of the matrix τ defined by

$$\tau_{n,m} = \psi_{1,n} \delta_{n,m} - \psi_{0,m} \chi_{n,m} \,. \tag{6.4.25}$$

In this expression $\chi_{n,m}$ was defined in (6.1.11), simplified in Sect. 6.1.2 and expressed in terms of analytic functions in (6.1.14). The electromagnetic problem has now been simplified to the inversion of a matrix whose components are analytic functions without a-priori assumption on the form of the electromagnetic field.

6.4.3 The Governing Equations

The motion of the electrons is determined by the longitudinal electric field averaged over the beam cross-section $[E(z)]$ as defined in (6.4.7). In this subsection we shall use (6.4.20) to simplify the relation between the normalized current density and $E(z)$. The longitudinal component of the electric field is related to the magnetic vector potential by

$$E_z(r,z;\omega) = \frac{c^2}{j\omega}\left[\frac{\omega^2}{c^2} + \frac{\partial^2}{\partial z^2}\right] A_z(r,z;\omega)\,, \tag{6.4.26}$$

which after substituting (6.4.9) reads

$$E_z(r,z;\omega) = \frac{c^2}{j\omega}\left[-\mu_0 J_z(r,z;\omega) - \frac{1}{r}\frac{\partial}{\partial r}r\frac{\partial}{\partial r}A_z(r,z;\omega)\right]. \tag{6.4.27}$$

Thus according to the definition of the effective electric field in (6.4.7), we have

$$E(z) = \frac{c^2}{j\omega}\left(-\mu_0 J(z) - \frac{2}{R_{\rm b}}\left[\frac{\partial}{\partial r}A_z(r,z;\omega)\right]_{r=R_{\rm b}}\right). \tag{6.4.28}$$

At this stage, we substitute the explicit expression for the magnetic vector potential in (6.4.11) and the result has two contributions: the space-charge term

$$\begin{aligned} E_{\rm SC}(z) = &-\frac{1}{j\omega\varepsilon_0}\int_{-\infty}^{\infty} {\rm d}z' J(z')\frac{1}{2\pi}\int_{-\infty}^{\infty} {\rm d}k\, {\rm e}^{-jk(z-z')} \\ &\times\left[1 - 2\frac{{\rm I}_1(\Delta_b)}{{\rm I}_0(\Delta)}\left({\rm I}_0(\Delta){\rm K}_1(\Delta_b) + {\rm K}_0(\Delta){\rm I}_1(\Delta_b)\right)\right], \end{aligned} \tag{6.4.29}$$

and the "pure" electromagnetic term

$$E_{\rm EM}(z) = \frac{-j\omega}{2\pi}\int_{-\infty}^{\infty} {\rm d}k\, {\rm e}^{-jkz}\frac{1}{{\rm I}_0(\Delta)}F(k)\sum_{n=1}^{N} D_n d_n \mathcal{L}_n(k)\psi_{0,n}\,, \tag{6.4.30}$$

where $F(k) \equiv 2{\rm I}_1(\Delta_b)/\Delta_b$ is the (continuous spectrum) filling factor and $\Delta_b = \Gamma R_{\rm b}$.

It can be seen that the grooves have no explicit effect on the space-charge term and we shall next take advantage of this fact to simplify (6.4.29). In order to evaluate the space-charge contribution in a uniform waveguide we start from Green's function associated with TM_{0s} modes in a cylindrical waveguide (2.4.34) and using the same method as in (6.4.26–28) we obtain

$$\begin{aligned} E(z) = &\frac{-1}{j\omega\varepsilon_0}\left[J(z) - \frac{1}{R_{\rm int}}\int_{-\infty}^{\infty} {\rm d}z' J(z')\right. \\ &\left.\times\sum_{s=1}^{\infty} {\rm e}^{-\Gamma_s|z-z'|}\frac{1}{2\Delta_s}\left(\frac{2{\rm J}_1(p_s R_{\rm b}/R_{\rm int})}{{\rm J}_1(p_s)}\right)^2\right], \end{aligned} \tag{6.4.31}$$

which can be simplified if all electromagnetic modes are below cutoff and in particular, for the case when the current density $|J(z)|$ varies much more slowly than $e^{-\Gamma_s|z-z'|}$. Subject to these assumptions, we can assume that the main contribution to the integral is from the region $z = z'$ and therefore $J(z)$ can be extracted from the integral. The result in this case reads

$$E(z) = \frac{-1}{j\omega\varepsilon_0} J(z) \left[1 - \sum_{s=1}^{\infty} \left(\frac{J_1(p_s R_b/R_{int})}{\Delta_s J_1(p_s)} \right)^2 \right], \tag{6.4.32}$$

or

$$E_{SC}(z;\omega) = -\frac{1}{j\omega\varepsilon_0} \xi_{SC} J(z), \tag{6.4.33}$$

where the space-charge coefficient ξ_{SC} is given by

$$\xi_{SC} = 1 - \sum_{s=1}^{\infty} \left[\frac{J_1(p_s R_b/R_{int})}{\Delta_s J_1(p_s)} \right]^2, \tag{6.4.34}$$

and is an approximation to the plasma frequency reduction factor.

It is possible to simplify the electromagnetic term by substituting the explicit expression for $\mathcal{L}_n(k)$ and using the Cauchy residue theorem. The result reads

$$E_{EM}(z) = \frac{\eta_0 I}{R_{int}} \sum_{n=1}^{N} f_n(z) \left[\sum_{m=1}^{N} \mathcal{T}_{n,m} a_m \right], \tag{6.4.35}$$

where

$$\mathcal{T}_{n,m} = \frac{j}{2\pi} \frac{d_n}{R_{int}} \psi_{0,n} \left[\tau^{-1} \right]_{n,m}, \tag{6.4.36}$$

and it can be considered as a "discrete" Green function of the system since a_m is the Fourier transform of the normalized current density with respect to the function $f_m(\zeta)$.

Now that the relation between the effective electric field acting on the particles and the current density has been established,

$$E(z) = E_{SC}(z) + E_{EM}(z), \tag{6.4.37}$$

we proceed to analysis of the beam-wave interaction. Substituting this effective field in the single particle energy conservation, defining $\bar{\xi}_{SC} = \xi_{SC} (R_{int}/R_b)^2 / \alpha\pi$ and $\bar{I} = \eta_0 I e/mc^2$, we obtain

$$\frac{d}{dz}\gamma_i = -\frac{\bar{I}}{2R_{int}} \Big\{ e^{j\omega\tau_i(z)} \Big[-j\bar{\xi}_{SC} \big\langle e^{-j\omega\tau_\nu(z)} \big\rangle_\nu + \sum_{n,m=1}^{N} \mathcal{T}_{n,m} a_m f_n(z) \Big] + \text{c.c.} \Big\}. \tag{6.4.38}$$

This is an integro-differential equation which describes the dynamics of the electrons. In order to determine γ_i at any given location it is necessary to know the Fourier transform of the normalized current density, a_n, which in turn requires to know the trajectories of all particles over the entire interaction region, as indicated in (6.4.22).

Before we proceed to actually presenting a solution of this set of equations it is important to make two comments which are evident from (6.4.38) and our prior definitions:
(i) Global energy conservation implies

$$\langle\gamma_i(\infty)\rangle - \langle\gamma_i(-\infty)\rangle = -\frac{1}{2}\bar{I} \sum_{n,m=1}^{N} a_n^* Z_{n,m} a_m \,, \tag{6.4.39}$$

where

$$Z_{n,m} \equiv \frac{1}{2}\left[\mathcal{T}_{n,m} + \mathcal{T}_{m,n}^*\right] \,, \tag{6.4.40}$$

is the *interaction impedance matrix*. This expression implies that in case of non-adiabatic changes from periodicity, as is the case in quasi-periodic structures, we can no longer refer to the interaction impedance as a scalar (and local) quantity but rather as a matrix and the interaction at a given location is affected by the geometry elsewhere. Furthermore, since the left-hand side of the global energy conservation [(6.4.39)] is proportional to the overall efficiency, it is evident that the latter is controlled by the interaction impedance matrix. In the example presented next, it will be shown that it is the largest eigen-value of this matrix which determines the efficiency of the interaction.
(ii) The space-charge term has no *explicit* effect on the global energy conservation. Furthermore, in the case of a very peaked distribution (e.g., single macro-particle) it has no effect on the equations of motion.

In order to solve the integro-differential equation in (6.4.38) for a large number of macro-particles (more than 30,000 were used), an iterative way was chosen. Typically a simple distribution is assumed, enabling the calculation of the zero iteration, $a_n^{(0)}$. With this quantity, the trajectories of all particles are calculated and in parallel, the "new" $a_n^{(1)}$ is evaluated; at the end of the iteration the two a_n's are compared. If the relative error is less than 1% the simulation is terminated. Otherwise we calculate the equations of motion again but this time using $a_n^{(1)}$ to determine the dynamics of the particles and calculate in parallel $a_n^{(2)}$. If the energy spread of the electrons at the input is not too large, then 3–4 iterations are sufficient for convergence.

Consider now a modulated beam which drives an output structure. The initial energy of the electrons is 850 keV and the structure should extract 70% of their kinetic energy; for the zero order design let us assume that in the interaction region there is only a single macro-particle at a time. Furthermore, the disk thickness is taken to be 1mm in order to ensure maximum group velocity. For the same reason the phase advance per cell is taken to be 90° . For the preliminary design a single macro-particle in one period of the wave and its velocity in the interaction region is assumed to satisfy

$$v(z) = v(0)/(1 + qz)\,; \tag{6.4.41}$$

q and the total interaction length $d_{\rm tot}$ are determined from the required efficiency and the condition that no two bunches will be present in the interaction region at a time. For simplicity, we also assume that the internal and external radius are the same in all cells. Their value is determined by *maximizing the largest eigen-value of the interaction impedance matrix* at 9 GHz – as illustrated in Fig. 6.13 where $R_{\rm int} = 9\,{\rm mm}$ and $R_{\rm ext} = 16.47\,{\rm mm}$; the other geometrical parameters are $d_1 = 6.5\,{\rm mm}$, $d_2 = 6.0\,{\rm mm}$, $d_3 = 5.7\,{\rm mm}$ and $d_4 = 5.4\,{\rm mm}$. Overlaid is also the efficiency assuming a single macro-particle injected into the system in one period of the wave. The dynamics of the particle is calculated numerically [(6.4.38)].

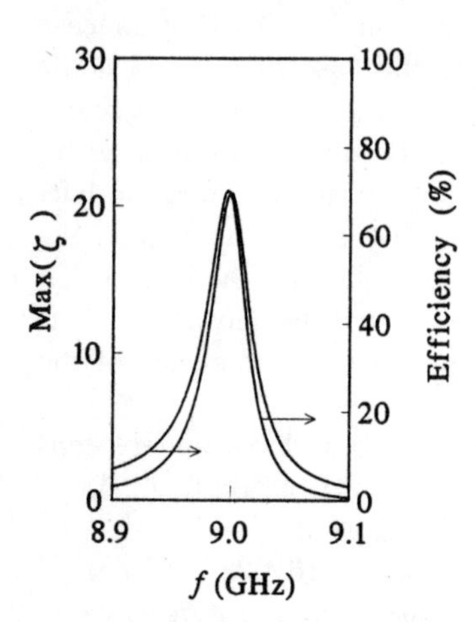

Fig. 6.13. The largest eigen-value of the interaction impedance and the efficiency as a function of the frequency using the resonant particle model

Figure 6.14 indicates that the efficiency of the electromagnetic energy conversion is strongly dependent on the phase-space distribution at the input; the phase here is defined as $\chi_i(z) \equiv \omega\tau_i(z)$. For a perfectly bunched beam the efficiency is as designed (for $I = 300A$). However, as the initial phase distribution increases to $-45° < \chi(0) < 45°$ the efficiency drops to 45% and to 25% for $-90° < \chi(0) < 90°$. It drops to virtually zero for a uniform distribution.

An interesting feature is revealed in Fig. 6.15 where we present the variation in space of the efficiency for two initial distributions: $-9° < \chi(0) < 9°$ and $-90° < \chi(0) < 90°$. We observe that the general pattern is virtually identical in both cases and only the spatial growth rate is smaller. The reduced efficiency is a result of energy transferred back to electrons which are

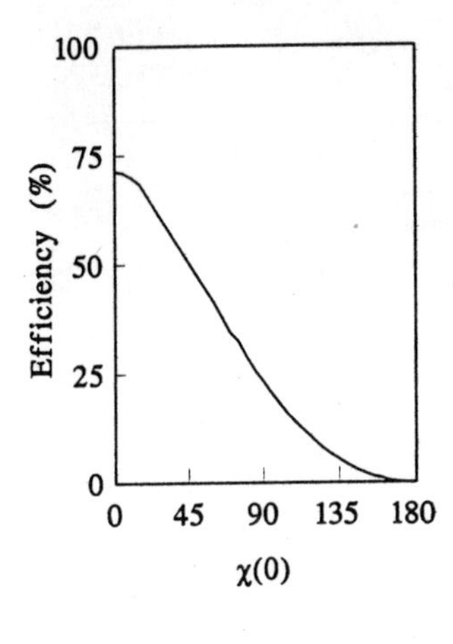

Fig. 6.14. The efficiency as a function of the modulation angle i.e., the fraction of wavelength in degrees occupied by electrons. The case $\chi(0) = 180°$ corresponds to a uniform beam

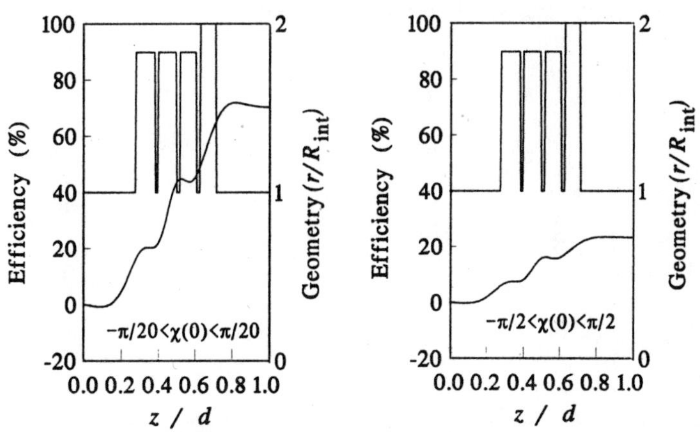

Fig. 6.15. The efficiency as a function of the location along the the interaction region for two different initially bunched beams

actually accelerated as illustrated in Fig. 6.16; clearly in the narrower initial phase-space distribution all the electrons are decelerated at the output, whereas in the case of broader distribution a substantial fraction of electrons is accelerated.

Finally, the efficiency is illustrated in Fig. 6.17 as a function of the frequency for $-15° < \chi(0) < 15°$. The curve is virtually identical to that of the single macro-particle case (Fig. 6.14). Overlaid, we present the energy spread ($\Delta\gamma \equiv \sqrt{\langle\gamma^2\rangle - \langle\gamma\rangle^2}$) at the output and we observe that up to a constant value, this quantity varies as the derivative of the efficiency with respect to the frequency. We shall discuss this effect in more detail with regard to the initial velocity of the electrons, in the next chapter in the context of Madey's theorem.

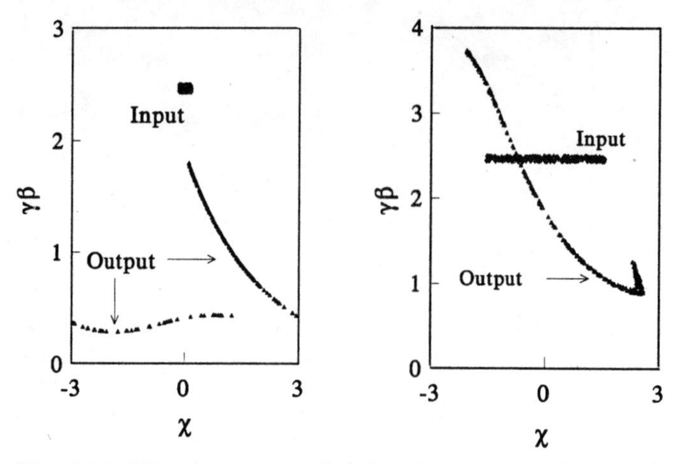

Fig. 6.16. The phase-space plots for the two cases illustrated in Fig. 6.15

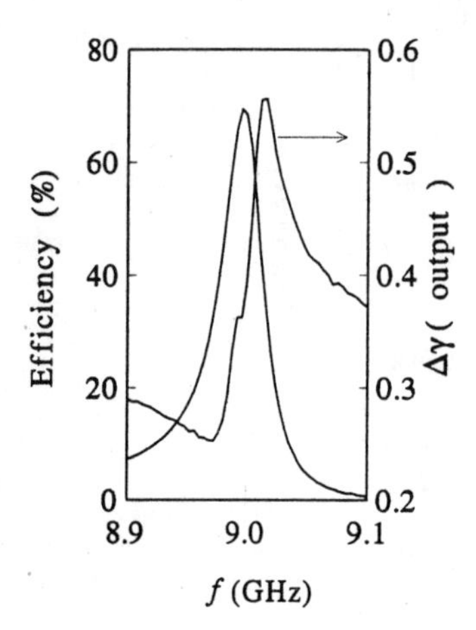

Fig. 6.17. The efficiency and the energy spread at the output as a function of the frequency

6.4.4 Qualitative Approach

The approach presented above provides us with a convenient 1D tool for calculating the dynamics of electrons in a quasi-periodic structure. Although this can be used as a design tool, it is usually convenient to apply more qualitative arguments for a zero order design which can be later improved with our model. Let us now follow such a qualitative argument. For this purpose, we shall assume an ideal bunch of electrons which carries a current I. The electrons are mono-energetic and they are initially accelerated by an initial voltage denoted by $\mathcal{E}$. If we require an extraction efficiency η in an interaction region of a length D then the average electric field experienced by the bunch is

$$E = \eta \frac{\mathcal{E}}{D}. \tag{6.4.42}$$

Based on the definition of the interaction impedance in (2.3.29) we conclude that the rf power in the system is

$$P_{\mathrm{rf}} = \frac{\pi R_{\mathrm{int}}^2 E^2}{2 Z_{\mathrm{int}}}, \tag{6.4.43}$$

and the power carried by the beam is

$$P_{\mathrm{beam}} = \mathcal{E} I. \tag{6.4.44}$$

These two are related since we assumed an efficiency η and energy conservation implies

$$P_{\mathrm{rf}} = \eta P_{\mathrm{beam}}. \tag{6.4.45}$$

From the expressions above we can determine the interaction impedance of the structure i.e.,

$$Z_{\mathrm{int}} = \frac{\pi}{2} \eta \left(\frac{R_{\mathrm{int}}}{D} \right)^2 \frac{\mathcal{E}}{I}, \tag{6.4.46}$$

and we observe that we should design the interaction impedance of the structure in conjunction with the effective impedance of the beam $(\mathcal{E}/I)$. In order to have a feeling as for the values of the impedance consider $\mathcal{E}$ =0.85 MV, I =0.5 kA, R_{int} =12 mm, total interaction length D =2.75 cm and efficiency of η =70%; for these parameters $Z_{\mathrm{int}} = 213\,\Omega$. It should be pointed out that here we tacitly assumed that E is constant thus the dynamics of the particles in space is different than the one prescribed in (6.4.41) and is given by

$$\gamma(z) = \gamma(0) - \frac{eEz}{mc^2} = \gamma(0)(1 - qz). \tag{6.4.47}$$

The length of the structure (D) and q can be determined exactly in the same way as prescribed at the end of the last subsection. Once D is determined and assuming that $\mathcal{E}$ and I are known, then (6.4.46) provides us with a simple relation between the internal radius and the interaction impedance. If we have in mind the disk-loaded structure then this relation in conjunction with the expression for the interaction impedance in (5.2.32) determine one constraint on the geometry of the structure. Thus out of the four geometric parameters ($R_{\mathrm{int}}, R_{\mathrm{ext}}, L$ and d in a periodic structure) we are left with three degrees of freedom. The resonance condition, the phase advance per cell and the group velocity (maximum gradient allowed) at resonance determine three additional constraints which in turn set the "local values" of these parameters. In other words, they roughly determine the geometry of the single cell which in turn is part of a quasi-periodic structure. Fine tuning of the design should be made following the approach in Sect (6.4.3).

Exercises

6.1 Formulate the electromagnetic problem as in Sect. 6.1 but with three modes in each groove and arm.
6.2 Formulate the electromagnetic problem as in Sect. 6.1 but for three arms which are not necessarily located at the ends of the structure.
6.3 Formulate the electromagnetic problem as in Sect. 6.1 but for a rectangular waveguide.
6.4 Formulate the electromagnetic problem as in Sect. 6.1 but for symmetric TE modes in a cylindrical waveguide.
6.5 Formulate the beam-wave interaction problem as in Sect. 6.4 for more than one arm.

7. Free-Electron Laser

In Chap. 1 we have shown that the interaction of electrons with an electromagnetic wave is possible even when the phase velocity of the latter is larger than c, provided that there is a way to conserve simultaneously both energy and momentum. In a free-electron laser (FEL) this is facilitated by the presence of a periodic magnetic field. In most cases the components of this field are transverse to the initial velocity of the electron. An electron injected in a periodic magnetic field (wiggler) oscillates and, as a result, it emits radiation. The highest frequency is emitted in the forward direction and in zero order it is determined by the periodicity of the wiggler, L, and the electron energy, γ. In Sect. 3.2.3 it was shown that for relativistic electrons ($\beta \sim 1$) this frequency is given by $\omega \simeq 2\gamma^2(2\pi c/L)$.

To the best of our knowledge the first analysis of the motion of an electron in a wiggler of this kind was performed in the early thirties by Kapitza and Dirac (1933). The question raised was whether it would be possible to observe stimulated scattered radiation from electrons moving in an electromagnetic wave. For this purpose the authors considered a low energy beam of electrons injected in a standing wave region and they estimated the number of scattered electrons due to the stimulated radiation. In the early fifties Motz (1951) investigated the radiation emitted by electrons as they move in a wiggler and later Phillips (1960) built the first coherent radiation source with a wiggler as its central component; it was called the Ubitron. In the late sixties Pantell (1968) suggested the same concept at much shorter wavelengths and Madey (1971) has proven that laser light can be amplified using this scheme but it was only later at Stanford that Elias *et. al.* (1976) demonstrated experimentally the amplification of a 10.6 μm laser beam and since then the name – free electron laser.

There are numerous books, review articles, proceedings and articles on free-electron lasers a small fraction of which will be mentioned in Sect. 7.5. However, for an introductory guide to the free-electron laser we find the article of Hasegawa (1978) as a good starting point. An excellent tutorial work on the theory of the free-electron laser is the article by Kroll, Morton and Rosenbluth (1981) which in fact inspired many of the topics presented in Sect. 7.4. An overview of the field is presented in an article by Roberson and Sprangle (1989) and among the books dedicated to FEL's, Marshall (1985) covers the basic theory and the early work done and more recently the book by Freund and Antonsen (1992) also covers advanced topics on free-electron lasers in addition to the basic theory.

In this chapter we present what we conceive as the basics of free-electron lasers. Alternative schemes of energy conversion from free electrons are briefly described in the last part of this chapter. Specifically, in the first section we consider the spontaneous emission as an electron traverses an ideal wiggler. This is followed by the investigation of coherent interaction in the low-gain Compton regime. Section 7.3 deals with the high-gain Compton regime which includes cold and warm beam operation. The macro-particle approach is presented in Sect. 7.4 and we conclude the chapter with a brief overview of the various alternative schemes of free-electron lasers.

7.1 Spontaneous Radiation

As an electron is injected into a periodic magnetic field it oscillates and emits spontaneous radiation. In this section we shall examine this process. For this purpose we consider a transverse periodic magnetic field which is uniform in the transverse direction - at least on the scale of the beam cross-section. This field can be derived from the following magnetic vector potential

$$\mathbf{A}_{\rm w} = -A_{\rm w}\Big[\mathbf{1}_x \cos(k_{\rm w}z) + \mathbf{1}_y \sin(k_{\rm w}z)\Big] , \tag{7.1.1}$$

and is given by

$$\mathbf{B}_{\rm w} = B_{\rm w}\Big[\mathbf{1}_x \cos(k_{\rm w}z) + \mathbf{1}_y \sin(k_{\rm w}z)\Big] ; \tag{7.1.2}$$

the two amplitudes $A_{\rm w}$ and $B_{\rm w}$ are related via $A_{\rm w} = B_{\rm w}/k_{\rm w}$ where $k_{\rm w} = 2\pi/L$ is the wiggler's wavenumber. An electron is injected along the z axis and we examine its motion in the absence of any radiation field. The relativistic Hamiltonian which describes the motion of an electron in the presence of an electromagnetic field was developed in Sect. 3.1 [(3.1.15)] and it is given by

$$\begin{aligned} H &= \sqrt{(\mathbf{p} + e\mathbf{A}_{\rm w})^2 c^2 + (mc^2)^2} \\ &= mc^2\gamma , \end{aligned} \tag{7.1.3}$$

where collective effects are ignored and since no external voltage is applied, the electrostatic potential is taken as zero. In addition, no boundaries are involved.

The canonical momentum $\mathbf{p}$ has two components: one which is parallel to the major velocity component of the electron and is denoted by $p_{\parallel}$ and the transverse one $p_{\perp}$. As indicated in Sect. 3.1, if the Hamiltonian is not explicitly dependent on the transverse coordinates then the transverse canonical momentum is conserved ($p_{\perp} = \text{const.}$). This canonical momentum has two contributions, the kinetic and the electromagnetic i.e., $p_\perp = m\gamma v_\perp - eA_\perp$. Assuming that the electron is born outside the magnetic field and its initial transverse motion is zero, we immediately conclude that $p_\perp = 0$, which implies

$$v_\perp = \frac{eA_\perp}{m\gamma} . \tag{7.1.4}$$

We also observe that this Hamiltonian does not explicitly depend on time therefore, energy is conserved i.e.,

$$\gamma = \text{const.} . \tag{7.1.5}$$

From the last two relations we can deduce the expressions which describe the motion of an electron in space, they read

$$\begin{aligned}
v_x(z) &= -\frac{eA_\text{w}}{\gamma m} \cos(k_\text{w} z) , \\
v_y(z) &= -\frac{eA_\text{w}}{\gamma m} \sin(k_\text{w} z) , \\
v_z(z) &= v_0 .
\end{aligned} \tag{7.1.6}$$

It is evident from these expressions that the particle undergoes a helical motion whose amplitude is determined by the amplitude of the wiggler B_w, its wavenumber k_w and the initial energy of the particle. This fact becomes clearer when realizing that in the x-y plane the particles undergo a circular motion as revealed by the first two equations of (7.1.6) which can be rewritten as

$$v_x^2 + v_y^2 = \left(\frac{eB_\text{w}}{k_\text{w} \gamma m} \right)^2 . \tag{7.1.7}$$

If we assume that the transverse motion is much slower than the longitudinal component we can assume that $z \simeq v_\| t$ and therefore,

$$\begin{aligned}
X(t) &= X(0) - \frac{eA_\text{w}}{\gamma m} \frac{1}{k_\text{w} v_\|} \sin(k_\text{w} v_\| t) , \\
Y(t) &= Y(0) + \frac{eA_\text{w}}{\gamma m} \frac{1}{k_\text{w} v_\|} \left[\cos(k_\text{w} v_\| t) - 1 \right] .
\end{aligned} \tag{7.1.8}$$

This trajectory of the particle implies that it will interact naturally with a circularly polarized plane wave. However, before we consider the radiation emitted, it is instructive to make one more observation. The energy factor, γ, is defined by

$$\gamma = \frac{1}{\sqrt{1 - \beta_\|^2 - \beta_\perp^2}} , \tag{7.1.9}$$

and we can also define a similar factor associated only with the longitudinal motion i.e.,

$$\gamma_{\|} \equiv \frac{1}{\sqrt{1-\beta_{\|}^2}}\,. \tag{7.1.10}$$

Since the transverse velocity, as determined in (7.1.4), is γ dependent, we find that

$$\gamma_{\|} = \frac{\gamma}{\sqrt{1+(eB_{\mathrm{w}}/mck_{\mathrm{w}})^2}}\,, \tag{7.1.11}$$

which indicates that the effective energy factor ($\gamma_{\|}$) can be substantially smaller than γ. For example, if $B_{\mathrm{w}} = 0.5\,\mathrm{T}$ and $L = 2\,\mathrm{cm}$ the longitudinal energy factor is about 70% of the original γ.

An electron which follows the trajectory described by (7.1.8) radiates. In order to calculate the emitted radiation we assume that the current density is given by

$$\begin{aligned}
J_x(\mathbf{r},t) &\simeq -ev_x(t)\delta(x)\delta(y)\delta(z-v_{\|}t)\,,\\
J_y(\mathbf{r},t) &\simeq -ev_y(t)\delta(x)\delta(y)\delta(z-v_{\|}t)\,,\\
J_z(\mathbf{r},t) &= -ev_{\|}\delta(x)\delta(y)\delta(z-v_{\|}t)\,,
\end{aligned} \tag{7.1.12}$$

where the transverse displacement of the electron was neglected. In free space the radiation generated by this current density is given by

$$\begin{aligned}
A_x(\mathbf{r},t) &= \mu_0 \int \mathrm{d}\omega \mathrm{e}^{j\omega t} \int \mathrm{d}\mathbf{r}' \frac{\mathrm{e}^{-j(\omega/c)|\mathbf{r}-\mathbf{r}'|}}{4\pi|\mathbf{r}-\mathbf{r}'|} J_x(\mathbf{r}',\omega)\,,\\
A_y(\mathbf{r},t) &= \mu_0 \int \mathrm{d}\omega \mathrm{e}^{j\omega t} \int \mathrm{d}\mathbf{r}' \frac{\mathrm{e}^{-j(\omega/c)|\mathbf{r}-\mathbf{r}'|}}{4\pi|\mathbf{r}-\mathbf{r}'|} J_y(\mathbf{r}',\omega)\,,\\
A_z(\mathbf{r},t) &= \mu_0 \int \mathrm{d}\omega \mathrm{e}^{j\omega t} \int \mathrm{d}\mathbf{r}' \frac{\mathrm{e}^{-j(\omega/c)|\mathbf{r}-\mathbf{r}'|}}{4\pi|\mathbf{r}-\mathbf{r}'|} J_z(\mathbf{r}',\omega)\,.
\end{aligned} \tag{7.1.13}$$

The time Fourier transform of the current density in (7.1.12), denoted above by $\mathbf{J}(\mathbf{r},\omega)$, is given by

$$\begin{aligned}
J_x(\mathbf{r},\omega) &= -\frac{e}{2\pi}\frac{v_x(t=z/v_{\|})}{v_{\|}}\delta(x)\delta(y)\mathrm{e}^{-j(\omega/v_{\|})z}\,,\\
J_y(\mathbf{r},\omega) &= -\frac{e}{2\pi}\frac{v_y(t=z/v_{\|})}{v_{\|}}\delta(x)\delta(y)\mathrm{e}^{-j(\omega/v_{\|})z}\,,\\
J_z(\mathbf{r},\omega) &= -\frac{e}{2\pi}\delta(x)\delta(y)\mathrm{e}^{-j(\omega/v_{\|})z}\,.
\end{aligned} \tag{7.1.14}$$

The integrals in (7.1.13) can be simplified for the case when the observer is far away from the wiggler $[(\omega/c)r \gg 1\,]$ in which case we have

$$A_x(\mathbf{r},t) = \mu_0 \mathrm{Re}\left[\int \mathrm{d}\omega \mathrm{e}^{j\omega t}\frac{\mathrm{e}^{-j(\omega/c)r}}{4\pi r}\int \mathrm{d}\mathbf{r}' \mathrm{e}^{j(\omega/c)z'\cos\theta} J_x(\mathbf{r}',\omega)\right],$$

$$A_y(\mathbf{r},t) = \mu_0 \mathrm{Re}\left[\int \mathrm{d}\omega \mathrm{e}^{j\omega t}\frac{\mathrm{e}^{-j(\omega/c)r}}{4\pi r}\int \mathrm{d}\mathbf{r}' \mathrm{e}^{j(\omega/c)z'\cos\theta} J_y(\mathbf{r}',\omega)\right],$$

$$A_z(\mathbf{r},t) = \mu_0 \mathrm{Re}\left[\int \mathrm{d}\omega \mathrm{e}^{j\omega t}\frac{\mathrm{e}^{-j(\omega/c)r}}{4\pi r}\int \mathrm{d}\mathbf{r}' \mathrm{e}^{j(\omega/c)z'\cos\theta} J_z(\mathbf{r}',\omega)\right]; \tag{7.1.15}$$

θ in these expressions is the angle between the vector which connects the center of the wiggler and the observer with the z axis. Substituting the explicit expression for the current densities allows us to evaluate the integral analytically. The result is presented next after terms of the type $(\omega/c)\cos\theta - k_\mathrm{w} - \omega/v_\|$ were neglected since they oscillate rapidly:

$$\begin{aligned}A_x(\mathbf{r},t) =& \frac{e\mu_0}{(4\pi)^2}\frac{eB_\mathrm{w}}{mk_\mathrm{w}\gamma v_\|}\frac{D}{r}\\ &\times \mathrm{Re}\left\{\int \mathrm{d}\omega\, \mathrm{e}^{j\omega(t-r/c)}\,\mathrm{sinc}\left[\left(\frac{\omega}{c}\cos\theta - \frac{\omega}{v_\|} + k_\mathrm{w}\right)\frac{D}{2}\right]\right\},\\ A_y(\mathbf{r},t) =& \frac{e\mu_0}{(4\pi)^2}\frac{eB_\mathrm{w}}{mk_\mathrm{w}\gamma v_\|}\frac{D}{r}\\ &\times \mathrm{Re}\left\{\int \mathrm{d}\omega\, \frac{1}{j}\mathrm{e}^{j\omega(t-r/c)}\,\mathrm{sinc}\left[\left(\frac{\omega}{c}\cos\theta - \frac{\omega}{v_\|} + k_\mathrm{w}\right)\frac{D}{2}\right]\right\},\\ A_z(\mathbf{r},t) \simeq& 0;\end{aligned} \tag{7.1.16}$$

where $\mathrm{sinc}(x) = \sin(x)/x$, the total length of the wiggler is denoted by D and it spans from $-D/2 < z < D/2$. The longitudinal component of the magnetic vector potential is negligible since it is proportional to $\mathrm{sinc}[(\cos\theta - 1/\beta_\|)(\omega D/2c)]$ and this function varies rapidly for $D \to \infty$.

The magnetic vector potential determines the electromagnetic field which in turn enables us to evaluate the power and energy emitted. The Poynting flux is given by

$$S_z(\mathbf{r},t) = E_x(\mathbf{r},t)H_y(\mathbf{r},t) - E_y(\mathbf{r},t)H_x(\mathbf{r},t), \tag{7.1.17}$$

and the energy emitted per unit area in this process is given by

$$W(\mathbf{r}) = \int \mathrm{d}t\, S_r(\mathbf{r},t), \tag{7.1.18}$$

where $S_z = S_r\cos\theta$. Substituting the explicit expressions for the components of the electromagnetic field, followed by the evaluation of the integral over t simplifies substantially the calculation since the resulting Dirac delta function can be utilized to evaluate the double integration

$$\begin{aligned}W(\mathbf{r}) =& \frac{4\pi}{\eta_0}\left[\frac{e\mu_0}{(4\pi)^2}\left(\frac{eB_\mathrm{w}}{mk_\mathrm{w}\gamma v_\|}\right)\frac{D}{r}\right]^2\\ &\times \int \mathrm{d}\omega\,\omega^2 \mathrm{sinc}^2\left[\left(\frac{\omega}{c}\cos\theta - \frac{\omega}{v_\|} + k_\mathrm{w}\right)\frac{D}{2}\right].\end{aligned} \tag{7.1.19}$$

This energy is emitted by a single electron. An ensemble of N electrons in the interaction region carry a current $I = eNv_{\|}/D$ and the energy in an angular interval $\theta \to \theta + d\theta$ is

$$\begin{aligned}\bar{W}(\theta) =& r^2 N W(\mathbf{r}) \\ =& \frac{eI\eta_0}{(4\pi)^3}\left(\frac{eB_{\mathrm{w}}}{mck_{\mathrm{w}}\gamma}\right)^2\left(\frac{D}{\beta_{\|}}\right)^3 \\ & \times \int \mathrm{d}\left(\frac{\omega}{c}\right)\frac{\omega^2}{c^2}\operatorname{sinc}^2\left[\left(\frac{\omega}{c}\cos\theta - \frac{\omega}{v_{\|}} + k_{\mathrm{w}}\right)\frac{D}{2}\right]. \end{aligned} \tag{7.1.20}$$

The term

$$g(\xi) \equiv \operatorname{sinc}^2(\xi), \tag{7.1.21}$$

represents the spontaneous emission line shape and we shall encounter it again when considering the power in the low-gain Compton regime. The argument of the sinc function is directly associated with the resonance condition and maximum power is emitted when this condition is satisfied i.e.,

$$\frac{\omega}{c}\cos\theta - \frac{\omega}{v_{\|}} + k_{\mathrm{w}} = 0. \tag{7.1.22}$$

It implies that the frequency emitted in the forward direction ($\theta = 0$) is given by

$$\omega = \omega_{\mathrm{res}} \equiv ck_{\mathrm{w}}\frac{\beta_{\|}(1+\beta_{\|})\gamma^2}{1+(eB_{\mathrm{w}}/mk_{\mathrm{w}}c)^2}, \tag{7.1.23}$$

which clearly depends on the strength of the wiggler. This can be considered the exact resonance condition and the expressions presented in the first and third chapters [(1.1.19) and (3.2.17)] are approximations which are valid in case of a weak wiggler field i.e. $eB_{\mathrm{w}}c/k_{\mathrm{w}} \ll mc^2$.

Rather than considering the whole spectrum of waves emitted in a given direction in space, it is instructive to present the energy emitted in a frequency interval $\omega \to \omega + d\omega$ in one period of the wave i.e.,

$$\begin{aligned}\frac{\omega}{2\pi}\frac{d\bar{W}(\theta)}{d\omega} =& \frac{eI\eta_0}{(4\pi)^4}\left(\frac{eB_{\mathrm{w}}}{mck_{\mathrm{w}}\gamma}\right)^2\left(\omega\frac{D}{v_{\|}}\right)^3 \\ & \times \operatorname{sinc}^2\left[\left(\frac{\omega}{c}\cos\theta - \frac{\omega}{v_{\|}} + k_{\mathrm{w}}\right)\frac{D}{2}\right]. \end{aligned} \tag{7.1.24}$$

Assuming operation at resonance we can substitute the explicit expression for the resonant frequency and obtain

$$\begin{aligned}\mathcal{W} \equiv& \left[\frac{\omega}{2\pi}\frac{\mathrm{d}\bar{W}(\theta=0)}{\mathrm{d}\omega}\right]_{\omega=\omega_{\mathrm{res}}} \\ =& \frac{eI\eta_0}{32\pi}\left[\frac{D}{L}(1+\beta_{\|})\right]^3\gamma^4\frac{\Omega_{\mathrm{w}}^2}{(1+\Omega_{\mathrm{w}}^2)^3}, \end{aligned} \tag{7.1.25}$$

where $\Omega_{\rm w} = eB_{\rm w}/mck_{\rm w}$. Note that as a function of $\Omega_{\rm w}$, the emitted power has a maximum at $\Omega_{\rm w} = 1/\sqrt{2}$ therefore

$$\mathcal{W}_{\max} = \frac{eI\eta_0}{216\pi}\left[\frac{D}{L}(1+\beta_{\parallel})\right]^3 \gamma^4 . \tag{7.1.26}$$

According to this result, a 10 MeV, 1 kA beam generates $\mathcal{W}_{\max} = 0.1\,\mu$J of energy in 10 periods of the structure. If the electron pulse duration corresponds to the radiation period 1/35 GHz, then the total energy carried by the beam is 0.28 J which is six orders of magnitude larger than the radiated power.

The energy lost by the electron as it traverses the periodic magnetic field, can be interpreted in terms of an effective gradient which decelerates the moving electron. In order to evaluate this gradient, we integrate (7.1.19) over the spherical envelope

$$\begin{aligned}\tilde{W} \equiv& 2\pi r^2 \int_0^{\pi} \mathrm{d}\theta \sin\theta\, W(r,\theta) \\ =& \frac{2(4\pi)^2}{\eta_0}\frac{(e\mu_0)^2}{(4\pi)^4}\left(\frac{eB_{\rm w}}{mck_{\rm w}\gamma\beta}\right)^2 D^2 \int_0^{\pi} \mathrm{d}\theta \sin\theta \int_{-\infty}^{\infty} \mathrm{d}\omega \\ & \times \omega^2 \mathrm{sinc}^2\left[\frac{D}{2}\left(\frac{\omega}{c}\cos\theta - \frac{\omega}{c}\frac{1}{\beta} + k_{\rm w}\right)\right] . \end{aligned} \tag{7.1.27}$$

For a long interaction region we use

$$\lim_{D\to\infty}\left\{\frac{D}{2}\mathrm{sinc}\left[\frac{D}{2}(k_1-k_2)\right]\right\} = 2\pi\delta(k_1-k_2) , \tag{7.1.28}$$

thus

$$\tilde{W} = \left(\frac{eB_{\rm w}}{2mck_{\rm w}\gamma\beta}\right)^2 D \frac{e^2}{4\pi\varepsilon_0/k_{\rm w}^2}\int_0^{\pi}\mathrm{d}\theta \frac{\sin\theta}{(1/\beta-\cos\theta)^3} . \tag{7.1.29}$$

The integral can be calculated analytically and, as we indicated, it is convenient to determine the effective gradient as $E_{\rm eff} \equiv \tilde{W}/eD$ which reads

$$E_{\rm eff} = \frac{1}{2}\beta\gamma^2 \frac{ek_{\rm w}^2}{4\pi\varepsilon_0}\left(\frac{eB_{\rm w}}{mck_{\rm w}}\right)^2 . \tag{7.1.30}$$

This is the decelerating gradient which acts on the particle. Note that for a relativistic particle it is quadratic in γ, the energy of the particle and it will become an important factor when discussing acceleration of electrons using the FEL scheme in Chap. 8.

7.2 Low-Gain Compton Regime

If an electromagnetic wave is injected parallel to the beam and its frequency matches the resonance condition, then stimulated radiation may occur. As a first stage, we examine the lowest order effect of the beam on the radiation field. To be more specific we look for the contribution to the radiated power of the first order in $\omega_{\rm p}^2$ term.

The wiggler is the same as in (7.1.1) and the injected wave is circularly polarized:

$$\mathbf{A}_{\rm rf}(\mathbf{r},t) = A_0\Big[\mathbf{1}_x\cos(\omega t - kz) + \mathbf{1}_y\sin(\omega t - kz)\Big]. \tag{7.2.1}$$

Ignoring space-charge effects and in the absence of boundaries it is justified to omit the electrostatic potential from the expression for the relativistic Hamiltonian [(3.1.15)] hence

$$H = \sqrt{(\mathbf{p} + e\mathbf{A}_{\rm w} + e\mathbf{A}_{\rm rf})^2c^2 + (mc^2)^2}. \tag{7.2.2}$$

Neither the wiggler nor the radiation field has components of $\mathbf{A}$ parallel to the beam and consequently,

$$H = \sqrt{(mc^2)^2 + c^2p_\parallel^2 + c^2(p_\perp + eA_{\rm w} + eA_{\rm rf})^2}. \tag{7.2.3}$$

As before, the conservation of the transverse canonical momentum ($p_\perp = 0$) implies

$$H = \sqrt{(mc^2)^2 + c^2p_\parallel^2 + c^2e^2(A_{\rm w} + A_{\rm rf})^2}, \tag{7.2.4}$$

whereas the linearization of the Hamiltonian in the radiation field reads

$$\begin{aligned} H &= H_0 + H_1 \\ &= mc^2\gamma + \frac{e^2}{m\gamma}\mathbf{A}_{\rm w}\cdot\mathbf{A}_{\rm rf} \\ &= mc^2\gamma - \frac{e^2A_0B_{\rm w}}{mk_{\rm w}\gamma}\cos[\omega t - (k + k_{\rm w})z]. \end{aligned} \tag{7.2.5}$$

From this expression we learn that the first order perturbation is proportional to the scalar product of the wiggler and radiation vector potentials. This part of the Hamiltonian determines the so-called pondermotive force (subscript p)

$$\begin{aligned} F_{\rm p} &= -\frac{\partial H_1}{\partial z} \\ &= \frac{e^2A_0B_{\rm w}}{m\gamma k_{\rm w}}(k + k_{\rm w})\sin[\omega t - (k + k_{\rm w})z]. \end{aligned} \tag{7.2.6}$$

For what follows it is convenient to adopt a phasor notation

$$F_{\rm p} \to \bar{F}_{\rm p} = -j\frac{e^2 A_0 B_{\rm w}}{m\gamma k_{\rm w}}(k+k_{\rm w})e^{j[\omega t-(k+k_{\rm w})z]}$$
$$= \tilde{F}_{\rm p} e^{j[\omega t-(k+k_{\rm w})z]} , \tag{7.2.7}$$

where

$$\tilde{F}_{\rm p} \equiv -je^2 A_0 B_{\rm w}(k+k_{\rm w})/(m\gamma k_{\rm w}) . \tag{7.2.8}$$

Next we linearize the Liouville equation i.e., assume that the distribution function f has the form

$$f(z,t;p) = f_0(p) + f_1(z,t;p) , \tag{7.2.9}$$

where f_0 is considered to be known and f_1 is linear in the pondermotive force hence

$$\left(j\omega + v\frac{\partial}{\partial z}\right) f_1 = -\tilde{F}_{\rm p} e^{j[\omega t-(k+k_{\rm w})z]} \frac{{\rm d}f_0}{{\rm d}p} . \tag{7.2.10}$$

A solution of this expression, assuming that the right hand side is known, can be formally written as

$$f_1(z,t) = -\frac{{\rm d}f_0}{{\rm d}p}\frac{\tilde{F}_{\rm p}}{v} e^{j\omega(t-z/v)} \frac{e^{j\delta k z} - e^{-j\delta k D/2}}{j\delta k} ,$$
$$= \frac{1}{p}\frac{{\rm d}f_0}{{\rm d}p}\frac{e^2 A_0 B_{\rm w}(k+k_{\rm w})}{k_{\rm w}} e^{j\omega(t-z/v)} \frac{e^{j\delta k z} - e^{-j\delta k D/2}}{\delta k} , \tag{7.2.11}$$

where D is the length of the interaction region which starts at $z = -D/2$ and $\delta k \equiv \omega/v - k - k_{\rm w}$. With this expression for the distribution function, we can define the macroscopic current density and in particular its transverse components read

$$J_\perp = -en_0 \int {\rm d}p v_\perp f_1$$
$$= \frac{1}{2} n_0 \left(\frac{e^2 B_{\rm w}}{k_{\rm w}}\right)^2 \frac{A_0(k+k_{\rm w})}{m} (\mathbf{1}_x - j\mathbf{1}_y)$$
$$\times \int {\rm d}p \frac{1}{\gamma p}\frac{{\rm d}f_0}{{\rm d}p} \left[e^{j\omega(t-z/v)+jk_{\rm w}z} \frac{e^{j\delta k z} - e^{-j\delta k D/2}}{\delta k} \right] ; \tag{7.2.12}$$

Here we used the explicit expression for the transverse velocity in (7.1.4) and ignored terms which vary rapidly in space; n_0 is the average density of the particles in the absence of the radiation. Since we calculated the current density generated by a known electric field the next step is to calculate the power

$$P = S_{\rm el} \int_{-D/2}^{D/2} {\rm d}z \frac{1}{2} {\rm Re}\,(E_\perp \cdot J_\perp^*) , \tag{7.2.13}$$

where $S_{\rm el}$ is the beam cross-section and

$$\begin{aligned} E_\perp &= -j\omega A_{\rm rf}\,, \\ &= -j\omega(\mathbf{1}_x - j\mathbf{1}_y)A_0 e^{j\omega t - jkz}\,. \end{aligned} \tag{7.2.14}$$

Note that it has been tacitly assumed here that the effect of the beam on the radiation field is negligible. Substituting in (7.2.13) we obtain

$$P = \frac{1}{4} n_0 \omega (k + k_{\rm w}) S_{\rm el} D^2 \left(\frac{e^2 B_{\rm w} A_0}{k_{\rm w}}\right)^2 \int {\rm d}p \frac{v}{p^2}\frac{{\rm d}f_0}{{\rm d}p} {\rm sinc}^2\left(\frac{1}{2}\delta k D\right). \tag{7.2.15}$$

At this point we can evaluate the last integral for two extreme regimes: *(i)* cold beam approximation and *(ii)* warm beam approximation. In the former case it is assumed that the initial distribution function f_0 is much sharper than the sinc function hence by integration by parts we get

$$P = \frac{1}{2\gamma_0^5 \beta_0^4} \frac{eI\eta_0}{mc^2} \left[\frac{(\omega A_0 D)^2}{2\eta_0}\right] \left(\frac{eB_{\rm w}}{mck_{\rm w}}\right)^2 (k + k_{\rm w}) D \left[-\frac{1}{2}\frac{\rm d}{{\rm d}\xi}{\rm sinc}^2 \xi\right]_{\xi = \delta k D/2}, \tag{7.2.16}$$

or

$$\begin{aligned} P = &\left[\left(\frac{\omega_{\rm p} D}{c}\right)^2 \frac{1}{(\gamma_0 \beta_0)^3}\right] \left[\frac{eB_{\rm w}}{mck_{\rm w}}\frac{1}{\gamma}\right]^2 \\ &\times \left[\frac{(\omega A_0)^2 S_{\rm el} D (k + k_{\rm w})}{4\eta_0}\right] \left[-\frac{1}{2}\frac{\rm d}{{\rm d}\xi}{\rm sinc}^2\xi\right]_{\xi = \delta k D/2}, \end{aligned} \tag{7.2.17}$$

where we used $f_0(p) = \delta(p - p_0)$. These expressions clearly indicate that the power is inversely proportional to the γ^5 out of which the γ^3 term is due to the longitudinal bunching, and γ^2 is due to the transverse oscillation in the wiggler. In addition, note that the power of the *coherent* radiation emitted is proportional to the derivative of the *spontaneous* emission line shape.

The second regime of interest is when the sinc function is much sharper than the distribution of particles and the power is

$$\begin{aligned} P &\propto \int {\rm d}p \frac{v}{p^2}\frac{{\rm d}f_0}{{\rm d}p}{\rm sinc}^2\left(\frac{1}{2}\delta k D\right), \\ &\propto \frac{v}{p^2}\int {\rm d}p \frac{{\rm d}f_0}{{\rm d}p}{\rm sinc}^2\left(\frac{1}{2}\delta k D\right), \\ &\propto \frac{v}{p^2}\left\{\int {\rm d}p \frac{\rm d}{{\rm d}p}\left[f_0 {\rm sinc}^2\left(\frac{1}{2}\delta k D\right)\right] - \int {\rm d}p f_0 \frac{\rm d}{{\rm d}p}{\rm sinc}^2\left(\frac{1}{2}\delta k D\right)\right\}, \\ &\propto -{\rm sinc}^2(\pi D/L)\,, \\ &\simeq 0\,, \end{aligned} \tag{7.2.18}$$

which indicates that in the low-gain Compton regime, "warm" electrons do not generate coherent radiation.

The coherent radiation which is generated in the course of the electrons' motion in the wiggler as revealed by (7.2.16,17) is illustrated in Fig. 7.1 where the normalized gain is $-\frac{1}{2}\frac{\mathrm{d}}{\mathrm{d}\xi}\mathrm{sinc}^2(\xi)$. We observe that when the velocity of the electrons is larger than the phase velocity of the pondermotive force i.e., $v > \omega/(k + k_{\mathrm{w}})$ meaning negative ξ, the normalized gain is negative thus energy is transferred from the electrons to the wave. And when the electrons are slower, they are accelerated by the pondermotive force. Maximum gain does not occur at resonance but for $|\xi| = 1.303$ in which case the absolute value of the normalized gain is 0.27.

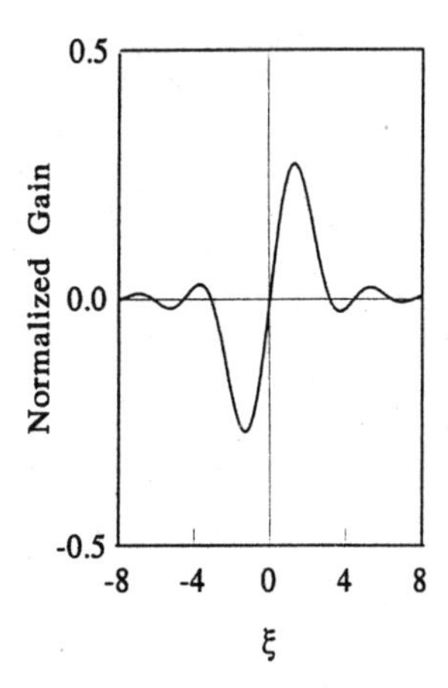

Fig. 7.1 Normalized gain

7.3 High-Gain Compton Regime

In the previous section the collective effect of the particles was neglected in the sense that the effect of the beam on the radiation field was ignored and the gain was a result of an ensemble of dipoles oscillating coherently due to the common excitation of the external field. This interpretation is supported by the expression for the current density in (7.2.12) which indicates that the electrons are organized in bunches. The amplitude of this modulation is proportional to the "dc" current. At low currents the effect of these bunches on the radiation field is indeed small but as the current is increased their effect becomes more and more significant. In parallel, as the modulation increases, the quasi-electrostatic forces between particles also increases and space-charge effects have to be accounted for. These will be represented here by a scalar electric potential Φ and the longitudinal component of the magnetic vector potential A_z. Following the same approach as previously, the dynamics of the distribution function satisfies

$$\left[j\omega + v\frac{\partial}{\partial z} + \left(F_{\mathrm{p}} + e\frac{\mathrm{d}\Phi}{\mathrm{d}z} + j\omega e A_z\right)\frac{\partial}{\partial p}\right] f = 0\,. \tag{7.3.1}$$

The electric scalar potential Φ is determined by the charge distribution via the non-homogeneous wave equation

$$\left[\frac{\mathrm{d}^2}{\mathrm{d}z^2} + \frac{\omega^2}{c^2}\right]\Phi = \frac{en_0}{\varepsilon_0}\left[\int \mathrm{d}p\, f - 1\right]. \tag{7.3.2}$$

This potential determines the longitudinal component of the magnetic vector potential since we have tacitly assumed the Lorentz gauge i.e.,

$$\frac{\mathrm{d}A_z}{\mathrm{d}z} + \frac{j\omega}{c^2}\Phi = 0. \tag{7.3.3}$$

Note that there is no magnetic field associated with this potential since it is dependent only on the z coordinate. In addition to these two potentials, the distribution of particles determines the transverse current density

$$J_\perp(z,\omega) = -en_0 \int \mathrm{d}p v_\perp f, \tag{7.3.4}$$

which in turn governs the magnetic vector potential via the non-homogeneous wave equation as in (2.1.39). In this case we ignore transverse effects therefore we integrate the wave equation over the transverse dimensions. Assuming that the effective area of the electromagnetic field is S_{em} and that of the electron beam is S_{el}, we define the filling factor term $F_{\mathrm{f}} \equiv S_{\mathrm{el}}/S_{\mathrm{em}}$ by whose means the 1D wave equation for the magnetic vector potential reads

$$\left[\frac{\mathrm{d}^2}{\mathrm{d}z^2} + \frac{\omega^2}{c^2}\right]A_\perp = -\mu_0 F_{\mathrm{f}} J_\perp; \tag{7.3.5}$$

this filling factor is assumed to be known.

This is the set of equations which describes the interaction in a free-electron laser in the high-gain Compton regime. Before we proceed to a solution of this set of equations it is instructive to examine the same set of equations when instead of the the Lorentz gauge we use the Coulomb gauge. In this case the equations read

$$\begin{aligned}
&\left[j\omega + v\frac{\partial}{\partial z} + \left(F_p + e\frac{\mathrm{d}\Phi}{\mathrm{d}z}\right)\frac{\partial}{\partial p}\right] f = 0, \\
&\frac{\mathrm{d}^2\Phi}{\mathrm{d}z^2} = \frac{en_0}{\varepsilon_0}\left[\int \mathrm{d}p f - 1\right], \\
&A_z = 0, \\
&J_\perp(z,\omega) = -en_0 \int \mathrm{d}p v_\perp f, \\
&\left[\frac{\mathrm{d}^2}{\mathrm{d}z^2} + \frac{\omega^2}{c^2}\right]A_\perp = -\mu_0 F_{\mathrm{f}} J_\perp.
\end{aligned} \tag{7.3.6}$$

In principle (classical) the physical result should not be affected by the gauge choice. From the point of view of the particles' dynamics what is important is the acting field and the latter is independent of the gauge choice. This is in particular easy to show in this 1D case: when using the Coulomb gauge, for a given source term $\varrho = -en$, assuming functional dependence of the form $e^{j\omega t - jkz}$, the Poisson equation dictates

$$\Phi = -\frac{en}{\varepsilon_0 k^2}, \tag{7.3.7}$$

and since the longitudinal component of magnetic vector potential vanishes the electric field reads

$$E_z = -\frac{d\Phi}{dz} = -jk\frac{en}{\varepsilon_0 k^2}. \tag{7.3.8}$$

When choosing a Lorentz gauge, the non-homogeneous wave equation dictates

$$\Phi = \frac{en/\varepsilon_0}{(\omega/c)^2 - k^2}, \tag{7.3.9}$$

and since Φ is related to A_z by $A_z = (\omega/c^2 k)\Phi$, the electric field reads

$$E_z = -j\omega A_z + jk\Phi = \frac{c^2}{j\omega}\left[\frac{\omega^2}{c^2} - k^2\right]\frac{\omega}{c^2 k}\Phi. \tag{7.3.10}$$

Substituting (7.3.9) in (7.3.10) we obtain the same expression for the electric field as in (7.3.8) i.e.,

$$\begin{aligned} E_z &= \frac{c^2}{j\omega}\left[\frac{\omega^2}{c^2} - k^2\right]\frac{\omega}{c^2 k}\Phi, \\ &= \frac{c^2}{j\omega}\left[\frac{\omega^2}{c^2} - k^2\right]\frac{\omega}{c^2 k}\frac{en/\varepsilon_0}{(\omega/c)^2 - k^2}, \\ &= -jk\frac{en}{\varepsilon_0 k^2}. \end{aligned} \tag{7.3.11}$$

In both cases, the magnetic field (associated with the space-charge) is zero: in the case of the Coulomb gauge it vanishes since A_z is zero and in the Lorentz gauge case, because A_z depends only on the z coordinate. Following the same procedure, it can be shown that the force term in Liouville equations [(7.3.1) and (7.3.6)] is identical in both cases.

7.3.1 The Dispersion Relation

The set of equations established previously will be analyzed in this subsection in order to quantify the energy exchange process. For this purpose it is convenient to adopt a phasor notation for all linearized quantities. According to (7.2.7) F_{p} is proportional to $\exp[j\omega t - j(k + k_{\mathrm{w}})z]$. However, since in this case the Hamiltonian is time dependent, γ is not conserved and therefore we redefine $\tilde{F}_{\mathrm{p}}$ such that it does not include the γ term i.e.,

$$F_{\mathrm{p}} = -j\tilde{F}_{\mathrm{p}}\frac{1}{\gamma}\mathrm{e}^{j\omega t - j(k+k_{\mathrm{w}})z},$$
$$\tilde{F}_{\mathrm{p}} = \frac{e^2 A_0 B_{\mathrm{w}}(k+k_{\mathrm{w}})}{mk_{\mathrm{w}}}. \tag{7.3.12}$$

Consequently, assuming that f_0 does not vary in time and in space, a similar dependence as F_{p} is anticipated for f_1. Furthermore, since according to (7.3.2) Φ is linear in f_1, a similar dependence is expected for Φ and A_z:

$$f_1 = \tilde{f}_1\mathrm{e}^{j\omega t - jKz},$$
$$\Phi = \tilde{\Phi}\mathrm{e}^{j\omega t - jKz},$$
$$\mathbf{A} = A_0(\mathbf{1}_x - j\mathbf{1}_y)\mathrm{e}^{j\omega t - jKz}, \tag{7.3.13}$$

where $K \equiv k + k_{\mathrm{w}}$. Substituting in (7.3.2) we obtain

$$\tilde{\Phi} = \frac{en_0/\varepsilon_0}{(\omega/c)^2 - K^2}\int \mathrm{d}p\tilde{f}_1, \tag{7.3.14}$$

and in a similar way we substitute in (7.3.1) to get

$$\tilde{f}_1 = \frac{1}{\delta\omega}\left[\frac{1}{\gamma}\tilde{F}_{\mathrm{p}} + \frac{e}{K}(K^2 - \frac{\omega^2}{c^2})\tilde{\Phi}\right]\frac{\mathrm{d}f_0}{\mathrm{d}p}, \tag{7.3.15}$$

where $\delta\omega \equiv \omega - v(k + k_{\mathrm{w}})$ is the resonance term. Substituting the latter into (7.3.14) we have, for the potential,

$$\tilde{\Phi} = \frac{m\omega_{\mathrm{p}}^2}{\varepsilon(\omega, K)}\frac{\tilde{F}_{\mathrm{p}}}{(\omega/c)^2 - K^2}\int \mathrm{d}p\frac{1}{\gamma\,\delta\omega}\frac{\mathrm{d}f_0}{\mathrm{d}p}, \tag{7.3.16}$$

where $\varepsilon(\omega, K)$ is the dielectric coefficient of the beam defined by

$$\varepsilon(\omega, K) = 1 + \frac{m\omega_{\mathrm{p}}^2}{K}\int \mathrm{d}p\frac{1}{\delta\omega}\frac{\mathrm{d}f_0}{\mathrm{d}p}. \tag{7.3.17}$$

The explicit expression for the amplitude of the scalar electric potential can be substituted into (7.3.15) which results in an expression for $\tilde{f}_1$:

$$\tilde{f}_1 = \frac{1}{\delta\omega}\left[\frac{1}{\gamma} - \frac{m\omega_{\mathrm{p}}^2}{K\varepsilon(\omega, K)}\left(\int \mathrm{d}p'\frac{1}{\gamma'\delta\omega'}\frac{\mathrm{d}}{\mathrm{d}p'}f_0(p')\right)\right]\frac{\mathrm{d}f_0}{\mathrm{d}p}\tilde{F}_{\mathrm{p}}. \tag{7.3.18}$$

The particles' density defines the current density via (7.3.4) whose linear term (in the radiation field) is $J_\perp = -en_0\int \mathrm{d}pv_\perp f_1$ or explicitly,

$$\mathbf{J}_\perp = \frac{1}{2}\omega_{\mathrm{p}}^2\varepsilon_0\frac{B_{\mathrm{w}}}{k_{\mathrm{w}}}\tilde{F}_{\mathrm{p}}(\mathbf{1}_x - j\mathbf{1}_y)\mathrm{e}^{j\omega t - jkz}$$
$$\times\int \mathrm{d}p\frac{1}{\delta\omega}\frac{\mathrm{d}f_0}{\mathrm{d}p}\left[\frac{1}{\gamma} - \frac{m\omega_{\mathrm{p}}^2}{K\varepsilon(\omega, K)}\int \mathrm{d}p'\frac{1}{\gamma'\,\delta\omega'}\frac{\mathrm{d}}{\mathrm{d}p'}f_0(p')\right], \tag{7.3.19}$$

where off-resonance terms of the form $\omega - v(k - k_{\rm w})$ were neglected and $\delta\omega' \equiv \omega - v'(k + k_{\rm w})$. The current distribution from the above is the source term to the wave equation in (7.3.5) which, after being substituted, gives the dispersion relation

$$\left(\frac{\omega}{c}\right)^2 - k^2 = -\frac{1}{2}\left(\Omega_{\rm w}\omega_{\rm p}\right)^2 F_{\rm f} K m \qquad (7.3.20)$$
$$\times \int {\rm d}p \frac{1}{\gamma\delta\omega}\frac{{\rm d}f_0}{{\rm d}p}\left[\frac{1}{\gamma} - \frac{m\omega_{\rm p}^2}{K\varepsilon(\omega,K)}\int {\rm d}p' \frac{1}{\gamma'\,\delta\omega'}\frac{\rm d}{{\rm d}p'}f_0(p')\right];$$

$\Omega_{\rm w}$ was defined in the context of (7.1.25). For a given initial distribution of particles, $f_0(p)$, this expression determines the relation between the ω and k in the system. In the remainder of this section we assume that the system operates as an amplifier which means that the frequency, ω, is set externally and the interaction determines the wavenumber k.

The integrals in the dispersion relation indicate that there are two critical functions: *(i)* the resonance term $\delta\omega^{-1}$ and *(ii)* the distribution function $f_0(p)$. At the simplest approximation one considers a distribution of particles which can be represented by the first two moments namely the average (longitudinal) momentum ($\langle p\rangle$) and its spread $\Delta p \equiv \sqrt{\langle p^2\rangle - \langle p\rangle^2}$. On the other hand, the "sharpness" of the resonance term is determined by the imaginary part of the wavenumber – which is basically the gain which a priori is not known. For solving the dispersion relation it is instructive to consider two extreme regimes: the first when the distribution of particles is much sharper than the resonance line. This will be referred to as the "cold beam operation" $[{\rm Im}(k)/|k| \gg \Delta v/\langle v\rangle]$. The other case corresponds to the opposite regime i.e., ${\rm Im}(k)/|k| \ll \Delta v/\langle v\rangle$ referred to as the "warm beam operation". The two are discussed in the following two subsections.

7.3.2 Cold Beam Operation

In the framework of the cold beam operation we shall consider the extreme case namely, a Dirac delta function distribution,

$$f_0(p) = \delta(p - p_0)\,. \qquad (7.3.21)$$

With this distribution in mind we can evaluate the three integrals in the dispersion relation. The integral

$$\int {\rm d}p \frac{1}{\gamma^2\delta\omega}\frac{{\rm d}f_0}{{\rm d}p}\,, \qquad (7.3.22)$$

can be simplified by integration by parts to read

$$\frac{1}{\gamma_0^2}\left[\int {\rm d}p \frac{\rm d}{{\rm d}p}\left(\frac{f_0}{\delta\omega}\right) - \int {\rm d}p f_0 \frac{\rm d}{{\rm d}p}\left(\frac{1}{\delta\omega}\right)\right]. \qquad (7.3.23)$$

In this expression it was assumed that γ^{-2} varies much more slowly than the other two functions. The first term in (7.3.23) is zero and in the second, the distribution function varies slower than the resonance term, thus

$$\int \mathrm{d}p \frac{1}{\gamma^2 \delta\omega} \frac{\mathrm{d}f_0}{\mathrm{d}p} \simeq -\frac{K}{(\delta\omega)^2} \frac{1}{m\gamma_0^5} . \tag{7.3.24}$$

In a similar way,

$$\int \mathrm{d}p \frac{1}{\gamma \delta\omega} \frac{\mathrm{d}f_0}{\mathrm{d}p} \simeq -\frac{K}{(\delta\omega)^2} \frac{1}{m\gamma_0^4} ,$$
$$\int \mathrm{d}p \frac{1}{\delta\omega} \frac{\mathrm{d}f_0}{\mathrm{d}p} \simeq -\frac{K}{(\delta\omega)^2} \frac{1}{m\gamma_0^3} . \tag{7.3.25}$$

With these results the dispersion relation reads

$$\left(\frac{\omega}{c}\right)^2 - k^2 = \frac{1}{2} \frac{\Omega_{\mathrm{w}}^2}{\gamma_0^2} \frac{\omega_{\mathrm{p}}^2 F_{\mathrm{f}}}{\gamma_0^3} \frac{K^2}{\varepsilon(\omega, K)} \frac{1}{(\delta\omega)^2} , \tag{7.3.26}$$

where

$$\varepsilon(\omega, K) = 1 - \frac{\omega_{\mathrm{p}}^2}{\gamma_0^3 (\delta\omega)^2} \tag{7.3.27}$$

and F_{f} is the filling factor defined in the context of (7.3.5). For a solution of this equation we shall examine two cases: firstly, when the space-charge effects are neglected and secondly, when their effect is taken into consideration.

No Space-Charge Effects. Firstly, we ignore the effect of the space-charge term in the dielectric coefficient of the beam. Following the same approach as in the case of the interaction in a slow-wave structure, we assume that the change, due to the interaction, of the vacuum solution is small and it is denoted by δk i.e., $\omega \gg c|\delta k|$. With this assumption the dispersion relation is identical with the one in a traveling-wave amplifier and it reads

$$\delta k \left(\delta k - \Delta k\right)^2 = -K_0^3 , \tag{7.3.28}$$

only that in this case

$$K_0^3 = \frac{1}{4} \left(\frac{\Omega_{\mathrm{w}} k_{\mathrm{w}}}{\gamma_0}\right)^2 \left(\frac{\omega_{\mathrm{p}}^2 F_{\mathrm{f}}}{c^2 \beta_0^2 \gamma_0^3}\right) \left(1 + \frac{\omega}{c k_{\mathrm{w}}}\right)^2 \left(\frac{\omega}{c}\right)^{-1} ,$$
$$\Delta k = \frac{\omega}{c} \frac{1}{\beta_{\|}} - \frac{\omega}{c} - k_{\mathrm{w}} . \tag{7.3.29}$$

Kroll (1978) was the first to point out the full equivalence between a free-electron laser and a traveling-wave amplifier. As in TWT, assuming that Δk and K_0 are independent, maximum gain occurs at resonance ($\Delta k = 0$) and it is given by

$$\mathrm{Im}(k) = \frac{\sqrt{3}}{2} K_0 \tag{7.3.30}$$

$$= \frac{\sqrt{3}}{2} \left[\frac{1}{4} \left(\frac{\Omega_\mathrm{w} k_\mathrm{w}}{\gamma_0} \right)^2 \left(\frac{\omega_\mathrm{p}^2 F_\mathrm{f}}{c^2 \beta_0^2 \gamma_0^3} \right) \left(1 + \frac{\omega}{c k_\mathrm{w}} \right)^2 \left(\frac{\omega}{c} \right)^{-1} \right]^{1/3} .$$

If we compare this result, as it stands, with the gain in a slow-wave structure we observe that the main difference is the fact that here K_0^3 is proportional to γ_0^{-5} and in the former it was proportional to γ_0^{-3}. However, for relativistic electrons, assuming that $\omega \gg c k_\mathrm{w}$ and bearing in mind that at resonance

$$\frac{\omega}{c} \simeq k_\mathrm{w} \frac{2\gamma_0^2}{1 + \Omega_\mathrm{w}^2} , \tag{7.3.31}$$

we find that the coupling wavenumber K_0 is

$$K_0^3 \simeq \frac{1}{2} \frac{\Omega_\mathrm{w}^2}{1 + \Omega_\mathrm{w}^2} \frac{F_\mathrm{f}}{\beta_0^2 \gamma_0^3} \frac{\omega_\mathrm{p}^2}{c^2} k_\mathrm{w} , \tag{7.3.32}$$

which for a strong wiggler, $\Omega_\mathrm{w} \gg 1$, reads

$$K_0^3 \simeq \frac{1}{2} \frac{\omega_\mathrm{p}^2 F_\mathrm{f}}{c^2 \beta_0^2 \gamma_0^3} k_\mathrm{w}$$

$$\simeq \frac{1}{2} \frac{e I \eta_0 F_\mathrm{f}}{m c^2} \frac{k_\mathrm{w}}{S_\mathrm{el}} \frac{1}{(\gamma_0 \beta_0)^3} . \tag{7.3.33}$$

This result indicates that for a given periodicity and strong wiggler the growth rate *scales* with particle's energy (γ_0) as in a traveling-wave amplifier but still their numerical value can differ quite substantially. A difference between this expression and the one in (4.1.18) is that ω/c was replaced here by k_w. However, in slow-wave structures driven by relativistic electrons, the two are of the same order of magnitude. A more important difference regards the interaction impedance: in this case it is simply

$$Z_\mathrm{int} = \eta_0 F_\mathrm{f} , \tag{7.3.34}$$

and since this might be substantially smaller than in a traveling-wave amplifier (based on metallic periodic structure) the gain per unit length in a FEL is typically smaller.

Before we consider the space-charge effect it is important to emphasize that the assumption $\Omega_\mathrm{w} \gg 1$ which leads to (7.3.33) should be considered only within the limited framework of the comparison with the traveling wave amplifier otherwise too large wiggler amplitude in a FEL has a detrimental effect on its performance which is clearly revealed when examining (7.1.23) since it lowers the operating frequency.

Space-Charge Effect. When the current density is high enough such that its effect on the dielectric coefficient $[\varepsilon(\omega, K)]$ of the beam is significant, we can simplify the dispersion for the forward propagating waves to read

$$\left(k - \frac{\omega}{c}\right)\left[\left(k + k_{\mathrm{w}} - \frac{\omega}{c}\frac{1}{\beta}\right)^2 - K_p^2\right] = -K_0^3 \,, \tag{7.3.35}$$

where $K_p^2 \equiv \omega_{\mathrm{p}}^2/v_0^2\gamma_0^3$. The space-charge waves in this case are characterized by

$$F(k) \equiv \left(k + k_{\mathrm{w}} - \frac{\omega}{c}\frac{1}{\beta}\right)^2 - K_p^2 = 0 \,. \tag{7.3.36}$$

Thus expanding this expression in conjunction with the FEL resonance condition, we obtain

$$\begin{aligned} F(k) &\simeq F\left(k = \frac{\omega}{c}\right) + \left(k - \frac{\omega}{c}\right)\left[\frac{\mathrm{d}}{\mathrm{d}k}F(k)\right]_{k=\omega/v_0 - k_{\mathrm{w}} + K_p} \\ &\simeq 2K_p\left(k - \frac{\omega}{c}\right) . \end{aligned} \tag{7.3.37}$$

This simplifies the dispersion relation to

$$\left(k - \frac{\omega}{c}\right)^2 \simeq -\frac{1}{2}\frac{K_0^3}{K_p} \,, \tag{7.3.38}$$

and the spatial growth rate is

$$\begin{aligned} \mathrm{Im}(k) &= \frac{\sqrt{2}}{2}\sqrt{\frac{K_0^3}{K_p}} \\ &= \frac{\sqrt{2}}{2}\left[\frac{1}{4}\Omega_{\mathrm{w}}^2\left(\frac{\omega_{\mathrm{p}}}{c}F_{\mathrm{f}}\right)\frac{\omega}{v_0}\frac{1}{\gamma_0^{7/2}}\right]^{1/2} \\ &= \frac{\sqrt{2}}{2}\left[\frac{1}{2}\frac{\Omega_{\mathrm{w}}^2}{1+\Omega_{\mathrm{w}}^2}\frac{F_{\mathrm{f}}}{\beta_0\gamma_0^{3/2}}\frac{\omega_{\mathrm{p}}}{c}k_{\mathrm{w}}\right]^{1/2} . \end{aligned} \tag{7.3.39}$$

The main difference between this regime and the former is that here the gain scales as $I^{1/4}$ compared to $I^{1/3}$. In addition, here the gain scales with energy of the electrons like $\gamma^{-7/4}$ compared to the $\gamma^{-5/3}$ in the former case.

7.3.3 Warm Beam Operation

So far we have investigated the dispersion relation in an FEL with a mono-energetic beam of electrons. In this subsection, we examine the operation of the FEL with a warm beam as defined in the context of (7.3.20). For this case we have to evaluate the integral

$$\int \mathrm{d}p \frac{1}{\delta\omega}\frac{1}{\gamma^2}\frac{\mathrm{d}}{\mathrm{d}p} f_0(p)\,, \tag{7.3.40}$$

only that the resonance term varies more rapidly than the distribution term. In the evaluation of the integral we assume that k is a complex quantity i.e.,

$$k = k_\mathrm{r} + jk_\mathrm{i}\,, \tag{7.3.41}$$

hence

$$\begin{aligned} &\int \mathrm{d}p \frac{\gamma^{-2}\,\mathrm{d}f_0(p)/\mathrm{d}p}{\omega - v(k_\mathrm{r}+k_\mathrm{w}) - jvk_\mathrm{i}} \\ &= \int \mathrm{d}p \frac{\gamma^{-2}(\mathrm{d}f_0/\mathrm{d}p)}{[\omega - v(k_\mathrm{r}+k_\mathrm{w})]^2 + [vk_\mathrm{i}]^2}\,[\omega - v(k_\mathrm{r}+k_\mathrm{w}) + jvk_\mathrm{i}]\,. \end{aligned} \tag{7.3.42}$$

The main contribution is from the region where the resonance term peaks i.e., $v = v_\mathrm{res} \equiv \omega/(k_\mathrm{r}+k_\mathrm{w})$. This allows us to extract the slow varying term out of the integral such that we are left with

$$\begin{aligned} &\left[\gamma^{-2}\frac{\mathrm{d}}{\mathrm{d}p}f_0(p)\right]_{v=v_\mathrm{res}} \int \mathrm{d}p \frac{\omega - v(k_\mathrm{r}+k_\mathrm{w})}{[\omega - v(k_\mathrm{r}+k_\mathrm{w})]^2 + [vk_\mathrm{i}]^2} \\ &+ \left[\gamma^{-2}\frac{\mathrm{d}}{\mathrm{d}p}f_0(p)\right]_{v=v_\mathrm{res}} \int \mathrm{d}p \frac{jvk_\mathrm{i}}{[\omega - v(k_\mathrm{r}+k_\mathrm{w})]^2 + [vk_\mathrm{i}]^2}\,. \end{aligned} \tag{7.3.43}$$

The contribution of the first term (near resonance) vanishes because of the asymmetry of the integrand relative to $v = v_\mathrm{res}$ and the second's can be evaluated analytically,

$$\int \mathrm{d}p \frac{1}{\delta\omega\gamma^2}\frac{\mathrm{d}f_0}{\mathrm{d}p} \simeq \left[\gamma\frac{\mathrm{d}f_0}{\mathrm{d}p}\right]_{v=v_\mathrm{res}} \frac{j\pi m}{k_\mathrm{r}+k_\mathrm{w}}\,. \tag{7.3.44}$$

In a similar way

$$\begin{aligned} &\int \mathrm{d}p \frac{1}{\delta\omega\gamma}\frac{\mathrm{d}f_0}{\mathrm{d}p} \simeq \left[\gamma^2\frac{\mathrm{d}f_0}{\mathrm{d}p}\right]_{v=v_\mathrm{res}} \frac{j\pi m}{k_\mathrm{r}+k_\mathrm{w}}\,, \\ &\int \mathrm{d}p \frac{1}{\delta\omega}\frac{\mathrm{d}f_0}{\mathrm{d}p} \simeq \left[\gamma^3\frac{\mathrm{d}f_0}{\mathrm{d}p}\right]_{v=v_\mathrm{res}} \frac{j\pi m}{k_\mathrm{r}+k_\mathrm{w}}\,. \end{aligned} \tag{7.3.45}$$

With these integrals the dispersion relation reads

$$\left(\frac{\omega}{c}\right)^2 - k^2 = -\frac{1}{2}\left(\Omega_{\rm w}\frac{\omega_{\rm p}}{c}\right)^2 F_{\rm f} \times \left\{ \frac{j\left[\pi(mc)^2\gamma({\rm d}f_0/{\rm d}p)\right]_{v=v_{\rm res}}}{1+j(\omega_{\rm p}/c)^2(\omega/c+k_{\rm w})^{-2}\left[\pi(mc)^2\gamma^3({\rm d}f_0/{\rm d}p)\right]_{v=v_{\rm res}}} \right\}. \tag{7.3.46}$$

As in the previous subsection the gain without space-charge effects is calculated neglecting the plasma frequency term in the denominator and it reads

$$\mathrm{Im}(k) \simeq \frac{1}{4}\left(\frac{\omega}{c}\right)^{-1}\left(\Omega_{\rm w}\frac{\omega_{\rm p}}{c}\right)^2 F_{\rm f}\left[\pi(mc)^2\gamma\frac{{\rm d}f_0}{{\rm d}p}\right]_{v=v_{\rm res}}. \tag{7.3.47}$$

When the space-charge effect is significant, the growth rate is given by

$$\mathrm{Im}(k) \simeq \frac{1}{4}\left(\frac{\omega}{c}\right)^{-1}\left\{\left(\Omega_{\rm w}\frac{\omega_{\rm p}}{c}\right)^2 F_{\rm f}\left[\pi(mc)^2\gamma\frac{{\rm d}f_0}{{\rm d}p}\right]_{v=v_{\rm res}}\right\} \times \left[1+\left(\frac{\omega_{\rm p}^2}{c^2K^2}\left[\pi(mc)^2\gamma^3\frac{{\rm d}f_0}{{\rm d}p}\right]_{v=v_{\rm res}}\right)^2\right]^{-1}, \tag{7.3.48}$$

which also determines a quantitative criterion for the regime when the space-charge effect is negligible i.e.,

$$\left(\frac{\omega_{\rm p}^2}{c^2K^2}\left[\pi(mc)^2\gamma^3\frac{{\rm d}f_0}{{\rm d}p}\right]_{v=v_{\rm res}}\right)^2 \ll 1. \tag{7.3.49}$$

Note that the spatial growth rate in these two cases is proportional to the current and if we consider Gaussian-like electrons' distribution i.e., $f_0(p) \simeq {\rm e}^{-(p-p_0)^2/\Delta p^2}$, then the gain vanishes when the resonance velocity corresponds to the peak value of the distribution function.

7.4 Macro-Particle Approach

Electrons which experience an electric field will have a momentum which is either larger or smaller than the average momentum of the beam. Since the system is designed to operate as an amplifier, the number of electrons which have energy below the average of the beam is larger than these which are faster and the energy difference is transferred to the electromagnetic field. In addition, exactly as the gain is associated with the imaginary part of the wavenumber, its real part changes the effective phase velocity of the wave and after a certain interaction length the electrons may be out of phase. Consequently, electrons which at the beginning of the interaction region were decelerated are now accelerated and vice versa. At the point in space where the slow electrons start to be accelerated because of the phase slip, they drain energy from the electromagnetic field whose growth saturates and beyond it, the gain decreases. In order to avoid this situation it is required to adjust the relative phase between the wave and the electrons. In the FEL this can be done by adjusting the wiggler period or amplitude (or both). Because of the large energy spread, fluid or kinetic approaches are inadequate and we then use the macro-particle approach which will be presented in this section. For free-electron lasers this approach was initially developed by Kroll, Morton and Rosenbluth (1981) and in this section we shall basically follow the essentials of their approach.

7.4.1 Basic Formulation

Assuming that space-charge effects are negligible, the scalar electric potential and the longitudinal magnetic vector potential can be omitted from the expression for the Hamiltonian, thus

$$\begin{aligned} H &= \sqrt{(p_\perp + eA_\perp)^2 c^2 + p_\parallel^2 c^2 + (mc^2)^2}\,, \\ &= mc^2\gamma \end{aligned} \tag{7.4.1}$$

The transverse magnetic vector potential has two components: the wiggler and radiation field i.e.,

$$\mathbf{A}_\perp = \mathbf{A}_\mathrm{w} + \mathbf{A}_\mathrm{rf}\,. \tag{7.4.2}$$

Since it will be necessary to adapt the wiggler parameters to the local conditions in order to keep the electron in resonance, we consider a wiggler with variable amplitude and wavenumber namely,

$$\mathbf{A}_\mathrm{w} = -A_\mathrm{w}(z)\left[\mathbf{1}_\mathrm{x}\cos\left(\int_0^z \mathrm{d}z' k_\mathrm{w}(z')\right) + \mathbf{1}_\mathrm{y}\sin\left(\int_0^z \mathrm{d}z' k_\mathrm{w}(z')\right)\right]\,; \tag{7.4.3}$$

in a similar way, the magnetic vector potential which describes the radiation field has an amplitude $A_0(z)$ which varies in space and so is its wavenumber:

$$\mathbf{A}_\mathrm{rf} = A_0(z)\left[\mathbf{1}_x\cos\left(\omega t - \int_0^z \mathrm{d}z' k(z')\right) + \mathbf{1}_y\sin\left(\omega t - \int_0^z \mathrm{d}z' k(z')\right)\right]\,. \tag{7.4.4}$$

The latter has two components, the wavenumber in vacuum (ω/c) and the effect of the interaction $[\theta(z)]$. Therefore $\int_0^z \mathrm{d}z' k(z') = (\omega/c)z + \theta(z)$; this is to say that θ is the phase accumulated by the wave due to the interaction. As in the previous sections the wiggler is assumed to be uniform in the transverse direction therefore the canonical momentum in these directions is conserved; for simplicity it will be assumed to be zero $(p_\perp = 0)$ hence

$$v_\perp = \frac{e}{m\gamma} A_\perp ,$$
$$\simeq \frac{e}{m\gamma} A_{\rm w} , \qquad (7.4.5)$$

where in the second expression it is assumed that the contribution of the radiation field to the transverse motion is negligible.

After substituting the expressions for the magnetic vector potentials into the Hamiltonian we obtain

$$H = mc^2 \sqrt{\mu^2 + \left(\frac{p_\parallel}{mc}\right)^2 - 2a_{\rm w}(z)a_{\rm rf}(z)\cos\psi}$$
$$= mc^2\gamma , \qquad (7.4.6)$$

where

$$a_{\rm w}(z) = \frac{eA_{\rm w}(z)}{mc} ,$$
$$a_{\rm rf}(z) = \frac{eA_{\rm rf}(z)}{mc} ,$$
$$\mu^2(z) = 1 + a_{\rm w}^2(z) + a_{\rm rf}^2(z) ,$$
$$K(z) = k(z) + k_{\rm w}(z) ,$$
$$\psi(z,t) = \omega t - \int_0^z {\rm d}z' K(z') . \qquad (7.4.7)$$

The last expression represents the phase between the wave and the particle, when the presence of the wiggler is accounted for. At the transition from the Hamiltonian in (7.4.1) to the latter no approximations are made other than $p_\perp = 0$. Note that μ^2 plays the role of a normalized effective mass of the electron which is z-dependent but not time dependent. This Hamiltonian enables us to determine a relatively simple expression for the longitudinal velocity of the particle; this is given by

$$v_\parallel = \frac{{\rm d}z}{{\rm d}t} = \frac{\partial H}{\partial p_\parallel}$$
$$= \frac{p_\parallel}{mc^2\sqrt{\mu^2 + \left(p_\parallel/mc\right)^2 - 2a_{\rm w}(z)a_{\rm rf}(z)\cos\psi}}$$
$$= \frac{c}{\gamma}\sqrt{\gamma^2 - \mu^2 + 2a_{\rm w}(z)a_{\rm rf}(z)\cos\psi} , \qquad (7.4.8)$$

where in the last expression we used (7.4.6) to express $p_\parallel$ in terms of γ.

Since we are interested in the operation of the system as an amplifier, it is assumed that the frequency is determined by the external source and only spatial variations are allowed. Consequently, we follow the particle in space and we consider the time it takes the ith particle to reach a point z starting from $z = 0$; this time interval is denoted by $\tau_i(z)$. Regarding the phase dynamics the situation seems at a first glance more complicated by the three dimensional motion of the electron (compared to 1D in the slow-wave structure). However, in practice we need only the projection of the motion along the wave propagation and this fact simplifies the calculation substantially as will be shown next.

The phase between the wave and the particle is given by $\psi_i(z) = \omega\tau_i(z) - \int_0^z \mathrm{d}z' K(z')$ thus the dynamics of the phase ψ in space reads

$$\frac{\mathrm{d}\psi_i(z)}{\mathrm{d}z} = \omega\frac{\mathrm{d}\tau_i}{\mathrm{d}z} - K(z) . \tag{7.4.9}$$

Now, the derivative of τ with respect to z is inversely proportional to the *longitudinal* component of the velocity as determined in (7.4.8) hence,

$$\frac{\mathrm{d}\psi_i(z)}{\mathrm{d}z} = \frac{\omega}{c}\frac{\gamma_i}{\sqrt{\gamma_i^2 - \mu^2 + 2a_{\mathrm{w}}a_{\mathrm{rf}}\cos\psi_i}} - K(z) . \tag{7.4.10}$$

As the velocity of the particle varies in space, so is its energy which satisfies

$$mc^2 v_{\|}\frac{\mathrm{d}\gamma_i}{\mathrm{d}z} = -e\mathbf{v}\cdot\mathbf{E} , \tag{7.4.11}$$

and since (in phasor notation) $E_\perp = -j\omega\mathbf{A}_{\mathrm{rf}}$ we can use (7.4.8) to write

$$\frac{\mathrm{d}\gamma_i}{\mathrm{d}z} = \frac{1}{2}\left[\left(-j\frac{\omega}{c}a_{\mathrm{rf}}\right)a_{\mathrm{w}}\frac{\mathrm{e}^{j\psi_i}}{\sqrt{\gamma_i - \mu^2 + 2a_{\mathrm{w}}a_{\mathrm{rf}}\cos\psi_i}} + c.c.\right] . \tag{7.4.12}$$

Next we determine the dynamics of the amplitude of the radiation field. The starting point is the non-homogeneous wave equation in (7.3.5). Its source term is the current density which in the framework of the present approach is

$$J_\perp(\mathbf{r},t) = -e\sum_i v_{\perp,i}\delta[x - x_i(t)]\delta[y - y_i(t)]\delta[z - z_i(t)] ; \tag{7.4.13}$$

in particular we can substitute the explicit expression for the magnetic vector potential and from the x component of the wave equation we obtain

$$\begin{aligned}\left[\frac{\partial^2}{\partial z^2} - \frac{1}{c^2}\frac{\partial^2}{\partial t^2}\right] A_0(z)\cos\psi(z,t) &= A_0(z)\cos\psi(z,t)\left[\frac{\omega^2}{c^2} - \left(\frac{\omega}{c} + \theta'\right)^2\right] \\ &+ 2A_0'(z)\sin\psi(z,t)\left(\frac{\omega}{c} + \theta'\right) . \end{aligned} \tag{7.4.14}$$

In a similar way we substitute the explicit expression for $v_\perp$ from (7.4.5) in the x component of the current density and write

$$J_x(z,t) = ec\sum_i \frac{a_{\mathrm{w}}}{\gamma_i}\cos\left[\int_0^z \mathrm{d}z' k_{\mathrm{w}}(z')\right]\frac{F_{\mathrm{f}}}{S_{\mathrm{el}}}\delta[z - z_i(t)] , \tag{7.4.15}$$

where we have already averaged out over the beam cross-section and the filling factor was included[see (7.3.5)]. Note that second derivatives of A_0 and θ were neglected in (7.4.14).

The coefficients of the trigonometric functions are time independent and therefore, we use the orthogonality of the trigonometric functions to average the wave equation over one period (T) of the wave. Firstly, we take advantage of the orthogonality of the $\cos(\omega t \cdots)$ function to obtain

$$-2\frac{\omega}{c}\theta' A_0 \frac{1}{2} = -\mu_0 F_\mathrm{f} \frac{1}{S_\mathrm{el}} \frac{1}{T} \int \mathrm{d}t \cos\left[\omega t - \int \mathrm{d}z' k(z')\right] \times \left\{ ec \sum_i \frac{a_\mathrm{w}}{\gamma_i} \cos\left[\int_0^z \mathrm{d}z' k_\mathrm{w}(z')\right] \delta\left[z - z_i(t)\right] \right\}, \tag{7.4.16}$$

and secondly, using the orthogonality of $\sin(\omega t \cdots)$ we have

$$2\frac{\omega}{c} A_0' \frac{1}{2} = -\mu_0 F_\mathrm{f} \frac{1}{S_\mathrm{el}} \frac{1}{T} \int \mathrm{d}t \sin\left[\omega t - \int \mathrm{d}z' k(z')\right] \times \left\{ ec \sum_i \frac{a_\mathrm{w}}{\gamma_i} \cos\left[\int_0^z \mathrm{d}z' k_\mathrm{w}(z')\right] \delta\left[z - z_i(t)\right] \right\}. \tag{7.4.17}$$

Assuming that electrons are not reflected backwards, the time integral can be readily evaluated using the Dirac delta function and if only slowly varying (resonant) terms are kept, then (7.4.16) reads

$$-\frac{\omega}{c}\theta' A_0 = -\frac{e^2 \mu_0}{cmT} \frac{F_\mathrm{f}}{S_\mathrm{el}} A_\mathrm{w} \frac{1}{2} \sum_i \frac{1}{\gamma_i \beta_{\|,i}} \cos\psi_i(z). \tag{7.4.18}$$

The summation in this case is over all particles in one period of the wave and assuming that there are N such particles, we can write $\sum_i \cdots = N\langle\cdots\rangle$. Since the average beam density is given by $n_0 = N/S_\mathrm{el}cT$, we can finally write

$$A_0 \frac{\mathrm{d}}{\mathrm{d}z}\theta = \frac{1}{2}\frac{\omega_\mathrm{p}^2}{c^2} F_\mathrm{f} A_\mathrm{w} \left(\frac{\omega}{c}\right)^{-1} \left\langle \frac{\cos\psi_i(z)}{\gamma_i \beta_{\|,i}} \right\rangle. \tag{7.4.19}$$

Following exactly the same procedure, we have for (7.4.17)

$$\frac{\mathrm{d}}{\mathrm{d}z} A_0 = -\frac{1}{2}\frac{\omega_\mathrm{p}^2}{c^2} F_\mathrm{f} A_\mathrm{w} \left(\frac{\omega}{c}\right)^{-1} \left\langle \frac{\sin\psi_i(z)}{\gamma_i \beta_{\|,i}} \right\rangle. \tag{7.4.20}$$

If d is the total length of the interaction region, it is convenient to use the following set of normalized quantities: $\zeta = z/d$, $\Omega = (\omega/c)d$, $K_\mathrm{w} = k_\mathrm{w} d$, $\bar{a}_\mathrm{rf} = a_\mathrm{rf} \mathrm{e}^{j\theta}$ and $\alpha = \frac{1}{2}(\omega_\mathrm{p} d/c)^2 F_\mathrm{f}$. With these definitions there are two equivalent ways to formulate the interaction: either in terms of complex variables

$$
\begin{aligned}
\frac{\mathrm{d}}{\mathrm{d}\zeta}\bar{a}_{\mathrm{rf}} &= -j\frac{\alpha}{\Omega}a_{\mathrm{w}}\langle\frac{\mathrm{e}^{-j\chi_i}}{\gamma_i\beta_{\|,i}}\rangle\,,\\
\frac{\mathrm{d}}{\mathrm{d}\zeta}\gamma_i &= \frac{1}{2}\left[(-j\Omega\bar{a}_{\mathrm{rf}})\,a_{\mathrm{w}}\frac{\mathrm{e}^{j\chi_i}}{\gamma_i\beta_{\|,i}} + c.\,c.\right],\\
\frac{\mathrm{d}}{\mathrm{d}\zeta}\chi_i &= \Omega\frac{1}{\beta_{\|,i}} - \Omega - K_{\mathrm{w}}\,,\\
\beta_{\|,i} &= \frac{1}{\gamma_i}\sqrt{\gamma_i^2-\mu^2+a_{\mathrm{w}}\left[\bar{a}_{\mathrm{rf}}\mathrm{e}^{j\chi_i}+c.c.\right]}\,,
\end{aligned}
\tag{7.4.21}
$$

or the alternative way, to formulate it in terms of the real variable $a \equiv |\bar{a}_{\mathrm{rf}}|$ and θ:

$$
\begin{aligned}
\frac{\mathrm{d}}{\mathrm{d}\zeta}a &= -\frac{\alpha a_{\mathrm{w}}}{\Omega}\langle\frac{\sin\psi_i}{\gamma_i\beta_{\|,i}}\rangle\,,\\
\frac{\mathrm{d}}{\mathrm{d}\zeta}\theta &= \frac{\alpha a_{\mathrm{w}}}{\Omega a}\langle\frac{\cos\psi_i}{\gamma_i\beta_{\|,i}}\rangle\,,\\
\frac{\mathrm{d}}{\mathrm{d}\zeta}\gamma_i &= \Omega a a_{\mathrm{w}}\frac{\sin\psi_i}{\gamma_i\beta_{\|,i}}\,,\\
\frac{\mathrm{d}}{\mathrm{d}\zeta}\psi_i &= \Omega\frac{1}{\beta_{\|,i}} - \Omega - K_{\mathrm{w}} + \frac{\mathrm{d}\theta}{\mathrm{d}\zeta}\,,\\
\beta_{\|,i} &= \frac{1}{\gamma_i}\sqrt{\gamma_i^2-\mu^2+2a_{\mathrm{w}}\,a\,\cos\psi_i}\,.
\end{aligned}
\tag{7.4.22}
$$

The proof of equivalence of the two formulations is left to the reader as an exercise. In both cases one can average over the equation of motion of the particles and substitute the amplitude equation to get

$$
\frac{\mathrm{d}}{\mathrm{d}\zeta}\left[\langle\gamma\rangle + \frac{1}{2\alpha}(\Omega a)^2\right] = 0\,,
\tag{7.4.23}
$$

which is the global energy conservation.

7.4.2 Resonant Particle Solution

Now that we have determined the equations which govern the dynamics of the electrons and the electromagnetic field in the presence of a quasi-periodic wiggler, we should be able to solve them provided that the initial conditions are known as well as the wiggler's parameters. However, we shall now make one step further and ask what the wiggler should be, for a given initial distribution of particles and electromagnetic field, that maximizes the energy extraction from the electrons. A general solution of this problem is difficult and practically impossible with analytical techniques. However, if the distribution of electrons occupies only a small region of the phase-space then the problem can be treated analytically.

For this purpose consider an ideally bunched beam such that we assume that all particles in one bunch move together forming a single macro-particle whose shape is preserved along the entire interaction region. Based on the equations of motion, the condition for maximum energy extraction is to keep it in correct phase with the wave i.e., maintain it in resonance. Assuming that at the input the macro-particle is in phase with the wave, the resonance along the interaction region will be defined as $\mathrm{d}\psi_{\mathrm{r}}/\mathrm{d}\zeta = 0$ and it translates into

$$\gamma_{\mathrm{r}}^2 \simeq \frac{1}{2}\frac{\Omega}{(K_{\mathrm{w}}+\theta')}\mu^2 , \tag{7.4.24}$$

where the subscript r indicates resonance conditions. This expression becomes exact if we choose the resonance phase to be $\psi_{\mathrm{r}} = \pm\pi/2$ since $d\theta/d\zeta = 0$. In this subsection we consider an amplifier configuration so we take $\psi_{\mathrm{r}} = -\pi/2$ in which case, $\beta_{\|,\mathrm{r}} = \sqrt{1-(\mu/\gamma_{\mathrm{r}})^2}$. Substituting in the equation for the amplitude we obtain

$$\frac{\mathrm{d}}{\mathrm{d}\zeta}a = \frac{\alpha}{\Omega}\frac{a_{\mathrm{w}}}{\sqrt{\gamma_{\mathrm{r}}^2 - 1 - a^2 - a_{\mathrm{w}}^2}} . \tag{7.4.25}$$

Bearing in mind that the total energy is conserved i.e., $\gamma_{\mathrm{r}}(\zeta)+[\Omega a(\zeta)]^2/2\alpha = \varepsilon \equiv \gamma_{\mathrm{r}}(0) + [\Omega a(0)]^2/2\alpha$, we substitute the expression for γ_{r} to get

$$\frac{\Omega}{\alpha}\frac{1}{a_{\mathrm{w}}}\int_{a(0)}^{a(\zeta)} \mathrm{d}x \sqrt{\left[\varepsilon - \frac{1}{2}\frac{\Omega^2}{\alpha}x^2\right]^2 - 1 - x^2 - a_{\mathrm{w}}^2} = \zeta , \tag{7.4.26}$$

This equation is solved numerically for a constant a_{w} (but variable k_{w}) assuming a 3 mm beam radius which carries 100 A current, the filling factor being $F_{\mathrm{f}} = 0.1$. The total interaction length is 5 m, at the entrance the wiggler period is $L = 2\,\mathrm{cm}$ and its amplitude is $B_{\mathrm{w}} = 0.2\,\mathrm{T}$. The initial energy of the electrons is 4.6 MeV $[\gamma(0) = 10]$ and the normalized amplitude of the radiation field is $a_{\mathrm{r}}(0) = 8.4 \times 10^{-5}$. The result is illustrated in Fig. 7.2: the upper left frame shows the way in which the amplitude should grow in space. The variation in space of γ (upper right frame) is calculated from the global energy conservation and the efficiency is illustrated in the lower left frame. The last frame illustrates the required variation in space of the period of the wiggler. It indicates that in order to achieve a 50% efficiency at 100 µm, the period of the wiggler has to be decreased from 2 cm to 0.6 cm and the intensity of the magnetic field increased to almost 0.7 T.

7.4.3 Buckets

In practice, any bunch has a finite spread in energy and phase. Let us denote the deviations from the resonant phase by $\delta\psi_i \equiv \psi_i - \psi_{\mathrm{r}}$ and from resonant energy by $\delta\gamma_i \equiv \gamma_i - \gamma_{\mathrm{r}}$. Based on the equations developed in the last subsection these two quantities satisfy

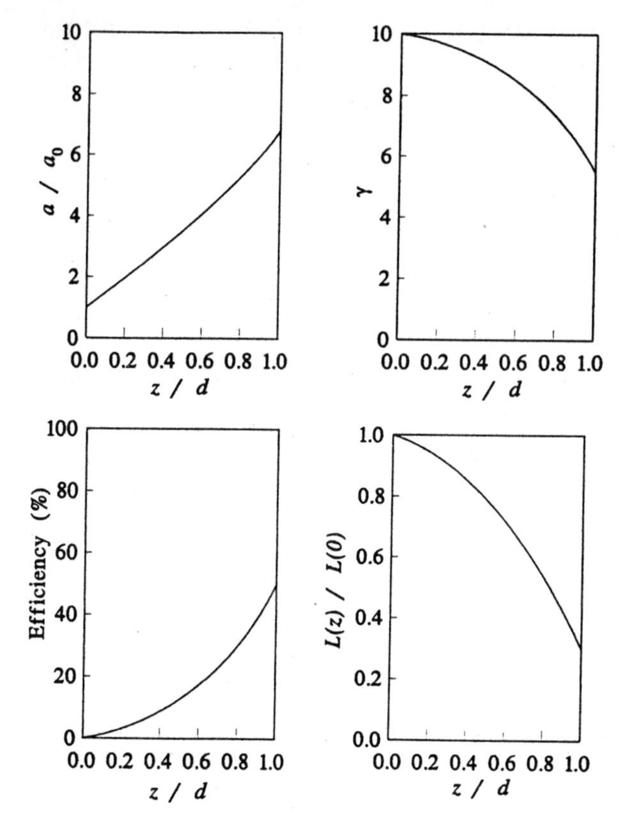

Fig. 7.2 Variation in space of the amplitude, energy, efficiency and period. All correspond to the resonant particle model

$$\frac{\mathrm{d}}{\mathrm{d}\zeta}\delta\psi_i = -\frac{(1+\beta_{\|,\mathrm{r}})K_\mathrm{w}}{\gamma_\mathrm{r}\beta_{\|,\mathrm{r}}}\delta\gamma_i\,, \tag{7.4.27}$$

and

$$\frac{\mathrm{d}}{\mathrm{d}\zeta}\delta\gamma_i = \Omega a_\mathrm{w} a_\mathrm{r}\frac{1}{\gamma_\mathrm{r}\beta_{\|,\mathrm{r}}}\cos\psi_\mathrm{r}\delta\psi_i\,, \tag{7.4.28}$$

provided that the deviations are small. For $\cos\psi_\mathrm{r} > 0$ the trajectories are stable and oscillate around the resonance point $(\psi_\mathrm{r}, \gamma_\mathrm{r})$ in the phase-space at a (spatial) "frequency"

$$\mathcal{O} = \frac{K_\mathrm{w}(1+\beta_{\|,\mathrm{r}})}{\mu\beta_{\|,\mathrm{r}}}\sqrt{a_\mathrm{r} a_\mathrm{w}\cos\psi_\mathrm{r}}\,. \tag{7.4.29}$$

These equations and the last result indicate that there is a whole range of trajectories around the resonance condition which are stable. However, the analysis was limited to small deviation from resonance. This might be justified in regard to the energy but it is a stringent constraint for the phase. For this reason we reformulate the dynamics for the case when large deviations of the phase (ψ_i) are permitted. The phase equation has a similar form as (7.4.27)

$$\frac{\mathrm{d}}{\mathrm{d}\zeta}\psi_i = -\frac{(1+\beta_{\parallel,\mathrm{r}})K_{\mathrm{w}}}{\gamma_{\mathrm{r}}\beta_{\parallel,\mathrm{r}}}\delta\gamma_i\,, \tag{7.4.30}$$

but the equation for $\delta\gamma_i$ reads

$$\frac{\mathrm{d}}{\mathrm{d}\zeta}\delta\gamma_i = \Omega a_{\mathrm{w}} a_{\mathrm{r}} \frac{1}{\gamma_{\mathrm{r}}\beta_{\parallel,\mathrm{r}}}\left[\sin\psi_i - \sin\psi_{\mathrm{r}}\right]\,. \tag{7.4.31}$$

It is convenient to redefine the phase as $\bar{\psi} \equiv -\psi$ and regard it as the canonical coordinate whereas $\delta\gamma_i$ is the canonical momentum. With these definitions, the Hamiltonian of the system reads

$$\begin{aligned} H &= \frac{1}{2}\frac{K_{\mathrm{w}}(1+\beta_{\parallel,\mathrm{r}})}{\gamma_{\mathrm{r}}\beta^2_{\parallel,\mathrm{r}}}\delta\gamma_i^2 - \Omega a_{\mathrm{w}} a_{\mathrm{r}}\frac{1}{\gamma_{\mathrm{r}}\beta_{\parallel,\mathrm{r}}}\left[\cos\bar{\psi}_i + \bar{\psi}_i \sin\bar{\psi}_{\mathrm{r}}\right] \\ &\equiv \frac{1}{2\mathcal{M}}\delta\gamma_i^2 - \frac{1}{2}\mathcal{K}\left[\cos\bar{\psi}_i + \bar{\psi}_i\sin\bar{\psi}_{\mathrm{r}}\right] \end{aligned} \tag{7.4.32}$$

and it corresponds to a particle whose mass, $\mathcal{M}^{-1} \equiv K_{\mathrm{w}}(1+\beta_{\parallel,\mathrm{r}})/\gamma_{\mathrm{r}}\beta^2_{\parallel,\mathrm{r}}$, is z dependent which moves in a potential $V(\bar{\psi})$,

$$V(\bar{\psi}) = -\mathcal{K}\left[\cos\bar{\psi} + \bar{\psi}\sin\bar{\psi}_{\mathrm{r}}\right]\,, \tag{7.4.33}$$

where $\mathcal{K} \equiv \Omega a_{\mathrm{w}} a_{\mathrm{r}}/\gamma_{\mathrm{r}}\beta_{\parallel,\mathrm{r}}$.

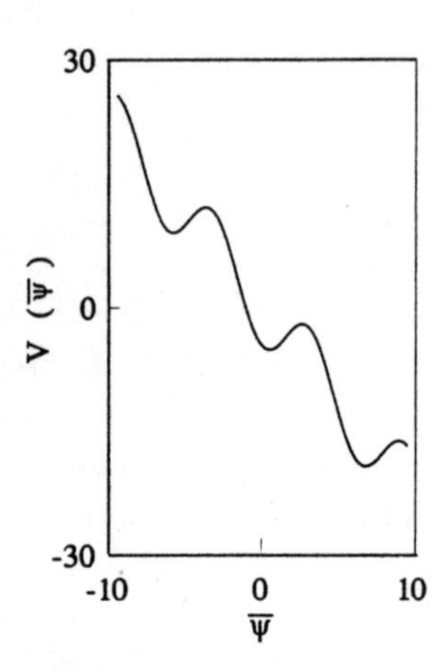

Fig. 7.3 Effective potential in whose minima electrons can be trapped

This potential is illustrated in Fig. 7.3 and it shows that particles can be trapped in its minima according to their initial conditions. The maximum stable trajectory of the particles is determined by the extrema of the potential in (7.4.33) and there are two sets of solutions: $\bar{\psi} = \bar{\psi}_r \pm 2\pi n$ or $\bar{\psi} = -\bar{\psi}_r + \pi \pm 2\pi n$. It is the latter which corresponds to the local maxima and thus represents the maximal value of a "bound state". Assuming that $\psi_r > 0$ and that at the extremum the (canonical) momentum is zero, we find that the maximum value of H, for which the trajectories are still expected to be stable is given by

$$H_{\max} = \frac{1}{2}\mathcal{K}\left[\cos\bar{\psi}_r + (\bar{\psi}_r - \pi)\sin\bar{\psi}_r\right] . \tag{7.4.34}$$

If ψ_r is negative, then π in this equation reverses its sign. Figure 7.4 illustrates the limits of the stable trajectories region (bucket) and a typical stable trajectory. Based on the maximal value of the Hamiltonian, one can also determine the maximal $\delta\gamma_{\max}$ permissible for stable trajectory. It occurs at $\bar{\psi} = \bar{\psi}_r$ and it is given by

$$\delta\gamma_{\max} = 2\sqrt{\mathcal{MK}}\sqrt{\cos\bar{\psi}_r + \left(\bar{\psi}_r - \frac{\pi}{2}\right)\sin\bar{\psi}_r} . \tag{7.4.35}$$

The two extreme phases of the bucket $(\bar{\psi}_1, \bar{\psi}_2)$ are determined in a similar way, only that in this case $\delta\gamma = 0$.

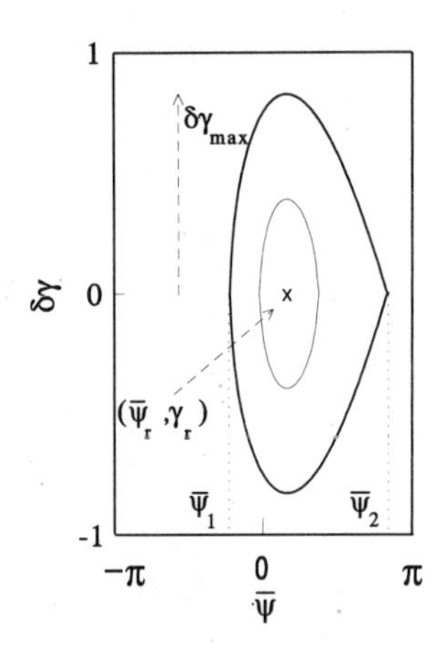

Fig. 7.4 The bucket limit (outer curve), a typical bucket shape (inner curve) and in the center the coordinates of the resonant particle are illustrated

The bucket method infers that there is an inherent limit on the efficiency of such a device since only those electrons which are trapped in the bucket can be decelerated. Furthermore, in the context of a traveling-wave amplifier, it was shown that in the interaction process the area of the phase-space increases and if the electrons are to be "recycled" (ring configuration FEL), there are two conditions to be satisfied: *(i)* increase to maximum the bucket at the entrance in order to capture the maximal number of electrons and *(ii)* minimize the energy spread, otherwise many electrons are lost in the next cycle. This kind of design was thoroughly investigated by Kroll, Morton and

Rosenbluth (1981). We shall not go any further into the details of the bucket method but rather investigate some additional aspects of the interaction and its manifestation in the phase-space.

7.4.4 Energy Spread

The set of equations as introduced in (7.4.21) can be reformulated in the case of a constant $a_{\rm w}$. From the equations of motion of the particles we conclude that the quantity $\tilde{a} \equiv j\Omega \bar{a}_{\rm rf} a_{\rm w}$ is the effective (normalized) longitudinal electric field which acts on a single particle. Substituting this definition in the amplitude equation it is only natural to redefine the normalized coupling coefficient α to read $\tilde{\alpha} \equiv \alpha a_{\rm w}^2$. With these definitions we have

$$\begin{aligned}
\frac{\rm d}{{\rm d}\zeta}\tilde{a} &= \tilde{\alpha}\left\langle \frac{{\rm e}^{-j\chi_i}}{\gamma_i \beta_{\|,i}} \right\rangle , \\
\frac{\rm d}{{\rm d}\zeta}\gamma_i &= -\frac{1}{2}\left[\tilde{a}\frac{{\rm e}^{-j\chi_i}}{\gamma_i \beta_{\|,i}} + c.\,c. \right] , \\
\frac{\rm d}{{\rm d}\zeta}\chi_i &= \Omega \frac{1}{\beta_{\|,i}} - \Omega - K_{\rm w} , \\
\beta_{\|,i} &= \frac{1}{\gamma_i}\sqrt{\gamma_i^2 - \mu^2 + \left[-j\tilde{a}{\rm e}^{j\chi_i} + c.c. \right]/\Omega} \, .
\end{aligned} \tag{7.4.38}$$

From the second equation we can develop the equation for the energy spread: $\Delta\gamma^2 \equiv \langle \gamma_i^2 \rangle - \langle \gamma_i \rangle^2$. This is done by firstly averaging over all particles

$$\frac{\rm d}{{\rm d}\zeta}\langle \gamma_i \rangle = -\frac{1}{2}\left[\tilde{a}\left\langle \frac{{\rm e}^{-j\chi_i}}{\gamma_i \beta_{\|,i}} \right\rangle + c.\,c. \right] ; \tag{7.4.39}$$

secondly, we multiply by the local average $\langle \gamma_i \rangle$ to get

$$\frac{1}{2}\frac{\rm d}{{\rm d}\zeta}\langle \gamma_i \rangle^2 = -\frac{1}{2}\left[\tilde{a}\left\langle \frac{{\rm e}^{-j\chi_i}}{\gamma_i \beta_{\|,i}} \right\rangle \langle \gamma_i \rangle + c.\,c. \right] . \tag{7.4.40}$$

We next repeat these steps but in the opposite order: we multiply the single particle equation of motion by γ_i and then average over the particles ensemble; the result is

$$\frac{1}{2}\frac{\rm d}{{\rm d}\zeta}\langle \gamma_i^2 \rangle = -\frac{1}{2}\left[\tilde{a}\langle \frac{{\rm e}^{-j\chi_i}}{\beta_{\|,i}} \rangle + c.\,c. \right] . \tag{7.4.41}$$

Subtracting from the last expression (7.4.40) and using the definition of the energy spread, we obtain

$$\frac{\rm d}{{\rm d}\zeta}\Delta\gamma^2 = -\left\{ \tilde{a}\left[\left\langle \frac{{\rm e}^{-j\chi_i}}{\gamma_i \beta_{\|,i}}\gamma_i \right\rangle - \left\langle \frac{{\rm e}^{-j\chi_i}}{\gamma_i \beta_{\|,i}} \right\rangle \langle \gamma_i \rangle \right] + c.\,c. \right\} . \tag{7.4.42}$$

We know from our analysis of traveling-wave amplifiers that the energy spread increases in the interaction process since part of the electrons are accelerated and others are decelerated. It is the same electromagnetic field which causes the average deceleration (in an amplifier) and at the same time it accelerates a fraction of the particles causing the energy spread at the output. It was Madey (1979) who initially showed that in the low-gain Compton regime the energy spread at the output is directly related to the gain; at the input the energy spread is assumed to be negligible. Here we quote a result which was revised by Kroll, Morton and Rosenbluth (1981) and it relates the gain $\langle\gamma_i(1)\rangle - \langle\gamma_i(0)\rangle$ to the energy spread at the output

$$\bar{\gamma}_1 - \bar{\gamma}_0 = \frac{1}{2}\frac{\mathrm{d}}{\mathrm{d}\bar{\gamma}_0}\Delta\gamma^2(1) \tag{7.4.43}$$

where $\bar{\gamma}_1 = \langle\gamma_i(1)\rangle$ and $\bar{\gamma}_0 = \langle\gamma_i(0)\rangle$. We will show now that Madey theorem as formulated above for the low-gain Compton regime is related to the equation which describes the energy spread [(7.4.42)].

Firstly, we note that in addition to $\Delta\gamma(\zeta)$, there are two other macroscopic quantities $\langle\gamma(\zeta)\rangle$ and $|\tilde{a}(\zeta)|$ which describe the system. Secondly, we bear in mind that these two are related via the energy conservation i.e., $\frac{\mathrm{d}}{\mathrm{d}\zeta}\left[\langle\gamma\rangle + |\tilde{a}|^2/2\tilde{\alpha}\right] = 0$, therefore we can use one of the two as an independent variable instead of ζ hence

$$\frac{\mathrm{d}}{\mathrm{d}\zeta}\Delta\gamma^2 = \left(\frac{\mathrm{d}\bar{\gamma}}{\mathrm{d}\zeta}\right)\frac{\mathrm{d}}{\mathrm{d}\bar{\gamma}}\Delta\gamma^2 . \tag{7.4.44}$$

Before we proceed there is one important comment: although ζ does not occur explicitly in the right-hand side of the equation, it is implicitly there since we consider the values of $\Delta\gamma$ and $\bar{\gamma} \equiv \langle\gamma(\zeta)\rangle$ at the same location ζ.

Using the energy conservation, we obtain

$$\frac{\mathrm{d}}{\mathrm{d}\zeta}\Delta\gamma^2 = -\frac{1}{\tilde{\alpha}}|\tilde{a}|\left(\frac{\mathrm{d}|\tilde{a}|}{\mathrm{d}\zeta}\right)\left(\frac{\mathrm{d}}{\mathrm{d}\bar{\gamma}}\Delta\gamma^2\right) , \tag{7.4.45}$$

and for simplicity we define $\tilde{a}e^{j\chi_i} \equiv |\tilde{a}|e^{j\tilde{\chi}_i}$. With this definition and using the amplitude equation in (7.4.38) we obtain

$$\begin{aligned}\frac{\mathrm{d}}{\mathrm{d}\zeta}\Delta\gamma^2 &= -|\tilde{a}|\left\langle\cos(\tilde{\chi}_i)/(\gamma_i\beta_{\|,i})\right\rangle\frac{\mathrm{d}}{\mathrm{d}\bar{\gamma}}\Delta\gamma^2 , \\ &= -2|\tilde{a}|\left[\left\langle\frac{\cos\tilde{\chi}_i}{\gamma_i\beta_{\|,i}}\gamma_i\right\rangle - \left\langle\frac{\cos\tilde{\chi}_i}{\gamma_i\beta_{\|,i}}\right\rangle\bar{\gamma}\right] ,\end{aligned} \tag{7.4.46}$$

where in the second expression we simply quote (7.4.42). From the two right-hand side expressions we conclude that

$$\frac{1}{2}\frac{\mathrm{d}}{\mathrm{d}\bar{\gamma}}\Delta\gamma^2 = \bar{\gamma} - \frac{\langle\cos(\tilde{\chi}_i)/\beta_{\|,i}\rangle}{\langle\cos(\tilde{\chi}_i)/\gamma_i\beta_{\|,i}\rangle} , \tag{7.4.47}$$

and the resemblance with Madey's theorem becomes apparent. However, in contrast with the latter this relation is exact at any point in the interaction region. Furthermore, it is also valid in the high-gain Compton regime.

Under the simplifying conditions of low-gain Compton regime it is possible to obtain (7.4.43). First we make a zero order estimate of the phase term [second term in the right-hand side of (7.4.47)] and for this purpose we assume that at the *input*, $\langle\gamma\rangle$ and $\Delta\gamma$ are independent, therefore the left-hand side of (7.4.47) is zero and we conclude that

$$\left[\frac{\langle\cos(\tilde{\chi}_i)/\beta_{\|,i}\rangle}{\langle\cos(\tilde{\chi}_i)/\gamma_i\beta_{\|,i}\rangle}\right]_{\zeta=0} = \bar{\gamma}(0)\,. \qquad (7.4.48)$$

We now substitute this estimate in (7.4.47) tacitly assuming that the term in the square brackets varies slowly along the interaction region, thus

$$\frac{1}{2}\frac{\mathrm{d}}{\mathrm{d}\bar{\gamma}(1)}\Delta\gamma^2(1) \simeq \bar{\gamma}(1) - \bar{\gamma}(0)\,. \qquad (7.4.49)$$

Bearing in mind that in the low-gain Compton regime the energy transfer is small, we can replace $\bar{\gamma}(1)$ on the left-hand side with $\bar{\gamma}(0)$ to obtain

$$\frac{1}{2}\frac{\mathrm{d}}{\mathrm{d}\bar{\gamma}(0)}\Delta\gamma^2(1) \simeq \bar{\gamma}(1) - \bar{\gamma}(0)\,, \qquad (7.4.50)$$

which is exactly Madey's theorem as formulated in (7.4.43).

Although the Madey theorem relates the moments of the electrons' distribution function it does not actually help us to calculate them and for this purpose we have to go back to the equations of motion i.e., (7.4.38). These were solved for a typical FEL set of parameters and the question we address now is how does the energy spread vary at the output of the device when the only other parameter which is changed is the energy spread at the input. The result of our simulation is presented in Fig. 7.5: the energy spread at the output $\Delta\gamma(1)$ as a function of the energy spread at the input, $\Delta\gamma(0)$ decreases for values of $\Delta\gamma(0)$ smaller than $\Delta\gamma(1)$. We observe that $\Delta\gamma(1)$ starts from a high value when the energy spread at the input is virtually zero. By increasing the latter we cause $\Delta\gamma(1)$ to decrease as does the gain. The latter vanishes when the energy spread at the output equals its value at the input. Any further increase of $\Delta\gamma(0)$ beyond this level does not change the gain and since the beam traverses the interaction region almost unaffected, $\Delta\gamma(1)$ increases linearly with $\Delta\gamma(0)$.

Another interesting aspect of the energy spread which we shall examine next is revealed when comparing the operation of an FEL and a TWT. In Sect. 7.3 it was shown that the dispersion relation of a free-electron laser is similar to that of a traveling wave tube but it was pointed out that the γ dependence of the gain is different in the two cases – a fact which may cause some differences in the operation of the two devices. In order to emphasize the similarities and the differences, we have summarized, in Table 7.1, the

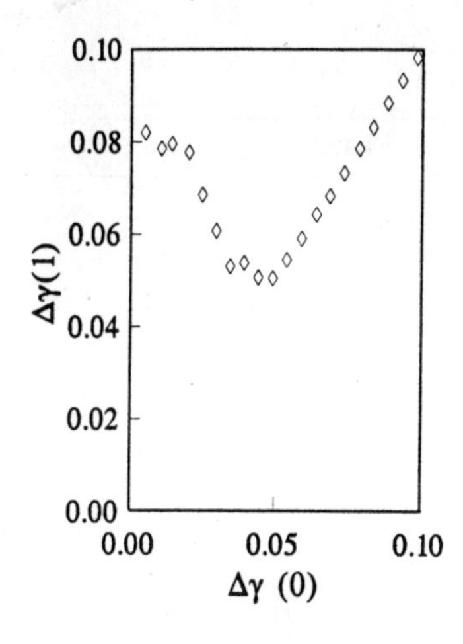

Fig. 7.5. Energy spread at the output as a function of the energy spread at the input

equations of the free-electron laser and the traveling-wave tube (TWT). In these equations each quantity which plays a similar role is denoted (intentionally) with the same notation: γ_i represents the energy of the ith particle and χ_i is its relative phase to the wave. In both cases a represents the normalized amplitude of the electromagnetic field; however in the TWT this is the longitudinal electric field ($a = eE_z d/mc^2$) whereas in the FEL it is the transverse field [$a = ja_{\rm w}a_{\rm rf}\omega d/c$ see also (7.4.7)]. The normalized coupling coefficient is denoted with α and it consists of different quantities: for the FEL α used here is $\alpha = \frac{1}{2}(\omega_{\rm p}d/c)^2 F_{\rm f} a_{\rm w}^2$ and for the TWT $\alpha = (eIZ_{\rm int}/mc^2)(d^2/\pi R^2)$. In the phase equation the longitudinal velocity is denoted by β_i and in the TWT case it is related to γ_i via $\beta_i = \sqrt{1 - 1/\gamma_i^2}$ whereas in the FEL case $\beta_i = \sqrt{1 - \mu^2/\gamma_i^2 + [-jae^{j\chi_i} + c.c.]/\Omega\gamma_i^2}$ and $\mu^2 = 1 + a_{\rm w}^2 + |a|^2/a_{\rm w}^2\Omega^2$. In addition $\Omega = \omega d/c$ and $K = kd$ in the TWT case and $K = (k + k_{\rm w})d$ in the FEL.

The general form of the equations is similar for both devices. In fact, the form of the global energy conservation law is identical. The major difference is the momentum term which occurs in the phase terms and which is not there in the TWT case. It was indicated previously that this term originates in the transverse oscillation that the wiggler forces the electrons to undergo and associated with that, is an "effective relativistic transverse mass" of the electron which is $m\gamma$ in contrast to $m\gamma^3$ associated with the longitudinal motion.

When comparing the TWT and FEL three parameters have to be the same: *(i)* the average energy of the electrons $\langle\gamma\rangle$, *(ii)* the electromagnetic energy per particle, $|a|^2/2\alpha$ and *(iii)* the total gain. Two cases have been examined. In the first the total length and the spatial growth rate were assumed to be the same but at the input $a_{\rm TWT} = a_{\rm FEL}/\langle\gamma\beta\rangle$ and $\alpha_{\rm TWT} = \alpha_{\rm FEL}/\langle\gamma\beta\rangle^2$ in order to satisfy the conditions above. In the linear regime, the two devices operated practically the same and Fig. 7.5 also represents the energy spread of the TWT.

Table 7.1.

	TWT	FEL
Amplitude dynamics	$\frac{d}{d\zeta}a = \alpha\langle e^{-j\chi_i}\rangle$	$\frac{d}{d\zeta}a = \alpha\langle \frac{e^{-j\chi_i}}{\gamma_i\beta_{\parallel,i}}\rangle$
Equation of motion	$\frac{d}{d\zeta}\gamma_i = -\mathrm{Re}[ae^{j\chi_i}]$	$\frac{d}{d\zeta}\gamma_i = -\mathrm{Re}[a\frac{e^{j\chi_i}}{\gamma_i\beta_{\parallel,i}}]$
Phase equation	$\frac{d}{d\zeta}\chi_i = \frac{\Omega}{\beta_i} - K$	$\frac{d}{d\zeta}\chi_i = \frac{\Omega}{\beta_i} - K$
Global energy conservation	$\frac{d}{d\zeta}[\langle\gamma\rangle + \frac{1}{2\alpha}\|a\|^2] = 0$	$\frac{d}{d\zeta}[\langle\gamma\rangle + \frac{1}{2\alpha}\|a\|^2] = 0$
Spatial growth rate	$q = \frac{\sqrt{3}}{2}\left[\frac{\alpha\Omega}{2}\langle\frac{1}{(\gamma_i\beta_i)^3}\rangle\right]^{\frac{1}{3}}$	$q = \frac{\sqrt{3}}{2}\left[\frac{\alpha\Omega}{2}\langle\frac{1}{(\gamma_i\beta_i)^5}\rangle\right]^{\frac{1}{3}}$

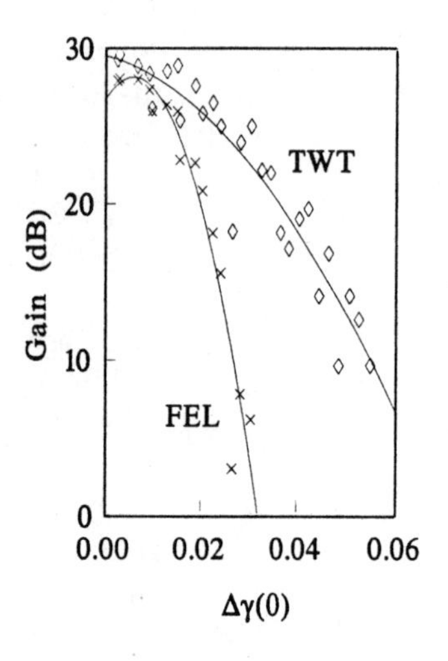

Fig. 7.6 Gain as a function of the energy spread at the input in a free-electron laser and in a traveling wave tube

In the second case examined it was assumed that a at the input is the same in both devices and consequently, from assumption *(ii)*, so is α. As a result, the spatial growth rate is smaller in the FEL by (roughly) a factor of $1/(\gamma\beta)^{2/3}$. Therefore, in order to satisfy the constraint in *(iii)*, we increase the length of the FEL by a factor of $(\gamma\beta)^2$ such that $\Omega_{\mathrm{FEL}} = \Omega_{\mathrm{TWT}}\langle\gamma\beta\rangle^2$ and $K_{\mathrm{FEL}} = K_{\mathrm{TWT}}\langle\gamma\beta\rangle^2$ the result is illustrated in Fig. 7.6 where we plotted the gain as a function of the energy spread at the input. The simulation reveals a clear sensitivity of the free-electron laser to initial energy spread compared to the TWT in the conditions determined above. The situation is even worse at higher energies.

7.5 Other FEL Schemes

One of the major advantages of free-electron lasers is the fact that no external means are required to confine the radiation. In fact Scharlemann, Wurtele and Sessler (1985) have shown that under certain conditions the beam acts like an optical fiber and guides the radiation. Later Sprangle (1987) formulated the three-dimensional problem introducing the source-dependent expansion technique. Beam guidance may become crucial in two cases: in the case of very high power radiation where the Ohm loss of walls makes the contact with the intense radiation field prohibitive. And in the case of very high frequency e.g., Ultra Violet or X-rays, where even if the contact of the radiation with a metallic surface is permissible, from the aspect of the power levels, the reaction of the surface is not as regular as at the low frequency (visible and below). That is to say, that the surface quality is poor since micro-perturbations are of the same size as the radiation wavelength. For the reasons mentioned above, the FEL has the potential to generate coherent and tunable radiation at short wavelengths such as UV and X-rays. There are, however, two major obstacles in its way: the beam quality which is a major limitation and the wiggler. As for the first it was shown in this chapter (Sects. 7.2 and 7.3) that the gain depends strongly on the temperature of the beam. This problem becomes acute at high frequencies. In addition, transverse beam effects (emittance), which were not discussed here, start to play an important role. The constraints imposed on the wiggler are also stringent. Obviously the shorter the period the better. But a short period implies a low intensity magnetic field which in turn implies a long interaction length. When a large number of magnets are involved, two problems occur: alignment and statistical errors in the intensity of each pole. While the first can be minimized the latter is unavoidable.

7.5.1 Gas Loaded FEL

In order to release some of the constraints on the beam and wiggler Pantell (1986) and Feinstein (1986) have shown that there are substantial advantages to slowing down the phase velocity of the wave by loading the FEL with gas since at the frequency of interest the refraction coefficient, n, is larger than unity. The resonance condition in this case is

$$\frac{\omega}{v_{\|}} = n(\omega)\frac{\omega}{c} + k_{\mathrm{w}}\,, \tag{7.5.1}$$

and the resonance frequency reads

$$\frac{\omega}{c} = k_{\mathrm{w}}\left[\frac{1}{\sqrt{1-\mu^2/\gamma^2}} - n\right]\,. \tag{7.5.2}$$

In order to emphasize the effect we shall examine two cases. Firstly we consider a vacuum system ($n = 1$) with $\mu = 1.2$ and $\gamma = 100$. If the period of the wiggler is 5 cm then the radiation wavelength is 3.6 μm. If the refraction coefficient of the gas is $n - 1 = 7 \times 10^{-5}$ then the radiation wavelength is 1000 Å. Therefore, the presence of the medium caused a frequency shift from infra red to UV.

7.5.2 Longitudinal Wiggler FEL

Another free-electron laser configuration which was considered by McMullin and Bekefi (1981,1982) consists of a longitudinal rather than a transverse wiggler. In this case the guiding magnetic field is rippled and it is approximately given by

$$\mathbf{B} = B_0 \mathbf{1}_z + B_1 \mathrm{I}_0(k_\mathrm{w} r) \sin(k_\mathrm{w} z) \mathbf{1}_z - B_1 \mathrm{I}_1(k_\mathrm{w} r) \cos(k_\mathrm{w} z) \mathbf{1}_r \,. \tag{7.5.3}$$

The transverse motion of the electrons is controlled here by both the wiggler and the guiding field and the resonance condition reads

$$\omega = \frac{\Omega_c}{\gamma} + (k + k_\mathrm{w}) v_\parallel \,, \tag{7.5.4}$$

where $\Omega_c = eB_0/m$ is the non-relativistic cyclotron frequency. Assuming that the interaction only slightly affects the TEM mode i.e., $k \simeq \omega/c$, the resonant frequency is

$$\omega = \gamma^2 \frac{1 + \beta_\parallel}{1 + (\gamma\beta_\perp)^2} \left(\Omega_c \frac{1}{\gamma} + k_\mathrm{w} c \beta_\parallel \right) \,, \tag{7.5.5}$$

and it can be readily seen that without the guiding field ($\Omega_c = 0$) the resonance corresponds to that of a transverse wiggler FEL. Therefore the guiding field causes an effective increase in the wave number of the wiggler by the factor $\Omega_c/c\gamma$ – which can be significant.

7.5.3 Rippled-Field Magnetron

The basic configuration of free-electron lasers discussed so far was co-linear in the sense that the dominant component of the electrons' velocity was in the longitudinal direction. Bekefi (1982) has suggested to construct a smooth bore magnetron where the interaction is facilitated by a wiggler rather than a slow-wave structure as is generally the case. The system consists of two cylindrical electrodes, an insulating magnetic field along the axis and a wiggler which is azimuthally periodic but its magnetic field is in the radial direction. A positive voltage V is applied on the anode. Electrons emitted from the cathode form a Brillouin flow around the cylinder provided that the intensity of the insulating magnetic field exceeds the critical value

$$B_z > B_{\rm cr} = \frac{mc}{e\delta R}\sqrt{\gamma^2 - 1}\,, \tag{7.5.6}$$

where $\gamma = 1 + eV/mc^2$ and δR is the anode-cathode gap. This equilibrium is altered by a wiggler which can be approximated by

$$B_r = B_0 \cos(N\phi)\,, \tag{7.5.7}$$

where N is the number of magnetic poles and ϕ is the azimuthal coordinate. Conceptually the interaction is similar to the co-linear case, however the cylindrical configuration complicates the detailed analysis. The concept was tested experimentally by Destler (1985) and good agreement with theoretical predictions was observed.

7.5.4 Wiggler and Guiding Magnetic Field

In many cases the electron beam is immersed in a guiding magnetic field even before it enters the wiggler field. It is therefore reasonable to calculate the trajectories of the electrons in a configuration which combines the two magnetic fields. Friedland (1980) calculated these trajectories for an idealized magnetic field and Freund (1983) has improved the model for a more realistic configuration. The guiding field causes an increase in the transverse velocity of the electrons but the effect on the gain is strongly dependent on the parameters of the wiggler according to detailed trajectory of the electrons – two types were initially emphasized. Conde and Bekefi (1991) discovered that by inverting the direction of the guiding field a substantial improvement of the efficiency can be achieved.

7.5.5 Electromagnetic Wiggler

In principle the magneto-static wiggler can be replaced by an intense electromagnetic wave which propagates anti-parallel to the beam. This was in fact the original concept considered by Kapitza and Dirac back in 1933, however they investigated the interaction with a standing wave (wavenumber was perpendicular to the beam).

In the process of interacting with the wave one gains a factor of 2 in the frequency of the emitted radiation. Since the wiggler field varies in time ($\omega_{\rm w}$) the resonance condition reads

$$\omega - \omega_{\rm w} = (k + k_{\rm w})v_{\parallel}\,. \tag{7.5.8}$$

Furthermore, both wiggler and emitted fields behave as free waves ($k \simeq \omega/c$ and $\omega_{\rm w} \simeq ck_{\rm w}$) therefore the resonance frequency reads

$$\omega = ck_{\rm w}\frac{1 + \sqrt{1 - \mu^2/\gamma^2}}{1 - \sqrt{1 - \mu^2/\gamma^2}} \simeq ck_{\rm w}\frac{4\gamma^2}{\mu^2}\,. \tag{7.5.9}$$

This concept was demonstrated by Carmel, Granatstein and Gover (1983) when a high power microwave pulse was generated with a backward-wave oscillator and used in a second stage as an electromagnetic wiggler for a free-electron laser.

7.5.6 Electrostatic Wiggler

One of the difficulties with magneto-static wigglers is that their period, for substantial field intensity, is of the order of cm's, therefore highly relativistic particles are required in order to achieve optical (or shorter) wavelengths. Even if the electrons with this energy are available, disregarding the cost of their acceleration, we still confront another problem which is: the scaling of the gain with γ. In a regular FEL, the electron undergoes a transverse motion under the influence of the transverse wiggler and since the coupling coefficient is quadratic in the wiggler field, this motion contributes a γ^{-2} term to this parameter. The pondermotive force modulates the beam in the longitudinal direction and this motion gives rise to an additional factor of $(\gamma\beta)^{-3}$ and therefore, the coupling coefficient in the high-gain Compton regime is proportional to $(\gamma\beta)^{-5}$. For comparison in TWT the coupling coefficient is proportional to $(\gamma\beta)^{-3}$.

It is relatively easy to make a short period electrostatic wiggler with say a period of a few microns and even shorter than that using photolithography techniques. However, the problem is that the gain is proportional to $\gamma^{-9}\beta^{-5}$. Originally, the calculation was done by Gover (1980) for the low-gain Compton regime but here we present the analysis of the high-gain Compton regime. In order to prove our previous statement let us consider an electrostatic potential of the following form

$$\phi(z, r \simeq 0) = \phi_0 \cos(k_{\mathrm{w}} z)\,, \tag{7.5.10}$$

and since all the discussion so far in this chapter considered only magnetic wigglers we shall present the analysis of this scheme in more detail.

The electron's motion has three components: the major one is the "dc", β, the second is due to this electrostatic potential (β_{w}) and the third is proportional to the radiation field, $\delta\beta$. If we ignore momentarily the radiation field then the motion of the electrons is longitudinal and it is given by

$$\beta_{\mathrm{w}} = \frac{e\phi_0}{mc^2}\frac{1}{\beta\gamma^3}\cos(k_{\mathrm{w}} z)\,. \tag{7.5.11}$$

As in the regular FEL we expect the "resonance" motion to be determined by the product of the wiggler induced motion and the magnetic vector potential of the radiation field. The Hamiltonian of the system can be approximated by

$$\begin{aligned} H &= H_0 + H_1 \\ &= mc^2\gamma + ec\beta_{\mathrm{w}} A_{\mathrm{rf}}\,, \end{aligned} \tag{7.5.12}$$

and the pondermotive force $F_{\rm p} = -\partial H_1/\partial z$. Assuming that the rf field is given by $A_{\rm rf} = A\cos(\omega t - kz)$, neglecting off resonance terms and using a phasor notation we have

$$H_1 \simeq h e^{j\omega t - jKz} \tag{7.5.13}$$

where $h = (ecA)(e\phi_0/2mc^2\beta\gamma^3)$ and $K = k + k_{\rm w}$. The phasor of the oscillatory motion is therefore

$$\delta\beta = \frac{Kc}{\omega - Kv}\frac{h}{mc^2\gamma^3}\,. \tag{7.5.14}$$

Next we use the continuity equation to determine the particles' density and, as above, keeping only terms which may contribute to the resonant process, we have

$$\delta n = n_0 \frac{Kc}{\omega - Kv}\delta\beta\,. \tag{7.5.15}$$

The current density is given by $J_z = -ec(n_{\rm w}\delta\beta + \beta_{\rm w}\delta n)$ and since the motion induced by the wiggler does not contribute to the net current we have $n_{\rm w} = -n_0\beta_{\rm w}/\beta$ which allows us to use the following approximation for the current density

$$\begin{aligned} J_z &= -ecn_0 \frac{Kc}{\omega - Kv}\beta_{\rm w}\delta\beta\,, \\ &\simeq -(ec)^2 n_0 \frac{(Kc)^2}{(\omega - Kv)^2}\left(\frac{e\phi_0}{2mc^2\beta\gamma^3}\right)^2 \frac{1}{mc^2\gamma^3} A e^{j\omega t - jkz}\,. \end{aligned} \tag{7.5.16}$$

This current density drives the magnetic vector potential which satisfies

$$\left[\frac{{\rm d}^2}{{\rm d}z^2} + k_0^2\right] A_{\rm rf} = -\mu_0 F_{\rm f} J_z\,, \tag{7.5.17}$$

where $F_{\rm f}$ is the filling factor and k_0 is the wavenumber of the interacting wave (harmonic) in the absence of the beam. Substituting the expression for the current density, assuming resonance and that $k = k_0 + \delta k$ we obtain

$$\begin{aligned} \delta k^3 &= -K_0^3 \\ &\equiv -\frac{F_{\rm f}}{8}\frac{\omega_{\rm p}^2}{c^2}\frac{(k_0 + k_{\rm w})^2}{k_0}\left(\frac{e\phi_0}{mc^2}\right)^2 \frac{1}{\beta^4\gamma^9} \\ &= -\frac{I\eta_0 F_{\rm f}}{8mc^2 S_{\rm el}}\frac{(k_0 + k_{\rm w})^2}{k_0}\left(\frac{e\phi_0}{mc^2}\right)^2 \frac{1}{\beta^5\gamma^9}\,. \end{aligned} \tag{7.5.18}$$

In the last expression I is the total current carried by the beam and $S_{\rm el}$ is the beam area. It clearly reveals what we indicated previously that the coupling coefficient is proportional to $\gamma^{-9}\beta^{-5}$. If we assume the same frequency and compare the coupling coefficient in this case with that of a regular FEL [see (7.3.29)] we get

$$\frac{K_{\text{elec}}^3}{K_{\text{mag}}^3} \propto \frac{(\gamma^5\beta^4)_{\text{mag}}}{(\gamma^9\beta^4)_{\text{elec}}} \left(\frac{e\phi_0 k_{\text{w}}}{eB_{\text{w}}c}\right)^2 . \tag{7.5.19}$$

If the two systems generate 5000 Å radiation in the magnetic wiggler case for a period of 10 cm the electrons must have $\gamma_{\text{mag}} \sim 3 \times 10^2$. In order to generate the same radiation with an electrostatic wiggler of 5 μm periodicity, the electrons must have $\gamma_{\text{elec}} = 2.4 (\beta = 0.9)$. If on the surface of a metallic surface the amplitude of the first harmonic is $\phi_0 \sim 50\,\text{V}$ and the intensity of the magnetic field is $B_{\text{w}} = 0.1\,\text{T}$, then the ratio of the two coupling coefficients is 10. However, in (7.5.19) the two filling factors were assumed to be the same and this is not generally the case. Only those electrons which are within $h \sim 5$ μm from the surface do interact therefore if the beam radius is $R_{\text{b}} \sim 2\,\text{mm}$ the filling factor is $h\sqrt{2hR_{\text{b}}}/\pi R_{\text{b}}^2 \sim 6 \times 10^{-5}$. Within 2–3 μm from the surface there is an exponential decay in the amplitude of the field by a factor of 10, thus a factor of 100 in the coupling coefficient. We may therefore expect the latter to be smaller by almost a factor of $K_{\text{elec}}^3/K_{\text{mag}}^3 \sim 10^{-6}$. In principle one can increase the amplitude of the electrostatic wiggler to 50 kV in which case the two growth rates are comparable. Unfortunately in this case we run into a breakdown problem since 50 kV applied on a structure of 5 μm period generate (dc) gradients of the order of 10 GV/m or higher. Once breakdown occurs the wiggler is short circuited and the gain vanishes. Consequently, the design of such a system is a trade-off between high voltage requirements dictated by the demand of a high gain and low voltage regime imposed by the need to avoid breakdown.

7.5.7 Channeling Radiation

All the wigglers mentioned so far, were on a the macroscopic scale i.e., order of cm's and down to the micron level. However, the lattice of a crystaline material forms a natural periodic electrostatic wiggler. If a beam of relativistic electrons is injected parallel into one of the symmetry planes of a lattice then it sees two periodicities. One in the longitudinal direction which is negligible since the intensity of the potential is too weak in order to bunch the beam, which is to say that the longitudinal momentum of the particle is many orders of magnitude larger than the quanta of lattice momentum. The other periodic system is in the transverse direction [Berry (1971)]. If the electrons have a small momentum in this direction, they are "reflected" by the lattice plane and they undergo an oscillatory motion [Kumakhov (1976)] – therefore they may emit radiation. The radiation is a direct result of the transverse momentum relative to the symmetry plane of the crystal. Terhune and Pantell (1977) and later Pantell and Alguard (1979) discussed the effect from the quantum mechanical perspective: the transverse potential of the lattice as seen by the moving electron consists of a set of "bound states". When the electrons are injected parallel to the plane of symmetry only the lowest state is populated. When the beam is tilted to this plane, higher states are populated. As in normal bound states, electrons can jump from a high state to a lower state emitting a photon in the process. It is interesting to note that from the point of view of the electron wave function propagation, it is completely analogous to the propagation of an electromagnetic wave in an optical fiber [Schächter (1988)]. Spontaneous emission of this process has been observed [Swent (1979)] but the condition imposed on the emittance of the beam is very stringent and to the best of our knowledge no stimulated radiation was observed so far. A review of the quantum picture of interaction of free electrons with radiation was given by Friedman (1988).

Exercises

7.1 Show that the equivalent to (7.4.47) in the case of a traveling-wave tube reads

$$\frac{1}{2}\frac{\mathrm{d}}{\mathrm{d}\bar{\gamma}}\Delta\gamma^2 = \bar{\gamma} - \frac{\langle \gamma_i \cos(\tilde{\chi}_i)\rangle}{\langle \cos(\tilde{\chi}_i)\rangle},$$

but the Madey theorem has the same form as in (7.4.43).

7.2 Use the formulation of the interaction in a tapered wiggler to calculate the power generated in the low-gain Compton regime in a uniform wiggler with stochastic errors in its parameters. Assume that the errors follow a Gaussian distribution.

7.3 Calculate the spontaneous emission emitted by a particle moving in a periodic electrostatic potential as in Sect. 7.5.6. Compare the spectrum with that of the Smith-Purcell effect. Calculate the effective decelerating force which acts on the particle and compare it with the result in (7.1.30) and with the decelerating force in the Smith-Purcell case.

7.4 Calculate the stable trajectories of a particle in a combination of wiggler and uniform magnetic field. Draw $\beta_{\|}$ as a function of B_{w}.

7.5 Based on Sects. 7.2 and 7.3 make a summarizing table which will include the gain and the condition (say on the current and beam temperature) that the system will operate in a particular regime. Discuss the transition from one regime to another.

8. Basic Acceleration Concepts

One of the important systems where beam-wave interaction in periodic structures plays a crucial role is the particle accelerator. The latter provides us with a unique tool to test on earth the different models which describe the constituents of matter. Accelerators have undergone a great deal of progress in the last fourty years and it seems that they still have a long way to go in order to meet the requirements necessary to test the theoretical models [Richter (1985)]. Over the years other disciplines learned to use the potential of accelerators and today they are widely used in chemistry, biology and medicine for generation of radiation.

There are two main types of particle accelerators: circular and linear machines. In this text we shall consider only the latter category – some fundamentals of circular machines were discussed by Collins (1983) and Ruth (1986). Linear accelerators can be divided according to the operation frequency: *(i)* dc linacs where charge is accumulated on a series of electrodes and the resulting field is used for acceleration of a dc current. *(ii)* Induction linacs rely on the Faraday law and the accelerating voltage is a result of a time varying magnetic field. Each unit can be conceived as a transformer in which high current and relatively low voltage form the primary whereas the accelerated beam, of relatively low current and high voltage, constitutes the secondary. *(iii)* At the other extreme of the dc machine stands the rf linac. It utilizes the energy carried by an electromagnetic wave to accelerate charged particles – in relatively short sections (1–2 m). The electrons are prepared in small bunches since only a small fraction of one wavelength can provide acceleration. Our treatment will be limited to this last category.

This chapter has basically two parts. In the first part (Sect. 8.1) we discuss the basics of linear accelerator (linac) concepts with particular emphasis on the beam-wave interaction [Loew and Tolman (1982)]. Our discussion will be limited to electron linear accelerators of the type *operational* today at SLAC (Stanford Linear Accelerator Center) whose basic concepts are applicable to what is today conceived as the Next Linear Collider. The second part (Sects. 8.2–8.6) is a collection of brief overviews of different alternative schemes of acceleration which are in their early stages of research. In these sections the discussion is in general limited to the basic concepts and the figure of merit which characterizes their application i.e. the achievable gradient.

For more detailed discussions the reader is referred to articles by the experts of each scheme.

8.1 Linear Accelerator Concepts

The field of linear accelerators is very broad and comprises many subfields which have been covered in books and presented in detail in summer schools [e.g. Lapostolle and Septier (1970)], and an entire set of proceedings of summer schools or meetings which will be referred to in what follows. When such a subject is presented in a single chapter, compromises have to be made as to what to include and what to leave out. Since the subject of this book is about periodic and quasi-periodic structures we shall concentrate on the acceleration structure itself. Therefore we ignore topics associated with rf generation, pulse compression, injection and extraction of the rf. Furthermore, we shall not discuss beam generation, cooling or focusing.

8.1.1 Constant Gradient and Constant Impedance Structures

The basic configuration of an acceleration section is conceptually identical to the coupled cavity structure as discussed in the context of closed periodic structure in Chap. 5 except that in this case the effect of the beam on the radiation field is small and the ohmic loss has to be considered since it causes amplitude and phase variations which affect the acceleration process. In Chap. 2 we defined several impedances which have been used throughout the text. At this point, we introduce an additional concept namely, the *shunt impedance*, and several other related quantities which are important to establish the dynamics of electrons in an acceleration structure. The shunt impedance ($Z_{\rm sh}$) is important in accelerators since large amounts of electromagnetic power flows in the system and any change in this power affects the electron dynamics. This quantity is a measure of the ohmic power loss in a unit length in terms of the electric field $[E(z)]$ which acts on the electrons. It is given by

$$Z_{\rm sh} \equiv \frac{|E(z)|^2}{-{\rm d}P/{\rm d}z}, \tag{8.1.1}$$

where $P(z)$ is the power which flows along the structure. According to this definition, the shunt impedance has units of impedance per unit length and it is related to the quality factor of a waveguide defined by

$$Q \equiv \frac{\omega {\rm W}}{-{\rm d}P/{\rm d}z}, \tag{8.1.2}$$

through

$$\frac{Z_{\rm sh}}{Q} = \frac{|E(z)|^2}{\omega \mathrm{W}}; \tag{8.1.3}$$

here W is the average electromagnetic energy per unit length. This quantity is also related to the spatial decay associated with the Ohmic loss

$$\alpha_{\rm ohm} = -\frac{\mathrm{d}P/\mathrm{d}z}{2P}, \tag{8.1.4}$$

which in turn is related to the skin resistance $R_{\rm s} \equiv (\sigma\delta)^{-1}$ where σ is the conductivity of the metal and δ is the skin-depth defined by

$$\delta = \sqrt{\frac{2}{\omega\sigma\mu_0}}. \tag{8.1.5}$$

For a cylindrical resonator of radius R and length d the shunt impedance is

$$Z_{\rm sh} = \eta_0 \left(\frac{\eta_0}{R_{\rm s}}\right) \frac{d}{\pi R(R+d)\mathrm{J}_1^2(p_1)}, \tag{8.1.6}$$

and

$$Q = \frac{\eta_0}{R_{\rm s}} \frac{dp_1}{2(R+d)}. \tag{8.1.7}$$

Based on (8.1.2,4) and the fact that the energy velocity is $v_{\rm en} \equiv P/W$, it can be shown that

$$\alpha_{\rm ohm} = \frac{\omega}{2v_{\rm en}Q}. \tag{8.1.8}$$

With the exception of the injection section, the electrons move in a typical acceleration section at almost c, therefore in what regards the phase velocity, there is no need to taper the structure. However, in order to have a feeling on the effect of the ohmic loss we can readily understand that part of the energy is absorbed by the walls and consequently, the field experienced by the electrons decreases in space. Therefore, if the motion is calculated for a lossless system, the variation in amplitude or phase due to lossy material may cause the electron to slip out of phase. Let us now calculate the energy transferred to the electron as a function of the shunt impedance and the electromagnetic power injected at the input. For this purpose we first assume that the electron is "riding on the crest of the wave" and it gains in a length D a kinetic energy

$$\delta E_{\rm kin} = e\int_0^D \mathrm{d}z E(z); \tag{8.1.9}$$

the effect of the phase will be considered in Sect. 8.1.3. The energy gain will be calculated for two different acceleration structures: *(i) constant impedance*, in which case the shunt impedance is constant and the gradient varies in space. And for a *(ii) constant gradient* structure in which case it is primarily the group velocity which varies and consequently, the interaction impedance changes.

Constant Impedance. The shunt impedance in this case is constant along the structure and this entails constant geometry and consequently uniform $\alpha_{\rm ohm}$. Bearing in mind that the power along the structure is given by $P(z) = P(0){\rm e}^{-2\alpha_{\rm ohm} z}$ and

$$E(z) = \sqrt{2\alpha_{\rm ohm} Z_{\rm sh} P(z)}\,, \tag{8.1.10}$$

we can readily conclude that when the beam loading effect is ignored the change in the kinetic energy is given by

$$\delta E_{\rm kin} = e\left[P(0) Z_{\rm sh} D\right]^{1/2} (2\xi)^{1/2} \left(\frac{1-{\rm e}^{-\xi}}{\xi}\right), \tag{8.1.11}$$

where $\xi \equiv \alpha_{\rm ohm} D$.

Constant Gradient. Here $\alpha_{\rm ohm}$ is tapered in such a way that the gradient remains constant along the structure. In order to avoid unnecessary complications we shall assume that we can ignore spatial variations in Q as well as in the shunt impedance and that the geometric variations affect primarily the group (energy) velocity and consequently, the interaction impedance. From the definition of the shunt impedance $E^2(z) = -Z_{\rm sh}({\rm d}P/{\rm d}z)$ we conclude that ${\rm d}P/{\rm d}z$ has to be constant, which implies

$$P(z) = A + Bz\,. \tag{8.1.12}$$

Since we know the power at the input $[P(0)]$ and at the output, $P(0){\rm e}^{-\zeta}$, where

$$\zeta \equiv \int_0^D {\rm d}z \alpha_{\rm ohm}(z)\,, \tag{8.1.13}$$

we conclude that the variation of the power in space is given by

$$P(z) = P(0)\left[1 - \frac{z}{D}\left(1-{\rm e}^{-\zeta}\right)\right]. \tag{8.1.14}$$

Using the definition of $\alpha_{\rm ohm}$, (8.1.4), we find that

$$\alpha_{\rm ohm} = \frac{1}{2D} \frac{1-{\rm e}^{-2\zeta}}{1-\left(1-{\rm e}^{-2\zeta}\right) z/D}\,, \tag{8.1.15}$$

which, subject to our assumptions, dictates the group velocity as defined by (8.1.8). Since the gradient is constant, the kinetic energy gained by one electron is $\delta E_{\rm kin} = eE(0)D$ which in terms of the input power (8.1.11,14) reads

$$\delta E_{\text{kin}} = e\left[P(0)Z_{\text{sh}}D\right]^{1/2}\left(1-\text{e}^{-2\zeta}\right)^{1/2} . \tag{8.1.16}$$

In either of the two cases, maximum energy is gained for a maximum shunt impedance. This conclusion leads us to the choice of the phase advance per cell. Recall that in Chap. 5, when solving the dispersion relation of a disk loaded structure, we assumed a certain phase advance per cell which together with the resonance condition determined the periodicity of the structure. No reason has been given for this particular choice. Now that we have concluded that for maximum energy gain, one has to maximize the shunt impedance, we can ask what is the number of disks in one period of the wave which satisfies this condition. It is intuitive that the larger the number of disks, the greater the total Ohmic loss and consequently for a given $E(z)$ the shunt impedance decreases. The same phenomenon occurs at the other extreme, since for a small number of disks, the electric field which acts on the electrons is expected to be small (for a given Ohmic loss). Simulations indicate that maximum shunt impedance occurs for three disks in one period of the wave and for this reason *traveling-wave* accelerating structures are designed with a phase advance per cell of $2\pi/3$.

Beam Loading. Up to this point the energy transferred to the beam was ignored. In order to consider the effect of the beam we can make a quasi quantitative argument as follows: in the absence of the beam and for *constant impedance* the electric field decays exponentially with α_{ohm} therefore it satisfies

$$\frac{\text{d}E}{\text{d}z} = -\alpha_{\text{ohm}}E . \tag{8.1.17}$$

Presence of the beam causes an additional change in the electric field $E \to E + Z_{\text{sh}}I$ whose variation in space is given by

$$\frac{\text{d}E}{\text{d}z} = -\alpha_{\text{ohm}}\left(E + Z_{\text{sh}}I\right) . \tag{8.1.18}$$

Here I represents the current carried by the beam in a narrow bunch. The solution of this equation is $E = A\text{e}^{\alpha_{\text{ohm}}z} - Z_{\text{sh}}I$ and since at the input the loading effect is expected to be zero we have, according to (8.1.10), $E(z = 0) = \sqrt{2\alpha_{\text{ohm}}Z_{\text{sh}}P(0)}$ hence

$$E(z) = \sqrt{2\alpha_{\text{ohm}}Z_{\text{sh}}P(0)}\,\text{e}^{-\alpha_{\text{ohm}}z} - Z_{\text{sh}}I\left(1-\text{e}^{-\alpha_{\text{ohm}}z}\right) . \tag{8.1.19}$$

Consequently, the kinetic energy gain of a single particle is given by

$$\delta E_{\text{kin}} = e\left[DZ_{\text{sh}}P(0)\right]^{1/2}(2\xi)^{1/2}\frac{1-\text{e}^{-\xi}}{\xi} - eZ_{\text{sh}}ID\left(1-\frac{1-\text{e}^{-\xi}}{\xi}\right) ; \tag{8.1.20}$$

clearly the second term represents the beam loading effect.

For a *constant gradient* structure, when the beam loading is ignored it implies that

$$\frac{\mathrm{d}E}{\mathrm{d}z} = 0 , \tag{8.1.21}$$

thus when the beam effect is included, in analogy with (8.1.18), we have

$$\frac{\mathrm{d}E}{\mathrm{d}z} = -\alpha_{\mathrm{ohm}} Z_{\mathrm{sh}} I . \tag{8.1.22}$$

Following the same procedure as previously, we have

$$\delta E_{\mathrm{kin}} = e\left[P(0) Z_{\mathrm{sh}} D\right]^{1/2} \left(1-\mathrm{e}^{-2\zeta}\right)^{1/2} - \frac{1}{2} e I Z_{\mathrm{sh}} D \left(1 - \frac{2\zeta \mathrm{e}^{-2\zeta}}{1-\mathrm{e}^{-2\zeta}}\right) ; \tag{8.1.23}$$

as in (8.1.20) the second term represents the beam loading effect. Note that subject to the present assumptions, we expressed the gain in kinetic energy of the bunch in terms of a few "global" parameters ($Z_{\mathrm{sh}}, \alpha_{\mathrm{ohm}}$ etc.) determined in turn by the geometric and electrical parameters which may vary from one structure to another.

8.1.2 Auxiliary Coupling

The design of an acceleration structure is a continuous process of trade-offs. In the previous subsection it was indicated that the phase advance per cell in a traveling-wave structure should be $2\pi/3$ in order to maximize the shunt impedance. There are, however, additional considerations which come into play. Strictly speaking from the electrons' point of view the best choice would be a π-mode i.e., *standing-wave* configuration, since the gradient for a given input power reaches its maximum. But a π-mode is unacceptable from the electromagnetic wave point of view since the group velocity at this point is zero, the filling time is long and if the structure is sufficiently long, the mode is unstable since near the crest of the dispersion relation the frequency separation of the modes is small. If we examine the same problem from the field aspect then the best choice would be a $\pi/2$-mode of operation since in this case the group velocity is the largest. Unfortunately, in this case half of the cavities do not contribute to acceleration. Knapp (1965) suggested a way to break this vicious circle. His basic idea is to satisfy both the electrons and the electromagnetic field: the former sees a π-mode and the electromagnetic wave sees a $\pi/2$-mode. This is possible because the beam occupies only a small fraction of the transverse dimension of the structure whereas the electromagnetic wave fills the entire volume. Implementation of this concept is possible by making each cell of two cavities: one cavity is the regular pill box cavity of a disk loaded waveguide while the second is recessed and its aperture is on the external wall coupling two adjacent pill box cavities. In this way, one can design the structure such that the electron sees a series of pill box cavities operating at the π-mode whereas the electromagnetic wave actually operates in a $\pi/2$-mode. Schematically this configuration is presented in Fig. 8.1.

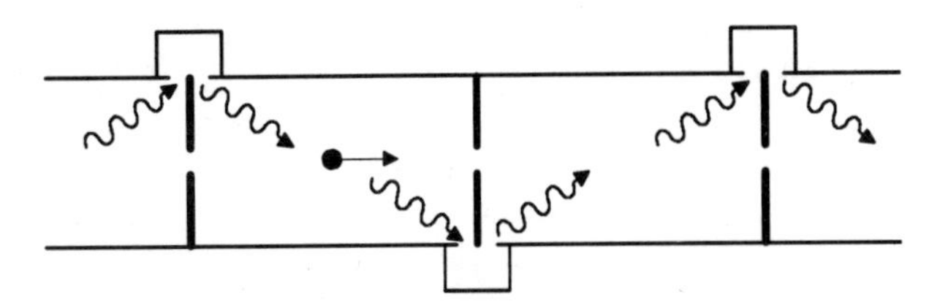

Fig. 8.1. Accelerating structure with side couplers. The electromagnetic wave undergoes a $\pi/2$ phase advance per cell but the accelerated electron sees a π phase advance

8.1.3 Phase Dynamics

In Sect. 8.1.1 we assumed that the particle "rides" on the crest of the wave and we examined the energy transfer assuming it stays on the crest along the entire interaction region. But this is not the case in general since as the particle is accelerated its relative phase varies. In this subsection we shall consider the phase dynamics assuming that the amplitude of the electric field is constant. For this purpose we now consider an accelerating structure which is designed for a phase velocity c. We ignore amplitude variation due to the interaction and disregard the transverse motion of the electrons. Subject to these conditions, the dynamics of the particles is given by

$$\frac{\mathrm{d}}{\mathrm{d}t}\gamma\beta = -\frac{eE_0}{mc}\cos\left[\omega t - \frac{\omega}{c}z(t)\right] . \tag{8.1.24}$$

It is convenient to define the phase of the particle relative to the wave as

$$\chi(t) = \chi(0) + \omega t - \frac{\omega}{c}z(t) , \tag{8.1.25}$$

and since β is always smaller than 1, the normalized velocity is expressed as

$$\beta = \cos\psi . \tag{8.1.26}$$

With these definitions we can write two equations which describe the dynamics of the particles as:

$$\begin{aligned} \frac{-1}{\sin^2\psi}\frac{\mathrm{d}}{\mathrm{d}t}\psi &= -\frac{eE_0}{mc}\cos\chi , \\ \frac{\mathrm{d}}{\mathrm{d}t}\chi &= \omega(1-\cos\psi) , \end{aligned} \tag{8.1.27}$$

and since we limited the motion to the longitudinal direction we can replace

$$\frac{\mathrm{d}\chi}{\mathrm{d}t} = \left(\frac{\partial \chi}{\partial \psi}\right) \frac{\mathrm{d}\psi}{\mathrm{d}t} . \tag{8.1.28}$$

Substituting, the two expressions in (8.1.27) we obtain,

$$\mathrm{d}\psi \frac{1 - \cos\psi}{\sin^2\psi} = \frac{eE_0}{mc\omega} \cos\chi \,\mathrm{d}\chi . \tag{8.1.29}$$

This expression can be integrated analytically and then re-arranged. Denoting with indexes "in" and "out" the corresponding values of the variables at the input of the interaction region and at its output; the result reads

$$\frac{\sin(\psi_{\mathrm{out}}/2)}{\cos(\psi_{\mathrm{out}}/2)} - \frac{\sin(\psi_{\mathrm{in}}/2)}{\cos(\psi_{\mathrm{in}}/2)} = \frac{eE_0}{mc\omega} \left[\sin\chi_{\mathrm{out}} - \sin\chi_{\mathrm{in}}\right] . \tag{8.1.30}$$

Finally this relation can be rewritten in terms of the familiar β and the phase χ as follows

$$\sin\chi_{\mathrm{out}} = \sin\chi_{\mathrm{in}} + \frac{m\omega c}{eE_0} \left(\sqrt{\frac{1-\beta_{\mathrm{out}}}{1+\beta_{\mathrm{out}}}} - \sqrt{\frac{1-\beta_{\mathrm{in}}}{1+\beta_{\mathrm{in}}}} \right) . \tag{8.1.31}$$

Expression (8.1.31) determines, for given initial conditions, the relation between the phase of the particle and its energy. Under certain circumstances, the particles are trapped as may be readily seen since once trapped they are accelerated and we may assume that they reach high γ. Explicitly if, $\beta_{\mathrm{out}} \sim 1$ we have

$$\sin\chi_{\mathrm{out}} = \sin\chi_{\mathrm{in}} - \frac{m\omega c}{eE_0} \sqrt{\frac{1-\beta_{\mathrm{in}}}{1+\beta_{\mathrm{in}}}} , \tag{8.1.32}$$

which indicates that the value of the phase at the output is determined only by the initial values and it is independent of the particle's energy at the output. Since the trigonometric functions together are of order 1, we conclude that the condition for particles to become trapped is

$$E_0 > E_{\mathrm{cr}} \equiv \frac{mc^2}{e} \frac{\omega}{c} \sqrt{\frac{1-\beta_{\mathrm{in}}}{1+\beta_{\mathrm{in}}}} . \tag{8.1.33}$$

For an initial energy of 300 keV the field intensity assuming operation at 10 GHz is 38 MV/m. Increasing the initial energy to 400 keV lowers the required field to 33 MV/m. Figure 8.2 illustrates the trapping process for $E_0 = 60\,\mathrm{MV/m}$, $f = 11.424\,\mathrm{GHz}$, $\chi_{\mathrm{in}} = 95°$ and the initial energy of the

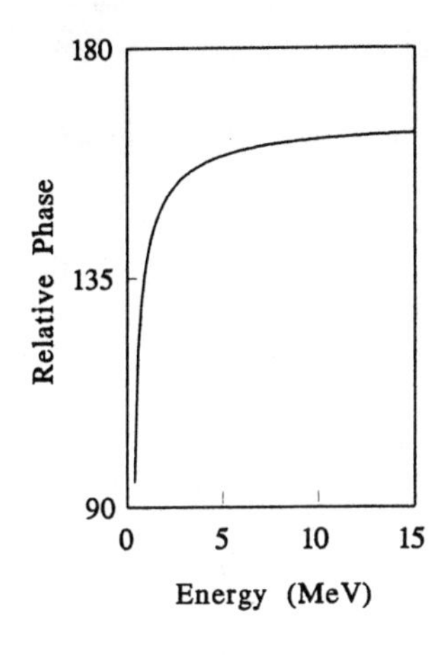

Fig. 8.2 Relative phase as a function of the energy. For an initial energy of 0.4 MeV the phase virtually remains constant when the electron's energy exceeds the 5 MeV level

particles is 400 keV: we observe that beyond 5 MeV, in practice, the phase does not change.

Another direct result of (8.1.31) is the fact that if the initial particle is also highly relativistic, implying that we can write

$$\begin{aligned}\sin\chi_{\text{out}} &= \sin\chi_{\text{in}} + \frac{m\omega c}{eE_0}\left(\frac{1}{\gamma_{\text{out}}(1+\beta_{\text{out}})} - \frac{1}{\gamma_{\text{in}}(1+\beta_{\text{in}})}\right),\\ &\simeq \sin\chi_{\text{in}} + \frac{m\omega c}{2eE_0}\left(\frac{1}{\gamma_{\text{out}}} - \frac{1}{\gamma_{\text{in}}}\right),\end{aligned} \tag{8.1.34}$$

and the change in γ is small relative to the initial value i.e., $\Delta\gamma = \gamma_{\text{out}} - \gamma_{\text{in}} \ll \gamma_{\text{in}}$ then the phase remains reasonably stable. In order to demonstrate this statement note that according to the single particle energy conservation, the energy change is

$$\Delta\gamma \simeq -\frac{eE_0 D}{mc^2}\cos\chi_{\text{in}}\,, \tag{8.1.35}$$

and therefore the shift in phase varies according to

$$\sin\chi_{\text{out}} - \sin\chi_{\text{in}} \simeq \frac{1}{\gamma_{\text{in}}^2}\left(\frac{\omega}{c}D\right)\cos\chi_{\text{in}}\,, \tag{8.1.36}$$

where D is the interaction length. For a 2 m structure, $f = 11.4$ GHz (SLAC) and $\gamma_{\text{in}} \sim 600$ the term in the right-hand side is on the order of 0.001.

Finally, (8.1.31) indicates that a bunch of finite (phase) spread is actually compressed in the acceleration process. For example consider a 20° bunch i.e., $130° > \chi_{\text{in}} > 110°$ while the other parameters are $\gamma_{\text{in}} = 2.0$, $f = 11.424$ GHz and $E_0 = 60$ MV/m. The asymptotic values ($\gamma \gg 1$) of the phase are $166° > \chi_{\text{out}} > 156°$ which is one half of the initial phase distribution. This calculation disregards the space-charge effect.

8.1.4 Transverse Effects: Panofsky-Wenzel Theorem

It was the term γ_{in} which was responsible for the fact that the bunch does not spread in phase and it is the same term which also helps us with regard to the transverse motion of the accelerated bunch. Until now we have considered only the longitudinal motion of the electrons assuming that the beam width is very small on the scale of the transverse variations of the electromagnetic field. However, its width is finite and the transverse components of the electromagnetic field may affect the bunch. Panofsky and Wenzel (1956) were the first to point out that the transverse kick on a relativistic bunch which traverses cavity is zero in the case of a symmetric TE mode and non-zero for a symmetric TM mode. In order to examine the effect we shall first adopt an intuitive approach [Palmer (1986)] followed by a generalized formulation of what is known as the Panofsky-Wenzel theorem.

Consider a particle moving with a velocity v_0 and a TM wave propagating at c. On axis the longitudinal electromagnetic wave is

$$E_z(r, z; \omega) = E_0 \mathrm{e}^{-j(\omega/c)z}, \tag{8.1.37}$$

and outside the bunch (but in its close vicinity) Maxwell equations imply

$$\frac{1}{r}\frac{\partial}{\partial r} r E_r(r, z; \omega) + \frac{\partial}{\partial z} E_z(r, z; \omega) = 0. \tag{8.1.38}$$

Solving for the radial component we obtain

$$E_r(r, z; \omega) = j\left(\frac{1}{2}\frac{\omega}{c} r\right) E_0 \mathrm{e}^{-j(\omega/c)z}, \tag{8.1.39}$$

and since $H_\phi = E_r/\eta_0$, the radial component of the force which acts on the envelope of the bunch is

$$\begin{aligned} F_r &= -e\Big\{E_r[r, z(t); t] - v_0 \mu_0 H_\phi[r, z(t); t]\Big\} \\ &= e\left(\frac{1}{2}\frac{\omega}{c} r\right) \frac{1}{(1+\beta)\gamma^2} E_0 \sin\chi, \end{aligned} \tag{8.1.40}$$

where $\chi = \chi_0 + \omega[t - z(t)/c]$. Expression (8.1.40) indicates that off-axis, the radial force is by a factor of γ^2 smaller than the longitudinal force. Let us, for the sake of simplicity, ignore variation in γ, that is to say that we examine the transverse motion in a relatively short section where the change in the energy of the electrons is small. For a relativistic particle, the radial motion is governed by

$$\left[\frac{\mathrm{d}^2}{\mathrm{d}t^2} - \frac{1}{\gamma^3}\frac{eE_0\omega \sin\chi}{4mc}\right] r = 0; \tag{8.1.41}$$

the azimuthal motion was neglected here. From this last expression we conclude that: *(i)* the radial motion scales as γ^{-3} therefore for highly relativistic particles in a symmetric TM mode propagating at c, the transverse motion is expected to be stable. *(ii)* If in the longitudinal direction the electromagnetic field is accelerating the particle i.e.,

$$F_z = -eE_0 \cos\chi \,, \tag{8.1.42}$$

is positive (say $\chi = 135°$), then the transverse motion is unstable. However for $\chi > 180°$ the sin function is negative and the motion is stable. Consequently, the phase domain where stable acceleration may occur is $180° < \chi < 270°$.

Now that we have a general feeling as to the transverse processes which occur when a bunch traverses an acceleration structure we shall introduce a systematic way to deal with the problem. In their original work, Panofsky and Wenzel (1956) defined the transverse momentum experienced by a relativistic bunch as it traverses an interaction of length D by

$$p_\perp = \frac{-e}{v_0}\int_0^D \mathrm{d}z\Big(\mathbf{E} + \mathbf{v_0}\times\mathbf{B}\Big)_{\perp, t=z/v_0} . \tag{8.1.43}$$

This definition takes into account the effect of the field generated by the particle on itself via the structure (the self field is obviously excluded). We have indicated that as a particle traverses a slow-wave structure or a cavity, it leaves behind a broad spectrum of electromagnetic waves – this is also referred to as electromagnetic wake-field [see Wilson (1988)]. Since this field may affect bunches trailing far behind the generating one, it is convenient to define the so-called transverse wake potential as

$$W_\perp(s,r) = -e\int_{-\infty}^{\infty} \mathrm{d}z\,[\mathbf{E} + \mathbf{v_0}\times\mathbf{B}]_\perp\,(r, z, t = (z+s)/v_0)\,, \tag{8.1.44}$$

which can be conceived as a generalization of (8.1.43). In an equivalent way one can define the longitudinal wake potential as

$$W_\mathrm{L}(s,r) = -e\int_{-\infty}^{\infty} \mathrm{d}z E_z\Big[r, z, t = (z+s)/v_0\Big] . \tag{8.1.45}$$

The Fourier transform of these two potentials

$$\begin{aligned} Z_\perp(\omega,r) &= \frac{j}{v_0 e^2}\int_{-\infty}^{\infty} \mathrm{d}s W_\perp(s,r)\mathrm{e}^{-j(\omega/v_0)s}\,, \\ Z_\mathrm{L}(\omega,r) &= \frac{1}{v_0 e^2}\int_{-\infty}^{\infty} \mathrm{d}s W_\mathrm{L}(s,r)\mathrm{e}^{-j(\omega/v_0)s}\,, \end{aligned} \tag{8.1.46}$$

determines the longitudinal and transverse impedances; several cases of interest were considered by Heifets and Kheifets (1990).

Since these two wake potentials previously introduced are determined by various components of the electromagnetic field, which are inter-dependent via Maxwell's equation, we may expect the two wake potentials to be also inter-dependent. The relation between the two can be shown based on Faraday law, assuming a steady state regime and having the symmetric TM mode in mind,

$$-\left[\frac{\partial}{\partial r}E_z - \frac{\partial}{\partial z}E_r\right] = -j\omega\mu_0 H_\phi\,. \tag{8.1.47}$$

We firstly calculate

$$\frac{\partial}{\partial r}W_{\rm L}(s,r) = -e\int_{-\infty}^{\infty} {\rm d}z \frac{\partial}{\partial r}E_z\Big[r,z,t=(z+s)/v_0\Big]\,, \tag{8.1.48}$$

and secondly,

$$\begin{aligned}\frac{\partial}{\partial s}W_\perp(s,r) &= -e\int_{-\infty}^{\infty} {\rm d}z \frac{\partial}{\partial s}\left[E_r - v_0\mu_0 H_\phi\right]\Big(r,z,t=(z+s)/v_0\Big)\,,\\ &= -e\int_{-\infty}^{\infty} {\rm d}z \left[j\frac{\omega}{c}\frac{1}{\beta}E_r(r,z,t) - j\omega\mu_0 H_\phi(r,z,t)\right]_{t=(z+s)/v_0}\,.\end{aligned} \tag{8.1.49}$$

Bearing in mind that

$$\begin{aligned}\frac{\partial}{\partial s}E_r[r,z,t=(z+s)/v_0] &= \left[\frac{\partial E_r}{\partial z}\right]_{t=(z+s)/v_0} + \frac{1}{c\beta}\left[\frac{\partial E_r}{\partial t}\right]_{t=(z+s)/v_0}\\ &= \left[\frac{\partial E_r}{\partial z}\right]_{t=(z+s)/v_0} + \left[\frac{1}{c\beta}j\omega E_r\right]_{t=(z+s)/v_0}\,,\end{aligned} \tag{8.1.50}$$

and assuming that for $z \to \pm\infty$ the transverse electric field vanishes, i.e., $E_r[r,z,t=(z+s)/v_0] \to 0$, we have

$$\frac{\partial}{\partial s}W_\perp(s,r) = -e\int_{-\infty}^{\infty} {\rm d}z \left[\frac{\partial E_r}{\partial z} - j\omega\mu_0 H_\phi\right]_{t=(z+s)/v_0}\,. \tag{8.1.51}$$

Adding (8.1.48,51) we obtain

$$\frac{\partial}{\partial r}W_{\rm L}(s,r) + \frac{\partial}{\partial s}W_\perp(s,r) = -e\int_{-\infty}^{\infty} {\rm d}z \left[\frac{\partial E_z}{\partial r} - \frac{\partial E_r}{\partial z} - j\omega\mu_0 H_\phi\right]_{t=(z+s)/v_0} \tag{8.1.52}$$

which by virtue of Faraday law (8.1.47) implies

$$\frac{\partial}{\partial s}W_\perp(s,r) = -\frac{\partial}{\partial r}W_{\rm L}(s,r)\,. \tag{8.1.53}$$

This relation is the formal notation of the Panofsky-Wenzel theorem phrased above. Equivalently, this theorem can be formulated in terms of the longitudinal and transverse impedances introduced in (8.1.46) as:

$$\frac{\partial}{\partial r} Z_{\mathrm{L}} = -\frac{\omega}{c} \frac{1}{\beta} Z_{\perp} . \tag{8.1.54}$$

Transverse gradients which may develop in accelerating structures may have a destructive effect on the beam dynamics. Already in the sixties it was observed at SLAC that the transmitted electron-pulse appears to shorten if the total current exceeds a certain threshold. This effect was attributed to a radial instability called beam break-up (BBU) which is due to the coherent interaction of the electrons with a hybrid mode, that is to say a mode which possesses properties of both TM and TE modes. In particular we can conceive a cavity in which the TM_{110} mode is excited. Longitudinal variations are ignored in this case ($\partial/\partial z \sim 0$) but we allow radial and azimuthal variations such that the non-zero component of the magnetic vector potential reads

$$A_z(r, \phi) = A J_1 \left(s_1 \frac{r}{R} \right) \cos \phi , \tag{8.1.55}$$

where s_1 is the first zero of the Bessel function of the first kind and first order i.e., $J_1(s_1) \equiv 0$. Consequently, the non-zero components of the electromagnetic field are

$$\begin{aligned} E_z(r, z) &= -j\omega A J_1 \left(s_1 \frac{r}{R} \right) \cos \phi , \\ B_r(r, z) &= -\frac{1}{r} A J_1 \left(s_1 \frac{r}{R} \right) \sin \phi , \\ B_\phi(r, z) &= -A \frac{s_1}{R} J_1' \left(s_1 \frac{r}{R} \right) \cos \phi , \end{aligned} \tag{8.1.56}$$

where $J_1'(u) = J_0(u) - J_1(u)/u$ and the eigen-frequency of this cavity is $\omega = s_1 c/R$. In the close vicinity of the axis the non-zero components are

$$\begin{aligned} E_z &\sim E \frac{x}{R} , \\ B_y &\sim \frac{1}{j\omega R} E . \end{aligned} \tag{8.1.57}$$

An electron which traverses this cavity experiences a deflection even if it is perfectly aligned because of the $\mathbf{v} \times \mathbf{B}$ term. Specifically, the change in the transverse momentum due to the excitation of this mode is

$$\Delta p_x = -e \int_0^D \mathrm{d}z \frac{1}{v_0} (-v_0 B_y) \sim \frac{e}{\omega} \frac{D}{R} \mathrm{Re}\,(-jE) . \tag{8.1.58}$$

Consistency with the prior assumptions forced us to assume a uniform field in the z direction – which is not the case in general but it is a reasonable approximation on the scale of a single cavity. Furthermore, from this expression we learn that the transverse deflection is proportional to the *longitudinal* electric field and therefore if we now consider a set of such coupled cavities, then the mode may grow in space and after a certain interaction region it dumps the beam to the wall. The corresponding mode is called hybrid electric magnetic mode and in our particular case it is HEM_{11}.

BBU can be divided into two different types: beam break-up which occurs in the scale of a single acceleration structure because of feedback (either due to backward-wave interaction or reflections) and then the condition for BBU occurs as the threshold condition is reached – as in an oscillator. This is called *regenerative* type of beam break-up. The other type, the *cumulative* beam break-up, is carried by the beam and it occurs on the scale of many acceleration sections (which are electromagnetically isolated). Panofsky and Bander (1968) developed based on Panofsky-Wenzel theorem a model which fits the basic features of long range BBU occurring on the scale of many acceleration sections. A good tutorial of the various BBU mechanisms has been given by Helm and Loew (1970) and more recently Lau (1989) has proposed a framework from which the various BBU regimes can be readily derived (see therein for reference to more recent work on BBU).

8.2 Advanced Accelerator Concepts: Brief Overview

The need for more powerful accelerators has triggered an extensive search for various other schemes to accelerate electrons. These schemes rely on either entirely new concepts as is the case in the plasma beat wave accelerator or on new technology which is the case in the laser wake-field accelerator. An intermediate approach was adopted by Henke (1994) of the Technische Universität of Berlin: one can employ ideas from radio frequency linacs in the millimeter wave regime $\sim$ 100GHz. Since the geometry is now below the millimeter range, the structures are built by means of techniques developed in the electronic industry for very large scale integrated (VLSI) circuits. At the bottom line all schemes are tested by their ability to generate a gradient in the longitudinal direction with the minimal possible transverse wake potential. Although energy is the major parameter of interest in accelerators, it is not the only one. Other parameters such as emittance, repetition rate and number of particles per bunch are also of great importance. Therefore the test of each one of the methods which will be discussed in the following sections is not only by the gradient which they are capable of generating but also in the potential of being incorporated in a large system taking into consideration all the other parameters mentioned above. At this stage of research the zero order parameter of comparison remains the accelerating gradient and this will be the basis for our comparison.

The various schemes can be divided according to several criteria. For example: is the accelerated electron in the close vicinity of a (metallic) structure or not? In the first case, the accelerating wave is evanescent and since the range of action of these waves is short this is called near-field scheme whereas the other is the far-field scheme. Another criterion is whether the rf accelerating field is continuous or comes in one pulse; as an example consider the various wake-field schemes vs. the long pulse machine like the present linacs or the inverse FEL, inverse Cerenkov etc.

An accelerator can be conceived as a transformer in the sense that high current – low voltage beams form the primary and low current - high voltage constitutes the secondary. Consequently, another criterion for examining the various acceleration schemes is the number of beams in the primary. In the present SLAC system there are many klystrons and correspondingly many beams however, in the proposed two beam acceleration scheme only one beam forms the primary. In the remainder of this chapter we shall briefly discuss the various advanced acceleration concepts.

8.3 Wake-Field Accelerator

The first group of advanced concepts we shall consider belongs to the "time domain" or short pulse schemes. It includes the *(i)* Dielectric Wake-Field Accelerator (DWFA), *(ii)* Periodic Structure Wake-Field Accelerator and *(iii)* Laser Wake-Field Accelerator. What is common to all is the fact that there is no one single frequency wave which accelerates the electron but rather an entire spectrum. Acceleration is achieved by "synchronizing" the radiation pulse and the accelerated bunch such that the latter trails behind and it sees an accelerating gradient.

8.3.1 Dielectric Wake-Field Accelerator

In order to understand the principles of the DWFA it is convenient to go back to Sect. 2.4.2 where we examined the Cerenkov radiation emitted by a point charge as it traverses a dielectric loaded waveguide along its axis. We found that provided the velocity of the particle is higher than the phase velocity of the plane wave in the medium, then the magnetic vector potential is given by

$$\begin{aligned} A_z(r,z,t) = &- \frac{q}{\pi\varepsilon_0 R^2}\frac{\beta^2}{n^2\beta^2-1}\sum_{s=1}^{\infty}\frac{J_0(p_s r/R)}{J_1^2(p_s)|\Omega_s|} \\ &\times \sin\Big[|\Omega_s|(t-z/v_0)\Big]h(t-z/v_0)\,; \end{aligned} \tag{8.3.1}$$

see (2.4.48) and the corresponding definitions. It is interesting to note that although the waveguide is dispersive, all the electromagnetic waves which belong to the wake travel at the particle's velocity v_0 – though they may trail far behind. It also explains why a "broad" spectrum signal can still provide net acceleration.

With this expression we can calculate the longitudinal electric field acting on a test particle which lags behind - see basic configuration in Fig. 8.3. On axis E_z is

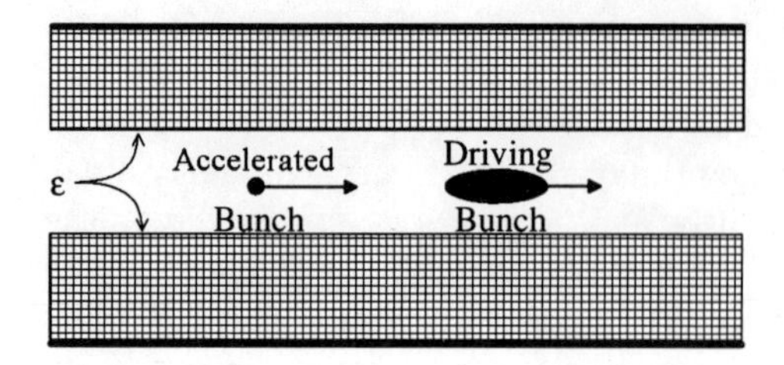

Fig. 8.3. Conceptual set up of the dielectric wake-field accelerator

$$E_z(r=0,z,t) = -\frac{q}{\pi\varepsilon_0 R^2}\sum_{s=1}^{\infty}\frac{1}{J_1^2(p_s)}\cos\Big[|\Omega_s|(t-z/v_0)\Big]h(t-z/v_0)\,, \tag{8.3.2}$$

and since in practice the dielectric coefficient is frequency dependent the summation is only on these modes for which the Cerenkov radiation is satisfied i.e., $\beta^2 > 1/\varepsilon(\omega = \Omega_s)$. Let us assume for simplicity that only the first two modes contribute, thus the normalized force which acts on a negative point charge e is

$$\begin{aligned}\mathcal{F}(\tau \equiv t - \frac{z}{v_0} > 0) &\equiv F_z\left[\frac{eq}{4\pi\varepsilon_0\varepsilon R^2}\right]^{-1}\\ &= \frac{4}{J_1^2(p_1)}\cos\Big(\frac{p_1 c\tau}{R\sqrt{\varepsilon-\beta^{-2}}}\Big)\\ &\quad+\frac{4}{J_1^2(p_2)}\cos\Big(\frac{p_2 c\tau}{R\sqrt{\varepsilon-\beta^{-2}}}\Big)\,,\end{aligned} \tag{8.3.3}$$

and it is plotted in Fig. 8.4. As anticipated, in the close vicinity of the particle, the force is decelerating since "naturally" the negative charge repels another negative charge. However, if the test particle is located adequately behind the leading bunch, the trailing one will be accelerated. This field distribution also helps us to envision the bunch *compression* which is a byproduct of this process. For this purpose consider a uniform distribution of particles which spread between the zero acceleration point and the crest of the wave. Electrons which experience zero acceleration preserve their relative location in the bunch while all the others are accelerated. Even if the accelerated electrons bypass the first group they immediately reach a deceleration region which pulls them back. Obviously, this process is limited by space charge effects which were disregarded in this discussion.

Conceptually, one can regard the system as a transformer with a low voltage and high current (say 10 MV, 1 kA) primary and the secondary is a high-voltage pulse of low current (say 1 GV, 1 A). DWFA was tested experimentally at Argonne National Laboratory by Gai (1988) and analyzed theoretically by Rosing (1990).

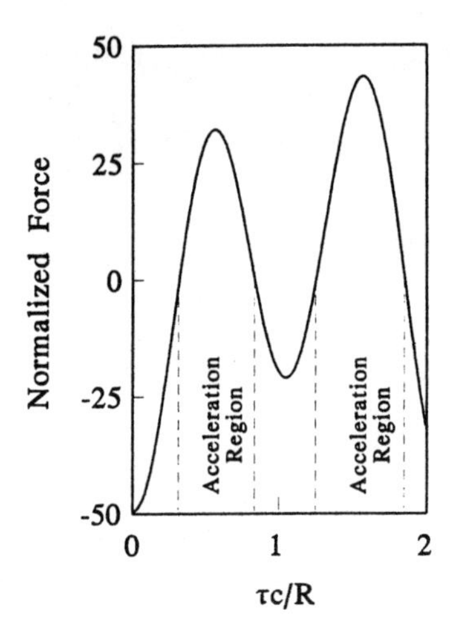

Fig. 8.4 Normalized force on a test particle which lags behind the driving bunch

8.3.2 Periodic Structure Wake-Field Accelerator

The concept in this case is very similar to the case of the dielectric structure namely, a driving pulse generates a wake in the periodic structure which in turn accelerates the trailing bunch. Mathematical complexity of the calculations involved is substantially higher because the boundary conditions, however the outcome is similar to the dielectric case. Voss and Weiland (1982) (at DESY, Germany) suggested an annular configuration that is to say that the driving beam forms a ring which excites a wake-field. The latter propagates toward the axis and in the process its amplitude increases. As it reaches the axis it accelerates a trailing bunch. Figure 8.5 illustrates this concept.

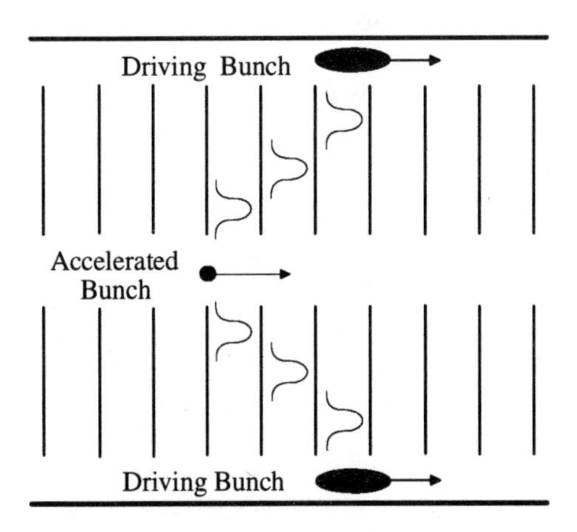

Fig. 8.5 Conceptual set up of the periodic structure wake-field accelerator

A different implementation of the same concept involves electro-optic switches: acceleration of a bunch requires a gradient at the (momentary)

location of the particle. The way this gradient is accomplished has no importance to the longitudinal motion however, the transverse wake may vary from one scheme to another. Progress in optical switching of semiconductor devices [Lee (1984)], facilitates the generation of fast voltage pulses which in turn can be used for acceleration. The essence of this concept is to optically switch a radial transmission line connected to a relatively high voltage source and benefit from the transformer effect as the voltage pulse propagates inwards to accelerate the electrons on axis. Recently, the concept was experimentally demonstrated by Bamber (1993) at the University of Rochester. More detailed reviews of periodic wake-field in periodic structures can be found in papers by Wilson (1988) and Cooper (1988).

8.3.3 Laser Wake-Field Accelerator

In the previous cases the wake-field was generated by a bunch of electrons injected in either a dielectric or a periodic structure and the accelerated bunch had to be injected with great precision at the location where the wake would accelerate it. In other words, the accelerated bunch sees a positive charge moving in its front and it is accelerated by it. It is possible to generate this situation in plasma. Consider a very intense laser pulse which propagates in the plasma. Its electric field is transverse to the direction of propagation, therefore it will tend to remove the electrons outwards leaving the positive charge in place since, on the time scale of the electrons' motion, the ions are motionless because of their large mass. Furthermore, the laser pulse virtually propagates at c, and the (relativistic) trailing bunch sees a continuous attractive force. The duration of the laser pulse is on the order of one period therefore it has to be very intense since it has to contain all the energy to be delivered to the accelerated bunch. Such lasers have been developed in the last decade [Strickland and Mourou (1985)] and intensities beyond 10^{18} W/cm^2 at $\lambda = 1$ μm for less than 1 psec are available. This energy flux translates into transverse electric field of

$$E_{\rm tr}\,[\mathrm{TV/m}] \sim 3\times 10^{-9}\sqrt{I\,[\mathrm{W/cm^2}]}\,, \tag{8.3.4}$$

therefore, at the intensity (I) mentioned above, the transverse electric field is of the order of 3 TV/m! This large gradient can not directly accelerate a bunch of electrons since it is transverse to their direction of motion. However, it is perfectly suited to generate the wake described above. For the parameters mentioned above Sprangle (1990) calculated the accelerating gradient in a plasma with $n_0 \sim 10^{23}\,\mathrm{m}^{-3}$ to be 12 GV/m which is less than 0.5% of the transverse field. Although this is a very large gradient, the energy gained by the electrons is limited (roughly) by the product gradient and interaction length. The latter is determined primarily by diffraction: $L_{\rm int} \simeq \pi D_{\rm R}$ where Rayleigh length is given by $D_{\rm R} = \pi r_{\rm L}^2/\lambda$. If we take the minimum laser beam waist ($r_{\rm L}$) to be 30μm, then the interaction length is about 9 mm and the gain of kinetic energy is 100 MeV which is comparable with what we would expect to achieve from one module of the Next Linear Collider (in 2 m). Sprangle (1993) has shown that the situation can be dramatically improved by optically guiding the laser beam. The essence of the mechanism is as follows: large transverse electric field brings the plasma electron from rest to relativistic velocities in a fraction of the laser pulse. This affects the density and consequently, the refractive coefficient: i.e., electrons are removed from regions of high gradients towards the envelope where they are denser. This relativistic effect may dominate the propagation characteristics of the laser beam provided that the laser power exceeds a critical value estimated to be $P_{\rm cr} \sim 17(\omega/\omega_{\rm p})^2$ GW. Using this method [Sprangle (1993)] it has been shown in simulations that electrons can be accelerated from 2 MeV to 1 GeV in 20 cm corresponding to an average gradient of 5 GV/m.

8.4 Plasma Beat-Wave Accelerator

Plasma beat-wave acceleration scheme was initially proposed by Tajima and Dawson (1979). It uses two laser pulses of different wavelength λ_1 and λ_2 to illuminate a plasma. The density of the latter is set such that its characteristic frequency ($\omega_{\rm p}$) equals the difference between the two laser lines i.e., $\omega_2 - \omega_1 = \omega_{\rm p}$. The two waves determine a space charge wave whose wavenumber is $k_{\rm p} = k_2 - k_1$ and which propagates at a phase velocity $v_{\rm ph} = \omega_{\rm p}/k_{\rm p} = c[1 - (\omega_{\rm p}/\omega)^2]^{1/2}$; this is very close to c. When electrons are injected with a velocity close to that of the space charge wave (beat wave) they can be trapped and therefore accelerated. Since no external walls are involved there is no problem of breakdown; however, it does not mean that the gradients are "infinite" since once the density modulation becomes significant (larger than 10%) non-linear effects dominate and the acceleration is altered. In order to have an idea as to the typical gradients achievable we quote here the numerical example presented by Joshi (1993): the longitudinal electric field which develops in a plasma as a result of a density modulation $\delta n/n_0$ where n_0 is the density of the unperturbed background plasma is $E \sim 100\sqrt{n_0}\delta n/n_0$ (V/m). If we take a 10% modulation on a background density of $10^{21} < n_0(m^{-3}) < 10^{23}$ then the achievable gradients are between 0.3 to 3 GV/m.

The number of plasma periods which will support acceleration is limited due to the occurrence of instabilities. Many of these can occur but if we limit the discussion only to ion instabilities then we can have a reasonable estimate of the number of periods along which the coherent process can be sustained. If we consider the case of hydrogen plasma, then the reaction time of the ions will be inversely proportional to their mass. Consequently, the number of periods will be as the ratio of the plasma frequency ratio i.e $\omega_{p,\rm elec}/\omega_{p,\rm ion} = \sqrt{M_{\rm ion}/m} \sim 43$. Fortunately, the electrons are synchronous with the beat-wave therefore the instabilities lag behind and the interaction length is actually higher.

Another difficulty which might be raised is the scattering of the accelerated electrons with the background plasma. Apparently the dominant scattering mechanism for over MeV electrons, is scattering by the plasma nuclei. Estimates made at UCLA [Katsouleas and Dawson (1989)] indicate that even at a density of 10^{26} m^{-3} the mean free path of a relativistic electron (few MeV) is 2 km and it increases with the energy of the accelerated electrons. Unfortunately, diffraction of laser beam limits the interaction length to many orders of magnitude shorter distances and the Rayleigh range as discussed by Joshi (1993) is of the order of 1 cm. This can be compensated by laser self-focusing in plasma and the price to be paid is a threshold in the laser power to be injected. According to Katsouleas and Dawson (1989) this is given by $P > P_{\mathrm{th}} \simeq 5(\omega/\omega_{\mathrm{p}})^2 \mathrm{GW}$.

8.5 Inverse of Radiation Effects

In all the effects where coherent radiation is generated by bunches of electrons, these are located such that the field which acts at their location is decelerating them. In principle, we can locate a similar bunch to be in anti-phase in which case the electrons are accelerated by an applied wave.

8.5.1 Inverse FEL

In principle the same mechanism which facilitates generation of coherent radiation in a periodic transverse wiggler as discussed in Chap. 7, allows to accelerate a bunch of electrons. In fact, Palmer (1972) suggested this concept several years before the renewed interest in the free electron laser as a radiation source [Elias (1976)]. An intense laser pulse interacts with a beam of electrons in the presence of a transverse and periodic magnetic field. As a result, electrons may be accelerated – see Courant, Pellegrini and Zakowicz (1985). The advantages and disadvantages of the free electron lasers as a radiation source, discussed in Chap. 7, apply also to its operation as an accelerator. In addition, Pellegrini (1982) pointed out that the decelerating gradient (E_{dec}) due to the emission of spontaneous radiation has to be smaller than the accelerating gradient associated with the laser field (E_{acc}). According to (7.1.30) and (7.2.6) this can be formulated as $E_{\mathrm{acc}} > E_{\mathrm{dec}}$ thus

$$\omega A_0 \frac{1}{\gamma} \frac{eB_{\mathrm{w}}}{mck_{\mathrm{w}}} > \frac{1}{2}\beta\gamma^2 \frac{e}{4\pi\varepsilon_0/k_{\mathrm{w}}^2} \left(\frac{eB_{\mathrm{w}}}{mck_{\mathrm{w}}}\right)^2 . \tag{8.5.1}$$

This expression determines a critical laser intensity I_{cr} which has to be exceeded in order to obtain net acceleration i.e.,

$$I \equiv \frac{1}{2\eta_0}|\omega A_0|^2$$
$$> I_{\rm cr} \equiv \left[\frac{1}{2\eta_0}\left(\frac{e}{4\pi\varepsilon_0 k_{\rm w}^{-2}}\right)^2\right]\left(\frac{eB_{\rm w}}{2mck_{\rm w}}\right)^2 \gamma^6 . \tag{8.5.2}$$

If we take $B_{\rm w} = 1{\rm T}$ and $2\pi/k_{\rm w} = 10\,{\rm cm}$ then $I_{\rm cr} \sim 10^{-16}\gamma^6{\rm W/cm}^2$ and for a typical γ on the order of 10^4 the required laser intensity is $10^8\,{\rm W/cm}^2$ as shown next for several other energies:

$$\begin{aligned}
\gamma &= 10^4 \rightarrow & I_{\rm cr} &= 10^8\,{\rm W/cm}^2 , \\
\gamma &= 10^5 \rightarrow & I_{\rm cr} &= 10^{14}\,{\rm W/cm}^2 , \\
\gamma &= 10^6 \rightarrow & I_{\rm cr} &= 10^{20}\,{\rm W/cm}^2 , \\
\gamma &= 10^7 \rightarrow & I_{\rm cr} &= 10^{26}\,{\rm W/cm}^2 .
\end{aligned} \tag{8.5.3}$$

This list indicates that with laser intensities of $10^{20}\,{\rm W/cm}^2$ one may accelerate electrons up to a few TeV but at least with the present technology, this seems to be the limit.

8.5.2 Inverse Cerenkov

Edighoffer (1981) at Stanford demonstrated experimentally the feasibility of the inverse Cerenkov effect for acceleration of electrons. The idea here is to illuminate an electron moving in a dielectric medium (gas) at the Cerenkov angle with a laser beam at the adequate frequency. Later Fontana and Pantell (1983) proposed an improved setup for the same purpose by ensuring an extended and symmetric interaction region with axicon lens – see Fig. 8.6. The lens generates a symmetric longitudinal electric field on axis and the gas slows down the electromagnetic wave which in turn, intersects the electron trajectory at an angle θ; the longitudinal wavenumber is $(\omega/c)n\cos\theta$ where n is the refraction coefficient. Consequently, the resonance occurs when the phase velocity equals the velocity of the particle i.e.,

$$v_{\rm ph} = \frac{c}{n\cos\theta} = v_0 . \tag{8.5.4}$$

Design of such a system is a trade-off between a (relatively) large refractive coefficient which requires high pressure and a long mean free path which implies low pressure. A possible solution, suggested in this context by Steinhauer and Kimura (1990) [previously suggested in the context of FEL by Feinstein (1986)], is to operate close to the resonant frequencies of the gas. This facilitates the required refraction coefficient but at low pressure. Recently, the concept was demonstrated experimentally [Kimura (1995)] when electrons accelerated by a CO_2 laser focused by an axicon lens have gained an energy which corresponds to a gradient of 31 MV/m in the interaction region.

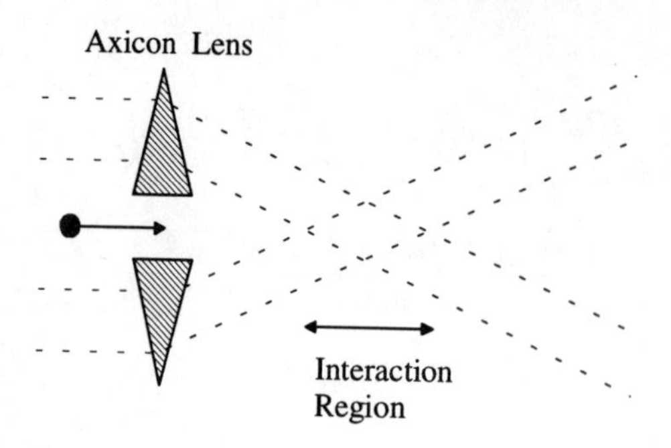

Fig. 8.6 Conceptual set-up of the inverse Cerenkov accelerator

8.5.3 Open Structure Accelerator

In Chap. 5 we discussed the electromagnetic characteristics of an open periodic structure and two main conclusions were emphasized: *(i)* the number of eigen-modes is finite and their number is controlled by the geometry of the structure. Each such mode consists of an infinite spectrum of harmonics whose phase velocity in the pass band is smaller than c; these harmonics correspond to evanescent waves and they do not carry power in the transverse direction. Another conclusion we reached was that *(ii)* a particle moving in the proximity of an open structure emits radiation (Smith-Purcell effect). In principle, one can use this effect to accelerate electrons by illuminating the grating at the adequate angle and wavelength. However, contrary to the inverse Cerenkov effect, where the use of the radiation field in the interaction region can be fairly efficient, in the grating case, the incident wave is scattered in a spectrum of harmonics, part of which are radiative therefore a substantial fraction of the energy is lost.

Alternatively, we may use the eigen-modes of an open structure to accelerate electrons [Palmer (1982) and Kroll (1985)]. However, one can immediately realize that a wave with phase velocity c is not supported by the kind of symmetric structure we discussed in Chap. 5 since if $k_z = \omega/c$ is parallel to the direction in which the electron moves then at least one harmonic of the mode does not decay exponentially. A simple solution of this problem is to enforce field variation in the third direction. Pickup (1985) has analyzed a grating which is periodic in the z direction, the y direction is perpendicular to the surface and in the x direction two metallic plates were placed at a distance D one from the other. Therefore, if the grating is designed such that in the z direction $k_z = \omega/c$ and in the x direction $k_x = \pi/D$ (lowest mode) then according to the homogeneous wave equation, $k_x^2 + k_y^2 + k_z^2 - (\omega/c)^2 = 0$, we have

$$\frac{\pi^2}{D^2} + k_y^2 + \frac{\omega^2}{c^2} - \frac{\omega^2}{c^2} = 0 \quad \rightarrow \quad k_y^2 = -\frac{\pi^2}{D^2} \,. \tag{8.5.5}$$

Consequently, the wave decays exponentially perpendicular to the grating.

Although the problem of the radiation confinement was solved there still is a problem of stability of the beam. To illustrate this problem let us consider the following electric field

$$\begin{aligned} E_x(\mathbf{r},t) &= 0\,, \\ E_y(\mathbf{r},t) &= E\frac{\omega}{cq}\sin\left[\omega(t-z/c)\right]\sin(qx)\mathrm{e}^{-qy}\,, \\ E_z(\mathbf{r},t) &= E\cos\left[\omega(t-z/c)\right]\sin(qx)\mathrm{e}^{-qy}\,, \end{aligned} \tag{8.5.6}$$

where $q = \pi/D$. We can substitute in Maxwell's equation and obtain the magnetic field:

$$\begin{aligned} H_x(\mathbf{r},t) &= \frac{q^2-(\omega/c)^2}{\omega\mu_0 q}E\sin\left[\omega(t-z/c)\right]\sin(qx)\mathrm{e}^{-qy}\,, \\ H_y(\mathbf{r},t) &= \frac{q}{\omega\mu_0}E\sin\left[\omega(t-z/c)\right]\cos(qx)\mathrm{e}^{-qy}\,, \\ H_z(\mathbf{r},t) &= \frac{1}{\eta_0}E\cos\left[\omega(t-z/c)\right]\cos(qx)\mathrm{e}^{-qy}\,. \end{aligned} \tag{8.5.7}$$

With these field components we can calculate the transverse force which acts on the particle. For this purpose we assume that the particle's motion is around $x \simeq D/2 + \delta x$ and $y = 0 + \delta y$ consequently,

$$\begin{aligned} &\left[\frac{\mathrm{d}^2}{\mathrm{d}t^2} + \frac{qeE\sin\phi}{m\gamma}\right]\delta x = 0\,, \\ &\left[\frac{\mathrm{d}^2}{\mathrm{d}t^2} - \frac{qeE\sin\phi}{m\gamma}\frac{qc}{\omega}\right]\delta y = -\frac{eE}{m}\frac{qc}{\omega}\sin\phi\,, \\ &\frac{\mathrm{d}}{\mathrm{d}t}\gamma = -\frac{eE}{mc}\cos\phi\,; \end{aligned} \tag{8.5.8}$$

here it was assumed that on the scale of the transverse motion variations in γ can be ignored, $\gamma \gg 1$ and $\phi = \omega[t - z(t)/c]$. The first two expressions indicate that even if, in the x direction, the motion is stable in the y direction the beam diverges and vice versa. A solution of this problem was suggested by Pickup (1985) and consists of rotating the orientation of the grating relative to the axis as illustrated in Fig. 8.7. Alternatively, the phase ϕ can be switched periodically as suggested by Kim and Kroll (1982).

8.5.4 PASER: Particle Acceleration by Stimulated Emission of Radiation

In Sect. 2.4.4 we calculated the decelerating force which acts on a single electron as it moves in a vacuum channel surrounded by lossy material. At the end of that section we indicated that if the conductivity of the material is negative which is to say that the medium is active, then the moving electron is accelerated [Schächter (1995)].

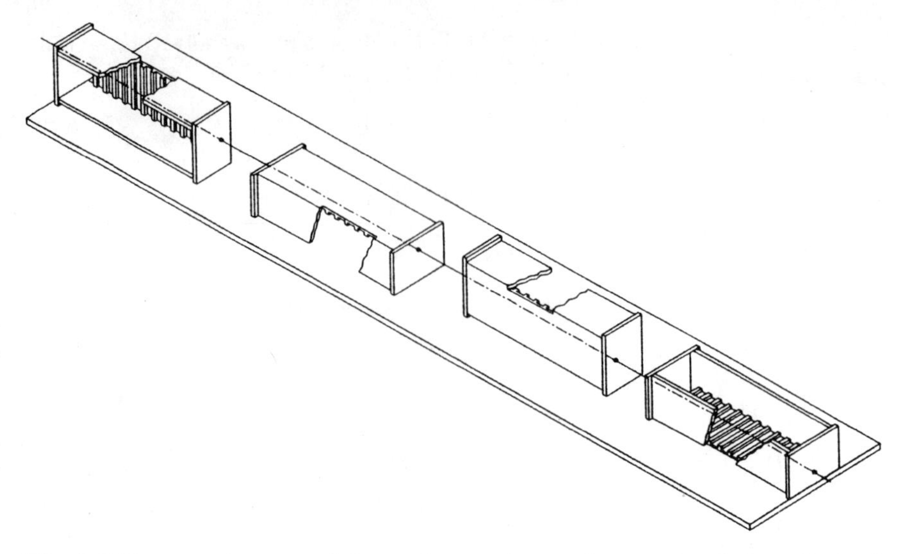

Fig. 8.7 Conceptual set up of the grating accelerator (inverse Smith-Purcell effect)

An additional way to examine this scheme is to consider the microscopic processes. Attached to a moving charge there is an infinite spectrum of evanescent waves; these can be viewed as a spectrum of virtual photons continuously emitted and absorbed by the electron. These photons impinge upon the excited atom which is conceived here as a two level system in its upper state. Since the spectrum of waves attached to this particle includes the resonance frequency of the medium, a photon with the adequate energy may stimulate the atom. As a result, two correlated photons are emitted: one is virtual as the initial one and the other is a real photon. Since the two are practically identical, the real photon is absorbed by the moving electron causing the latter's acceleration. The inverse process is also possible: if the virtual photon encounters an atom in the ground state and excites it, the moving electron loses energy – thus it is decelerated. We may expect net acceleration only if the number of atoms in the excited state is larger than those in the lower state i.e., the population is inverted. From the description above the acceleration force is a result of stimulated radiation; therefore this scheme was called PASER which stands for Particle Acceleration by Stimulated Emission of Radiation.

Interaction of a moving "macro-particle" with a stationary two-state quantum system which consists of either atoms or molecules can be considered within the framework of the macroscopic (and scalar) dielectric coefficient. This coefficient is given by

$$\varepsilon(\omega) = 1 - \chi \frac{(\omega - \omega_0)T_2 + j}{1 + \xi^2 + (\omega - \omega_0)^2 T_2^2}, \tag{8.5.9}$$

and it is tacitly assumed that the transients at the microscopic level are negligible. The macro-particle moves along a vacuum channel "bored" in an otherwise infinite dielectric medium of radius R_c and which is much larger than the radius of the macro-particle. In the expression above $\omega_0/2\pi$ is the resonant frequency of the medium, T_2 is the line width, $\chi = \mu^2 \Delta N T_2/\varepsilon_0 \hbar$ is the normalized population inversion (and it is negative in this case); μ is the atom's dipole moment, $\Delta N \equiv N_1 - N_2$ is the density of the population difference – subscript 1 represents the lower energy state and subscript 2 the higher one. Changes in the population difference due to energy transfer is considered here through the saturation term $\xi^2 = (E/E_{\rm sat})^2$; E is the amplitude of the acting electric field and the saturation field is given by $E_{\rm sat} = \hbar/\mu\sqrt{\tau T_2}$ where τ is the characteristic time in which the population reaches its equilibrium state.

Based on the model presented above the electromagnetic field in the entire system is calculated following the same approach presented in Sects. 2.4.3–4. Firstly it is convenient to simplify this model further. The dielectric coefficient in (8.5.9) is unity at resonance and the deviation of the real part, $[\mathrm{Re}(\varepsilon_{\rm r}-1)]$, is anti-symmetric relative to resonance; furthermore, it vanishes far away from this point. Consequently we approximate the dielectric coefficient with one whose real part is unity at all frequencies and its imaginary part is constant in a window of frequencies around resonance. The width of this window is determined by T_2 and is determined such that the area of this window is identical with that calculated from (8.5.9). Explicitly $\varepsilon_{\rm r}(\omega) \simeq 1 - j\bar{\sigma}(\omega)$ and

$$\bar{\sigma}(\omega) \equiv \bar{\sigma}_0 \begin{cases} 1 & \text{for } |\omega - \omega_0| T_2 < \pi/2\ , \\ 0 & \text{for } |\omega - \omega_0| T_2 > \pi/2\ , \end{cases} \tag{8.5.10}$$

where $\bar{\sigma}_0 = \chi/(1+\xi^2)$.

With this definition the normalized impedance reads

$$\zeta(x) = \frac{1}{1 - j\bar{\sigma}(x)} \sqrt{1 + j(\gamma\beta)^2 \bar{\sigma}(x)} \frac{\mathrm{K}_0\left[x\sqrt{1 + j(\gamma\beta)^2 \bar{\sigma}(x)}\right]}{\mathrm{K}_1\left[x\sqrt{1 + j(\gamma\beta)^2 \bar{\sigma}(x)}\right]}\ . \tag{8.5.11}$$

We consider the relativistic case such that $(\gamma\beta)^2|\sigma_0| \gg 1$ moreover at the typical frequencies of interest we assume that $\omega_0 R_{\rm c}|\sigma_0|^{1/2}/c \ll 1$. Consequently, the normalized impedance function is

$$\zeta(x) \simeq jx\gamma^2 \bar{\sigma}(x)\,. \tag{8.5.12}$$

This implies that $|\zeta| = \gamma^2|\bar{\sigma}|$ and the phase of ζ is given by

$$\psi = \begin{cases} \pi/2 & \text{for } \bar{\sigma}_0 > 0\,, \\ 0 & \text{for } \bar{\sigma}_0 = 0\ , \\ -\pi/2 & \text{for } \bar{\sigma}_0 < 0\,, \end{cases} \tag{8.5.13}$$

where we assumed that $|\bar{\sigma}_0| < 1$ thus $\psi = (\pi/2)\mathrm{sgn}(\bar{\sigma})$. The normalized gradient according to (2.4.78) is

$$\mathcal{E} = \frac{2}{\pi}\int_0^\infty dx \frac{x\gamma^2\bar{\sigma}(x)}{\mathrm{I}_0^2(x) + x^2\gamma^4\bar{\sigma}^2(x)\mathrm{I}_1^2(x)}. \tag{8.5.14}$$

In practice, for highly relativistic particles or explicitly for $\omega_0 R_c/c \ll \gamma$, the main cotribution to the integral is from the small values of x thus the argument of the modified Bessel functions is small and consequently we can approximate $\mathrm{I}_0(x) \simeq 1$ and $\mathrm{I}_1(x) \simeq x/2$, thus

$$\mathcal{E} = \frac{2}{\pi}\int_{x_-}^{x_+} \frac{dx\, x\gamma^2\bar{\sigma}_0}{1 + (x^2\gamma^2\bar{\sigma}_0/2)^2}, \tag{8.5.15}$$

where $x_\pm = R_c(\omega_0 \pm \pi/2T_2)/c\beta\gamma$. Analytic evaluation of this integral is possible and the result is

$$\mathcal{E} = \frac{2}{\pi}\left[\mathrm{atan}\left(\frac{1}{2}\bar{\sigma}_0\gamma^2 x_+^2\right) - \mathrm{atan}\left(\frac{1}{2}\bar{\sigma}_0\gamma^2 x_-^2\right)\right]. \tag{8.5.16}$$

Substituting the explicit expression for $x_\pm$ and assuming that $\bar{\sigma}_0(\omega_0 R_c/c)^2 \ll 1$, we can approximate the trigonometric functions with their arguments thus

$$\begin{aligned}\mathcal{E} &\simeq \frac{2}{\pi}\left(\frac{1}{2}\gamma^2\bar{\sigma}\right)\left(\frac{R_c}{c\beta\gamma}\right)^2\left[\left(\omega_0 + \frac{\pi}{2T_2}\right)^2 - \left(\omega_0 - \frac{\pi}{2T_2}\right)^2\right], \\ &\simeq 2\bar{\sigma}\left(\frac{\omega_0}{c}R_c\right)\frac{R_c}{cT_2}.\end{aligned} \tag{8.5.17}$$

Using the explicit expressions for $\mathcal{E}$ and χ we have

$$\begin{aligned}E_z(z = v_0 t, t) &= \frac{e}{2\pi\varepsilon_0}\frac{\omega_0}{c}\frac{1}{cT_2}\chi \\ &= \frac{e}{2\pi\varepsilon_0}\frac{\omega_0}{c}\frac{\mu^2}{c\varepsilon_0\hbar}\Delta N.\end{aligned} \tag{8.5.18}$$

This expression reveals that the particle is accelerated in case of population inversion ($\Delta N < 0$), it is decelerated by a passive medium ($\Delta N > 0 \rightarrow$ e.g., Cerenkov effect) and in vacuum ($\Delta N = 0$ – see (8.5.9)) the particle moves without being affected. To realize the potential of this effect we shall now give a rough estimate of the gradients which may develop in the system. For this purpose consider a medium of Ar^+ which lases close to $\lambda \simeq 0.5\,\mu$m and we shall assume $|\Delta N| \simeq 3 \times 10^{19}\mathrm{cm}^{-3}$ which is achievable at high pressure. The dipole moment (μ) is determined at the atomic level therefore it is anticipated to be of the order of $\mu \simeq (1.6 \times 10^{-19}\,\mathrm{C})\times(1\,\text{Å}) = 1.6 \times 10^{-29}\mathrm{C\,m}$. The gradient which develops, if the bunch consists of 10^6 electrons, is on the order of 1 GV/m.

8.6 Two Beam Accelerator

Conceptually the linear accelerator consists of many sections of acceleration structures each one fed by one or more klystrons. Each klystron in turn is driven by an electron beam generated separately therefore thousands of beams form the primary and they accelerate a single beam which is the secondary of a large "transformer". There might be several advantages if all these discrete beams are replaced by a *single driving beam* carrying all the required energy. Sessler (1982) has initially suggested this concept. It came in parallel to the substantial progress in understanding the operation of the free electron laser. In particular, the fact that electrons could be trapped [see Sect. 7.4 as originally shown by Kroll, Morton and Rosenbluth (1981)] and their energy extracted without substantial energy spread, suggested that after extraction in an FEL, the electrons could be re-accelerated. In addition, the operation of a klystron beyond X-band becomes problematic because of the small structure required whereas the free electron laser can generate high power levels at high frequencies without inherent structure limitations but with substantial constraint upon the beam quality.

This original approach of the so-called Two-Beam Accelerator (TBA) as proposed in the United States, see Fig. 8.8, was to start with a medium-energy (3 MeV) high-current (1 kA) beam, extract power in each segment and compensate the driving beam for the lost energy in a re-acceleration unit. Thus each section consists of three units: the acceleration unit, extraction unit and re-acceleration unit [Hübner (1993)].

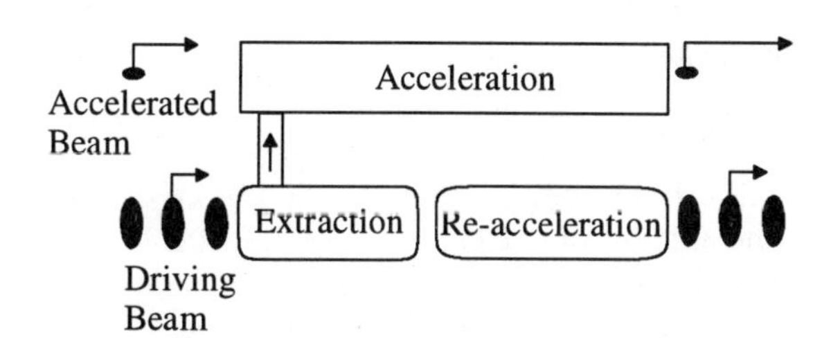

Fig. 8.8 Conceptual set up of the two-beam accelerator as proposed by Sessler (1982)

In Europe, at CERN, the approach is somewhat different [Schnell (1991)]: the initial energy of the electrons is three orders of magnitude higher, in the GeV range, and at least in the preliminary experimental stages (2×250 GeV) no re-acceleration is planned as illustrated in Fig. 8.9. In the final system (2 TeV) a few superconducting re-acceleration cavities are included in the design. Traveling-wave structures are planned to extract electromagnetic power on the order of 40 MW from a prebunched beam at 30 GHz. In order to have the correct perspective of the performance of each section we should note that the 40 MW of power generated at 30 GHz produce almost the same gradient as 400 MW at 11.4 GHz as is the case in the Choppertron [Haimson

(1992)]. In spite of the clear advantage of operation at high frequency with regard to the acceleration gradient, the wake-fields are correspondingly high.

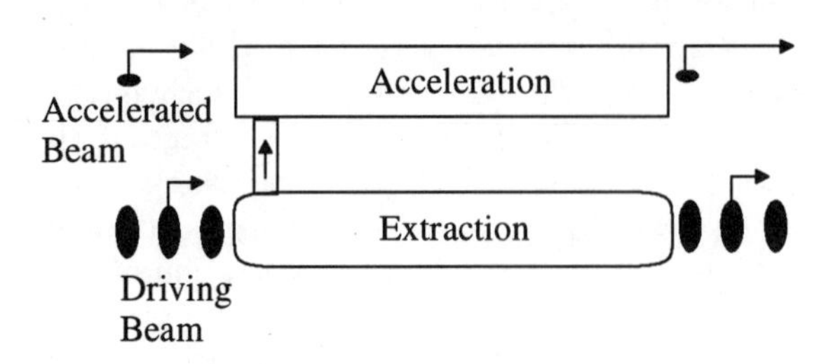

Fig. 8.9 Conceptual set up of the two-beam accelerator being designed at CERN

If, in order to reduce wake effects, the frequency is reduced from 30 to 24 GHz the longitudinal wake effects are reduced to about 60% and the transverse wakes to 50% but this comes at the expense of the accelerating gradient which is lowered from 80 to 50 MV/m. The scaling law behind this result can be readily retrieved bearing in mind that in a uniform waveguide the relation between the power carried by a single TM_{01} mode and the electric field on axis is

$$\frac{P}{|E|^2} \sim \beta_{gr} R^2 \left(\frac{\omega}{c} \frac{R}{p_1} \right)^2 . \tag{8.6.1}$$

If we require a single mode operation, $\omega R/c$ is limited by the cutoff of the second mode i.e., $\omega < p_2 c/R$. Consequently, for a given group velocity $\beta_{gr} c$ and a given gradient, the power (in the accelerating structure) is expected to be inversely proportional to the frequency i.e.,

$$P \times \omega^2 \sim \text{const.} \tag{8.6.2}$$

Although the beam energy in both schemes is larger than in the klystrons of the "conventional" scheme, the amount of current required to provide the power goes up too. Consequently, the amount of charge which propagates is larger and this may deteriorate the bunching via space-charge forces.

Exercises

8.1 Calculate the two wake potentials, (8.1.45–46), as an electron traverses two parallel plates separated by a distance D. Consider only the region between the plates.

8.2 Calculate the two wake potentials, (8.1.45–46), as an electron traverses a pill-box cavity of radius R and length D; the electron moves along the axis. Consider only the internal region. Compare your result with that in Exercise 8.1.

8.3 Consider a uniform cylindrical waveguide of radius R which is infinitely long; ignore walls loss. Between $z = 0$ and $z = D$ the waveguide is filled with a dielectric material ε which is frequency independent; otherwise the waveguide is empty. Calculate the two wake potentials, (8.1.45–46), as an electron traverses this system along its axis. Compare your result with that in Exercises 8.1 and 8.2.

References

Abramowitz M. and Stegun I.A. (1968): *Handbook of Mathematical Functions*, Fifth edition, Dover Publications, New York.

Alyokhin B.V., Dubinov A.E., Selemir V.D., Shamro O.A., Shibalko O.A., Stepanov N.V. and Vatrunin V.E. (1994): "Theoretical and Experimental Studies of Virtual Cathode Microwave Devices", IEEE Trans. **PS-22**, p. 945.

Ashcroft N.W. and Mermin N.D. (1976): *Solid State Physics*, Saunders College, Philadelphia.

Baird J.M. (1987): "Gyrotron Theory", in *High Power Microwave Sources*, edited by V.L. Granatstein and I. Alexeff, Artech House, Boston, p. 103.

Bamber C., Donaldson W.R., Lincke E. and Melissinos (1983): "A Pulsed Power Electron Accelerator Using Laser Driven Photoconductive Switches", in AIP conference proceedings 279, *Advanced Accelerator Concepts*, Port Jefferson, NY 1992, Editor Jonathan S. Wurtele, p. 802.

Beck A.H.W. (1958): *Space Charge Waves and Slow Electromagnetic Waves*, Pergamon Press, Inc., New York.

Bekefi G. (1982): "Rippled-Field Magnetron", Appl. Phys. Lett. **40**(7), p. 578.

Berry M.V. (1971): "Diffraction in Crystals at High Energies", J. Phys. C: Solid St. Phys. **4**, p. 697.

Born M. and Wolf E. (1984): *Principles of Optics*, 6th Edition, Pergamon Press, Oxford, Ch.8.

Brillouin L. (1948): "Wave Guides for Slow Waves", J. Appl. Phys. **19**, p. 1023.

Brillouin L. (1953): *Periodic Structures: Electric Filters and Crystal Lattices*, 2nd edition, Dover, New York.

Bugaev S.P., Cherpenin V.A., Kanavets V.I., Klimov A.I., Kopenkin A.D., Koshelev V.I., Popov V.A. and Slepov A.I. (1990): "Relativistic Multiwave Cerenkov Generator", IEEE Trans. **PS-18**, p. 525.

Carmel Y. and Nation J.A. (1973): "Instability of an Unneutralized Relativistic Electron Beam", Phys. Rev. Lett. **31**, p. 286.

Carmel Y., Granatstein V.L and Gover A. (1983): "Demonstration of Two-stage Backward-wave-oscillator Free Electron Laser", Phys. Rev. Lett.**51**, p. 566.

Carmel Y., Minami K., Kehs R.A., Destler W.W., Granatstein V.L., Abe D. and Lou W.L. (1989): "Demonstration of Efficiency Enhancement in a High Power Backward Wave Oscillator by Plasma Injection". Phys. Rev. Lett. **62**, p. 2389.

Caryotakis G. (1994): "High Power Microwave Tubes: In the Laboratory and Online", IEEE Trans. **PS-22**, p. 683.

Chang D.B. and McDaniel J.C. (1989): "Compact Short-Wavelength Free-Electron Laser". Phys. Rev. Lett. **63**, p. 1066.

Chodorow M. and Susskind C. (1964): *Fundamentals of Microwave Electronics.* McGraw-Hill Book Company, New York.

Chu J.L. and Jackson J.D. (1948): "Field Theory of Traveling Wave Tubes", Proc. IRE **36**, p. 853.

Collins T. (1983): "Concepts in the Design of Circular Accelerators", AIP Conference Proceeding 105, Physics of High Energy Particle Acceleration, SLAC Summer School 1982, Edited by M. Month., AIP New York, p. 93.

Conde M.E. and Bekefi G. (1991): "Experimental Study of a 33.3GHz Free Electron Laser Amplifier with a Reversal Axial Guide Magnetic Field". Phys. Rev. Lett. **67**, p. 3082.

Cooper R.K (1988): "Wake Fields: Limitations and Possibilities". NATO ASI Series Vol. 178, *High Brightness Accelerators*, edited by Hyder A. K., Rose M.F. and Guenther A.H. Plenum Press, New York, p. 157.

Courant E.D., Pellegrini C. and Zakowicz W. (1985): "High Energy Inverse Free Electron Laser Accelerator". Phys. Rev A. **32**, p. 2813.

Davis T.J., Nation J.A. and Schächter L. (1994). "Results from an X-Band Coaxial Extended Length Cavity". IEEE-Trans. Plasma Science **22** p. 504.

DeSanto J.A. (1971): "Scattering from a Periodic Corrugated Structure: Thin Comb with Soft Boundary Conditions", J. Math. Phys. **12**, p. 1913.

DeSanto J.A. (1972): "Scattering from a Periodic Corrugated Structure: Thin Comb with Hard Boundary Conditions", J. Math. Phys. **13**, p. 337.

Destler W.W., Aghimir F.M., Boyd D.A., Bekefi G., Shefer R.E. and Yin Y.Z. (1985): "Experimental Study of Millimeter Wave Radiation from a Rotating Electron Beam in a Rippled Magnetic Field", Phys. Fluids **28**(6), p. 1962.

Doucas G., Mulvey J.H., Omori M., Walsh J.E. and Kimmitt M.F. (1992): "First Observation of Smith–Purcell Radiation from Relativistic Electrons", Phys. Rev. Lett. **69**, p. 1761.

Edighoffer J.A., Kimura W.D., Pantell R.H., Piestrup M.A. and Wang D.Y. (1981): "Observation of Inverse Cerenkov Interaction between Free Electrons and Laser Light". Phys. Rev. A. **23**, p. 1848.

Elachi, C. (1976): "Waves in Active and Passive Periodic Structures: a Review", Proceedings of IEEE, Vol. 64, p. 1666.

Elias L.R., Fairbank W.M., Madey J.M.J., Schwettman H.A. and Smith T.I. (1976): "Observation of Stimulated Emission of Radiation by Relativistic Electrons in a Spatially Periodic Transverse Magnetic Field", Phys. Rev. Lett. **36**, p. 717.

Feinstein J., Pantell R.H. and Fauchet A.M. (1986): Prospects for Visible and VUV Free Electron Lasers Using Dielectric Resonance". IEEE Trans. Quantum Electr. **QE-22**, p. 587.

Flaygin V.A., Gaponov A.V., Petelin M., and Yulpatov V.K. (1977): "The Gyrotron", IEEE Trans. **MTT-25**, p. 514.

Freund H.P., Johnston S. and Sprangle P. (1983): "Three-dimensional Theory of Free Electron Lasers with an Axial Guide Field", IEEE J. Quant. Electron., **QE-19**, p. 322.

Freund H.P and Antonsen T. (1992): *Principles of Free Electron Lasers*, Chapman and Hall, London.

Friedland L., (1980): "Electron Beam Dynamics in Combined Guide and Pump Magnetic Fields for Free Electron Laser Applications", Phys. Fluids **23**, p. 2376. See also Friedland L. and Hirshfield J.L. (1980):"Free Electron Laser with a Strong Axial Magnetic Field", Phys. Rev. Lett. **44**, p. 1456.

Friedman A., Gover A., Kurizki G., Ruschin S. and Yariv A. (1988): "Spontaneous and Stimulated Emission from Quasi-free Electrons", Rev. Mod. Phys. **60**(2), p. 471.

Friedman M. and Serlin V. (1985): "Modulation of Intense Relativistic Electron Beams by an External Microwave Source", Phys. Rev. Lett. **55**, p. 2860 .

Gai W., Schoessow P., Cole B., Konecney R., Norem J., Rosenzweig J. and Simpson J. (1988): "Experimental Demonstration of Wake-field Effects in Dielectric Structures", Phys. Rev. Lett. **61**, p. 2756.

Gaponov A.V. (1959): (Translation of) "Interaction of Nonlinear Electron Beams with Electromagnetic Waves in Transmission Lines". Soviet Radiophysics Vol. 2, p. 837.

Gilmour A.S. Jr. (1986): *Microwave Tubes*, Artech House, Norwood.

Goldstein H. (1950): *Classical Mechanics*, Addison-Wesley Publishing Company Inc., Reading Massachusetts.

Gover A. (1980): "An Analysis of Stimulated Longitudinal Electrostatic Bremsstrahlung in a Free Electron Laser Structure", Appl. Phys. **23**, p. 295.

Granatstein V.L. (1987): "Gyrotron Experimental Studies", in *High Power Microwave Sources*, edited by V.L. Granatstein and I. Alexeff, Artech House, Boston, p. 103.

Haimson J. (1992): "Suppression of Beam-induced Pulse-shortening Modes in High Power RF Generator TW Output Structures", SPIE Proceedings Vol. 1629, *Intense Microwave and Particle Beams III*, Edited by H.E. Brandt, Los Angeles California, p. 209.

Hasegawa A. (1978): "Free Electron Laser", The Bell System Technical Journal, **57**, p. 3069.

Haus H.A. (1959): "Signal and Noise Propagation along Electron Beams", in *Noise in Electron Devices*, Edited Smullin L.D. and Haus H.A., John Wiley & Sons, New York, p. 77.

Haus H. A. and Melcher J.R. (1989): *Electromagnetic Fields and Energy*, Prentice Hall, Englewood Cliffs, New Jersey.

Heifets S. and Kheifets S. (1990): "Coupling Impedance in Modern Accelerators", SLAC-PUB-5297. See also: Heifets S. (1992): "Broadband Impedances of Accelerating Structures: Perturbation Theory". SLAC-PUB-5792.

Helm R.H. and Loew G.A. (1970): "Beam Breakup". in *Linear Accelerators* Edited by P.M. Lapostolle and A.L. Septier, North-Holland Pub. Company - Amsterdam, p. 173.

Henke H. (1994): "MM-Wave Linac and Wiggler Structures", Technische Universität Berlin, Institute für Theoretische Elektronik, TET-94/04.

Hirshfield J. and Granatstein V.L. (1977): "The Electron Cyclotron Maser - An Hystorical Survey", IEEE Trans. **MTT-25**, p. 522.

Hübner K. (1993): "Two–beam Linear Colliders", HEACC'92, XV Int. Conference on High Energy Accelerators, Hamburg, Germany. Int. J. Mod. Phys. A 2B, Editor J. Rossbach, World Scientific, River Edge, New Jersey, p. 791.

Hutter R.G.E. (1960): *Beam and Wave Electronics in Microwave Tubes*, D. Van Nostrand Company Inc., Princeton New Jersey.

Ivers J.D., Advani R., Kerslick G.S., Nation J.A. and Schächter L. (1993): "Electron Beam Using Ferroelectric Cathodes". J. Appl. Phys. **73**, p. 2667.

Joshi C., Clayton C. E., Marsh K.A., Dyson A., Everett M., Lal A., Leemans W.P., Williams R., Katsouleas T. and Mori W.B. (1993): "Acceleration of Injected Electrons by the Plasma Beat Wave Accelerator", AIP conference proceedings 279, *Advanced Accelerator Concepts*, Port Jefferson, NY 1992, Editor Jonathan S. Wurtele, p. 379.

Kapitza P.L. and Dirac P.A.M. (1933): "The Reflection of Electrons from Standing Light Waves", Proc. Cambridge Phil. Soc. **29**, p. 297.

Katsouleas T. and Dawson J.M. (1989): "Plasma Acceleration of Particle Beams", AIP Conference proceedings 184, *Physics of Particle Accelerators*, Editors M. Month and M. Dienes, AIP, New York, p. 1799.

Kim Kang-Je and Kroll N.M. (1982): "Some Effects of the Transverse Stability Requirement on the design of a Grating Linac", AIP Conference Proceedings 91, *Laser Acceleration of Particles*, Los Alamos 1982, Edited by P.J. Channell, AIP New York, p. 190.

Kimura W.D., Kim J.H., Romea R.D., Steinhauer L.C.,Pogoreisky, I.V., Kusche K.P., Fernow R.C., Wano X. and Liu Y. (1995): "Laser Acceleration of Relativistic Electrons Using the Inverse Cerenkov Effect". Phys. Rev. Lett. **74**, p. 546.

Kittel C. (1956): *Introduction to Solid State Physics*, 2nd edition, John Wiley & Sons, New York.

Knapp B.C., Knapp E. A., Lucas G.L. and Potter J.M. (1965): "Accelerating Structures for High Current Proton Linacs", IEEE Trans. Nucl. Sci. **NS-12**, p. 159.

Kroll N.M. (1978): "The Free Electron Laser as a Traveling Wave Amplifier", Physics of Quantum Electronics, *Novel Sources of Coherent Radiation*, Vol. 5 edited by S.F. Jacobs, M. Sargent and M.O. Scully. Addison Wesley Pub. Corp. Massachusettes, p. 115.

Kroll N.M., Morton P.L. and Rosenbluth M.N. (1981): "Free–electron Lasers with Variable Wigglers", IEEE Quantum Electronics **QE-17**, p. 1436.

Kroll N.M. (1985): "General Features of Accelerating Modes in Open Structures", AIP Conference Proceedings 130, *Laser Acceleration of Particles*, Malibu 1985, Edited by C. Joshi and T. Katsouleas, AIP New York, p. 253.

Kuang E., Davis T.J., Kerslick G.S., Nation J.A. and Schächter L. (1993): "Transit Time Isolation of a High Power Microwave TWT". Phys. Rev. Lett. **71**, p. 2666.

Kumakhov M.A. (1976): "On the Theory of Electromagnetic Radiation of Charged Particles in a Crystal", Phys. Lett. A, **57**, p. 17.

Landau, L.D. and Lifshitz, E.M. (1960): *Mechanics*, Pergamon Press, Oxford.

Lapostolle P.M. and Septier A.L. Editors (1970): *Linear Accelerators.* North-Holland Pub. Company - Amsterdam.

Lapostolle P.M. (1971): "Possible Emittance Increase Through Filamentation due to Space-charge Effects", IEEE Trans. Nucl. Sci. **NS-18**, p. 1101.

Lawson J.D. (1988): *The Physics of Charged Particles Beams*, Second Edition, Clarendon Press, Oxford.

Lau Y.Y. (1989): "Classification of Beam Breakup Instabilities in Linear Accelerators", Phys. Rev. Lett. **63**, p. 1141.

Lau Y.Y., Friedman M., Krall J. and Serlin V. (1990): "Relativistic Klystron Amplifiers Driven by Modulated Intense Beams", IEEE Trans. **PS-18**, p. 553.

Lee C.H. (Editor) (1984): *Picosecond Optoelectronic Devices*, Academic Press, New York. In particular see Chapt. 7 by Mourou G., Knox W.H. and Williamson S.

Lewin L. (1975): *Theory of Waveguides*, John Wiley & Sons, New York.

Loew G. A. and Tolman R. (1983): "Lectures on the Elementary Principles of Linear Accelerators", AIP Conference Proceeding 105, Physics of High Energy Particle Acceleration, SLAC Summer School 1982, Edited by M. Month., AIP New York, p. 1.

Madey J.M.J. (1971): "Stimulated Emission of Bremsstrahlung in a Periodic Magnetic Field", J. Appl. Phys. **42**, p. 1906.

Madey J.M.J. (1979): "Relationship between Mean Radiated Energy, Mean Square Radiated Energy and Spontaneous Power Spectrumin a Power Series Expansion of the Equations of Motion in a Free Electron Laser", Il Nuovo Cimento **50B**(1), p. 64.

Marshall T.C. (1985): *Free Electron Lasers.* Macmillan Pub. Comp. New York.

McMullin W.A. and Bekefi G. (1981): "Coherent Radiation from a Relativistic Electron Beam in a Longitudinal Periodic Magnetic Field", Appl. Phys. Lett. **39**(10), p. 845.

McMullin W.A. and Bekefi G. (1982): "Stimulated Emission from Relativistic Electrons Passing through a Spatially Periodic Longitudinal Magnetic Field", Phys. Rev. A **(25)**(4), p. 1826.

Miller, R.B. (1982): *An Introduction to the Physics of Intense Charged Particle Beams*, Plenum Press, New York.

Mittra R. and Lee S.W. (1971): *Analytical Techniques in the Theory of Guided Waves*, Macmillan, New York.

Motz H. (1951): "Applications of the Radiation from Fast Electron Beams", J. Appl. Phys. **22**, p. 527.

Nation J.A. (1970): "On the Coupling of a High-current Relativistic Beam to a Slow Wave Structure". Appl. Phys. Lett. **17**, p. 491.

Naqvi S.A., Nation J.A., Kerslick G.S. and Schächter L. (1996). "Resonance Shift in Relativistic Traveling Wave Amplifiers" Phys. Rev. E **53**(4) p. 4229.

Orzechowsky T.J., Anderson B.R., Fawley W.M., Prosnitz D., Scharlemann E.T., Yarema S.M., Hopkins D., Paul A.C., Sessler A.M. and Wurtele J. (1985). "Microwave Radiation from a High Gain FEL Amplifier", Phys. Rev. Lett. **54**, p. 889.

Palmer R.B., (1972): "Interaction of Relativistic Particles and Free Electromagnetic Waves in the Presence of a Static Helical Magnet", J. Appl. Phys. **43**, p. 3014.

Palmer R.B. (1982): "Near Field Accelerators", AIP Conference Proceedings 91, *Laser Acceleration of Particles*, Los Alamos 1982, Edited by P.J. Channel, AIP New York, p. 179. See also Palmer R.B., Baggett N., Claus J., Fernow R., Stumer I., Figueroa H., Kroll N.M., Funk W., Lee-Whiting G., Pickup M., Goldstone P., Lee K., Corkum P. and Himel T. (1985): "Report of Near Field Group", AIP Conference Proceedings 130, *Laser Acceleration of Particles*, Malibu 1985, Edited by C. Joshi and T. Katsouleas, AIP New York, p. 234.

Palmer R.B. (1986): "An Introduction to Acceleration Mechanisms", in *Frontiers of Particle Beams*, Edited by M. Month and S. Turner, Springer-Verlag, Berlin, p. 607.

Panofsky W.K.H. and Wenzel W.A. (1956): "Some Considerations Concerning the Transverse Deflection of Charged Particles in Radio Frequency Fields", Rev. Sci. Instr. **27**, p. 967.

Pantell R.H., Soncini G. and Puthoff H.E. (1968). "Stimulated Photon-Electron Scattering", J. Quant. Elect. **QE-4**, p. 905.

Pantell R.H. and Alguard M.J. (1979): "Radiation Characteristics of Planar Channeled Positrons", J. Appl. Phys. **50**, p. 798.

Pantell R.H. (1981): "Interactions between Electromagnetic Fields and Electrons", Physics of High Energy Particle Accelerators, Fermilab Summer School July 13-24 1981, Editors Carrigan R.A., Huson F.R. and Month M., AIP Conference Proceedings No. 87, p. 864.

Pantell R.H., Feinstein J., Fisher A.L., Deloney T.L., Reid M.B. and Grossman W.M. (1986): "Benefits and Costs of the Gas-loaded Free Electron Laser", Nucl. Instr. Methods Phys. Res., **A250**, p. 312.

Pauli W. (1958): *Theory of Relativity*, Pergamon Press, London.

Petit R. (1980): *Electromagnetic Theory of Gratings*, Springer-Verlag, Berlin.

Phillips R.M. (1960): "The Ubitron, a High Power Traveling Wave Tube Based on a Periodic Beam Interaction in Unloaded Waveguide", Trans. IRE Elec. Dev. **7**, p. 231.

Pickup M. (1985): "A Grating Linac at Microwave Frequencies", AIP Conference Proceedings 130, *Laser Acceleration of Particles*, Malibu 1985, Edited by C. Joshi and T. Katsouleas, AIP New York, p. 281.

Pierce J.R. (1947): "Theory of the Beam-Type Traveling Wave Tube", Proc. IRE, p. 111.

Pierce J.R. (1950): *Traveling-Wave Tubes*, D. van Nostrand Company Inc., Princeton, New Jersey.

Ramo S., Whinnery J. R. and Van Duzer T. (1965): *Fields and Waves in Communication Electronics*, John Wiley & Sons, New York.

Richter B. (1985): "Requirements for Very High Energy Accelerators", AIP Conference Proceedings 130, *Laser Acceleration of Particles*, Malibu California Edited by C. Joshi and T. Katsouleas, AIP New York, p. 8.

Riege H. (1993): "Electron Emission from Ferroelectrics - A Review". Nucl. Instrum. and Meth. **340**, p. 80.

Roberson C.W. and Sprangle P. (1989): "A Review of Free–Electron Lasers", Phys. Fluids B**1**, p. 3.

Rosing M. and Gai W. (1990): "Longitudinal and Transverse Wake Field Effects in Dielectric Structures, Phys. Rev. D. **42**, p. 1829.

Rubin L. D. (1992): "CESR Status", HEACC'92 XVth International Conference on High Energy Accelerators, Hamburg July 20-24 1992, Editor J. Rossbach, Int. J. Mod. Phys. A. (Proc. Suppl.) 2A (1993), Vol. 1, p. 78.

Ruth R.D. (1986): "Single Particle Dynamics and Nonlinear Resonances in Circular Accelerators", in Lecture Notes in Physics, 247, *Non-linear Dynamics Aspects of Particle Acceleration*, edited by Jowett J.M., Month M. and Turner S., Springer-Verlag, Berlin, p. 37.

Salisbury W.W. (1970): "Generation of Light from Free Electrons", J. Opt. Soc. Am. **60**, p. 1279.

Schächter L. (1988): "Remarks on Channeling Radiation", J. Appl. Phys. **63**, p. 712.

Schächter L., Nation J.A. and Shiffler D. A. (1991): "Theoretical Studies of High Power Cerenkov Amplifiers", J. Appl. Phys. **70**, p. 114.

Schächter L. and Nation J.A. (1992): Slow Wave Amplifiers and Oscillators: a Unified Study", Phys. Rev. A. **45**, p. 8820.

Schächter L. (1995): "PASER: Particle Acceleration by Stimulated Emission of Radiation", Phys. Lett. A. 205, p. 355.

Scharlemann E.T, Sessler A.M. and Wurtele J.S. (1985): "Optical Guiding in Free Electron Laser", Phys. Rev. Lett. **54**(17), p. 1925.

Schieber D. (1986): *Electromagnetic Induction Phenomena*, Springer-Verlag, Berlin.

Schneider J. (1959): "Stimulated Emission of Radiation by Relativistic Electrons in a Magnetic Field", Phys. Rev. Lett.**2**, p. 504.

Schnell W. (1991): "The CERN Study of a Linear Collider in the TeV Range", CERN Div. Rep. SL/91-49.

Sessler A.M. (1982): "The FEL as a Power Source for a High Gradient Accelerating Structure", AIP Conference Proceedings 91, Editor P.J. Channel, p. 154.

Serlin V. and Friedman M. (1994): "Development and Optimization of the Relativistic Klystron Amplifier", IEEE Tran. **PS-22**, p. 692.

Shiffler D.A., Nation J.A. and Wharton C. B. (1989): "High–Power Traveling Wave Amplifier", Appl. Phys. Lett. **54**, p. 674.

Shiffler D. A., Ivers J.D., Kerslick G.S., Nation J.A. and Schächter L. (1991): "A High Power Two Stage Traveling Wave Amplifier", J. Appl. Phys. **70**, p. 106.

Slater J.C. (1950): *Microwave Electronics*, D. Van Nostrand Company Inc., New York.

Smith S.J. and Purcell E.M. (1953): "Visible Light from Localized Charges Moving Across a Grating", Phys. Rev. **92**, p. 1069.

Sprangle P., Esarey E. and Ting A. (1990): "Nonlinear Interaction of Intense Laser Pulses in Plasmas". Phys. Rev. A. **41**, p. 4463.

Sprangle P., Ting A. and Tang C.M. (1987): "Radiation Focusing and Guiding with Application to the Free Electron Laser", Phys. Rev. Lett. **59**, p. 202.

Sprangle P., Esarey E., Krall J., Joyce G. and Ting A. (1993): "Electron Acceleration and Optical Guiding in the Laser Wake Field Accelerator", AIP conference proceedings 279, *Advanced Accelerator Concepts*, Port Jefferson, New York 1992, Editor J. S. Wurtele, p. 490.

Steinhauer L.C. and Kimura W.D. (1990): "High-γ Inverse Cerenkov Acceleration in Resonant Media". J. Appl. Phys. **68**, p. 4929.
Strickland D. and Moureau G. (1985): "Compression of Amplified Chirped Optical Pulses", Opt. Commun. **56**, p. 219.
Sullivan D.J., Walsh J.E. and Coutsias E.A. (1987): "Virtual Cathode Oscillator Theory", in *High-Power Microwave Sources*, edited by V.L. Granatstein and I. Alexeff, Artech House, Boston, p. 441.
Swent R.L., Pantell R.H., Alguard M.J., Berman B.L., Bloom S.D. and Datz S. (1979): "Observation of Channeling Radiation from Relativistic Electrons", Phys. Rev. Lett. **43**, p. 1723.
Tajima T. and Dawson J.M. (1979): "Laser Electron Accelerator", Phys. Rev. Lett. **43**, p. 267.
Terhune R.W. and Pantell R.H. (1977): "X-ray and γ-ray Emission from Channeled Relativistic Electrons and Positrons". Appl. Phys. Lett. **30**, p. 265.
Toraldo di Francia, G. (1960): "On the Theory of some Cerenkovian Effects", Il Nuovo Cimento, **16**, p. 61.
Twiss R. O. (1958): "Radiation Transfer and the Possibility of Negative Absorption in Radio Astronomy". Australian J. Phys. **11**, p. 564.
Van Bladel J. (1984): *Relativity and Engineering*, Springer Verlag, Berlin.
Van den Berg P.M. (1973): "Smith-Purcell Radiation from a Line Charge Moving Parallel to a Reflection Grating", J. Opt. Soc. Am. **63**, p. 689. See also, Van den Berg P.M. (1973): "Smith-Purcell Radiation from a Point Charge Moving Parallel to a Reflection Grating", J. Opt. Soc. Am. **62**, p. 1588.
Voss G.A. and Weiland T. (1982): "The Wake Field Acceleration Mechanism", DESY 82-074. See also, Voss G.A. and Weiland T. (1982): "Particle Acceleration by Wake Fields", DESY 82-10.
Walsh J.E. (1987): "Cerenkov Masers: Experiment", in *High Power Microwave Sources*, Edited by Granatstein V.L. and Alexeff I., Artech House, Boston, p. 421.
Wheeler J.A. and Feynman R.P. (1945): "Interaction with the Absorber as the Mechanism of Radiation", Rev. Mod. Phys. **17**, p. 157. See also Wheeler J.A. and Feynman R.P. (1949): "Classical Electrodynamics in terms of Direct Interparticle Action", Rev. Mod. Phys. **21**, p. 425.
Wilson P. (1988): "Wake Field Accelerators: Concepts and Machines", NATO ASI Series Vol. 178, *High Brightness Accelerators*, edited by Hyder A. K., Rose M.F. and Guenther A.H. Plenum Press, New York, p. 129.

Subject Index